Soil Carbon, Nitrogen Sequestration and Greenhouse Gas Mitigation under Global Change

Soil Carbon, Nitrogen Sequestration and Greenhouse Gas Mitigation under Global Change

Editor

Ling Zhang

MDPI • Basel • Beijing • Wuhan • Barcelona • Belgrade • Manchester • Tokyo • Cluj • Tianjin

Editor
Ling Zhang
College of Forestry
Jiangxi Agricultural University
Nanchang
China

Editorial Office
MDPI
St. Alban-Anlage 66
4052 Basel, Switzerland

This is a reprint of articles from the Special Issue published online in the open access journal *Life* (ISSN 2075-1729) (available at: www.mdpi.com/journal/life/special_issues/Greenhouse_Gas_Mitigation).

For citation purposes, cite each article independently as indicated on the article page online and as indicated below:

LastName, A.A.; LastName, B.B.; LastName, C.C. Article Title. *Journal Name* **Year**, *Volume Number*, Page Range.

ISBN 978-3-0365-7345-8 (Hbk)
ISBN 978-3-0365-7344-1 (PDF)

Cover image courtesy of Ling Zhang

Contents

Preface to "Soil Carbon, Nitrogen Sequestration and Greenhouse Gas Mitigation under Global Change"

Global-change-induced extreme climate events are becoming more common than ever. Soil carbon and nitrogen pools correlate significantly with the changes in atmospheric greenhouse gas levels. A large increase in atmospheric greenhouse gases, namely carbon dioxide, nitrous oxide, and methane, can accelerate atmospheric heating, which is generally followed by global warming. The mitigation of greenhouse gas emissions via various strategies, such as the sequestration of carbon and nitrogen in soil, plant, or ecosystems; the efficient management of agricultural and forestry ecosystems; the mitigation of ecosystem carbon and nitrogen leaching; the mitigation of greenhouse gas emissions from all kinds of sources; etc. will therefore be crucial in the mitigation of global climate change.

By gathering the latest case studies and methodologies, including but not limited to the measurement and mitigation strategies of carbon and nitrogen pools, and greenhouse gas emissions, this reprint will substantially improve our understanding of the potential, ability, and capacity of ecosystems in mitigating greenhouse gas emissions and, hence, global climate change.

This reprint can be used by colleagues working on global climate change, ecology, agriculture, forestry, and policy making associated with global change. The articles included in this reprint were contributed by colleagues from China, Egypt, Italy, Jordan, Mexico, Pakistan, Saudi Arabia, Turkey, and more. I appreciate their substantial contributions to this reprint and the cooperation of my colleagues at Jiangxi Agricultural University.

Ling Zhang
Editor

Review

Carbon Pool in Mexican Wetland Soils: Importance of the Environmental Service

Sergio Zamora [1], Irma Zitácuaro-Contreras [2], Erick Arturo Betanzo-Torres [3], Luis Carlos Sandoval Herazo [3], Mayerlin Sandoval-Herazo [4], Monserrat Vidal-Álvarez [2,*] and José Luis Marín-Muñiz [2,*]

[1] Facultad de Ingeniería, Construcción y Habitad, Universidad Veracruzana, Bv. Adolfo Ruíz Cortines 455, Costa Verde, Boca del Rio 94294, Veracruz, Mexico; szamora@uv.mx
[2] Academy of Sustainable Regional Development, El Colegio de Veracruz, Xalapa 91000, Veracruz, Mexico; izitacuaro@yahoo.com
[3] Wetlands and Environmental Sustainability Laboratory, Division of Graduate Studies and Research, Tecnológico Nacional de México/Instituto Tecnológico de Misantla, Veracruz, Km 1.8 Carretera a Loma del Cojolite, Misantla 93821, Veracruz, Mexico; eabetanzot@itsm.edu.mx (E.A.B.-T.); lcsandovalh@gmail.com (L.C.S.H.)
[4] Department of Business Management Engineering, Tecnológico Nacional de México/Instituto Tecnológico de Misantla, Veracruz, Km 1.8 Carretera a Loma del Cojolite, Misantla 93821, Veracruz, Mexico; mayerli.sandoval24@gmail.com
* Correspondence: monserrat.vidal@gmail.com (M.V.-Á.); soydrew@hotmail.com (J.L.M.-M.); Tel.: +52-2-281-261-814 (M.V.-Á.); +52-2-281-624-680 (J.L.M.-M.)

Citation: Zamora, S.; Zitácuaro-Contreras, I.; Betanzo-Torres, E.A.; Herazo, L.C.S.; Sandoval-Herazo, M.; Vidal-Álvarez, M.; Marín-Muñiz, J.L. Carbon Pool in Mexican Wetland Soils: Importance of the Environmental Service. *Life* **2022**, *12*, 1032. https://doi.org/10.3390/life12071032

Academic Editor: Ling Zhang

Received: 12 June 2022
Accepted: 7 July 2022
Published: 11 July 2022

Publisher's Note: MDPI stays neutral with regard to jurisdictional claims in published maps and institutional affiliations.

Abstract: Mexican wetlands are not included in Earth system models around the world, despite being an important carbon store in the wetland soils in the tropics. In this review, five different types of wetlands were observed (marshes, swamps, flooded grasslands, flooded palms and mangroves) in which their carbon pool/carbon sequestrations in Mexican zones were studied. In addition, it was shown that swamps (forested freshwater wetlands) sequestered more carbon in the soil (86.17 ± 35.9 Kg C m^{-2}) than other types of wetlands ($p = 0.011$); however, these ecosystems are not taken into consideration by the Mexican laws on protection compared with mangroves (34.1 ± 5.2 Kg C m^{-2}). The carbon pool detected for mangrove was statistically similar ($p > 0.05$) to data of carbon observed in marshes (34.1 ± 5.2 Kg C m^{-2}) and flooded grassland (28.57 ± 1.04 Kg C m^{-2}) ecosystems. The value of carbon in flooded palms (8.0 ± 4.2 Kg C m^{-2}) was lower compared to the other wetland types, but no significant differences were found compared with flooded grasslands ($p = 0.99$). Thus, the carbon deposits detected in the different wetland types should be taken into account by policy makers and agents of change when making laws for environmental protection, as systematic data on carbon dynamics in tropical wetlands is needed in order to allow their incorporation into global carbon budgets.

Keywords: wetlands; environmental services; carbon soil sequestration; carbon budgets

1. Introduction

Atmospheric concentrations of greenhouse gases (GHGs) are at levels unprecedented in at least 800,000 years. Concentrations of carbon dioxide (CO_2), methane (CH_4) and nitrous oxide (N_2O) have all shown large increases since 1750 (40%, 150% and 20%, respectively) [1]. Thus, wetlands are an option to help to mitigate the impact of climate change as they can regulate, capture and store GHGs [2,3] due to their dense vegetation, microbial activity and soil conditions.

Wetlands are transitional zones between terrestrial and aquatic ecosystems, areas of water saturated with soil that covers about 5 to 8% of the land surface of Earth [2]. They include floodplains with forest or herbaceous vegetation. Wetlands play an important role in the global carbon cycle; they have the best capacity of any ecosystem to retain carbon in the soil, as these ecosystems have a high capacity to limit the availability of oxygen to soil microbes and the decomposition of organic matter [4,5].

Decomposition of organic matter within wetlands involves aerobic and anaerobic processes. Organic matter decomposition (shrub residues, detritus, etc.) is often incomplete under anaerobic conditions due to the lack of oxygen [5]. The carbon accumulation formed over time is vulnerable when the wetland soil is affected by five factors, which includes pollution, biological resource use, natural system modification, agriculture and aquaculture [6]. These changes may have important repercussions on global warming. However, despite the fact that approximately 30% of the world's wetlands are found in the tropics [2,7], only a few studies [7,8] regarding carbon sequestration in wetlands have been conducted and taken into account for the global carbon budget in the tropical regional zone of Mexico.

Of the total storage of carbon in the earth's soils (1400–2300PgC (Pg = 10^{15} g of carbon)), 20–30% is stored in wetlands [2]. Some studies around the world on carbon sequestration have shown the importance of natural soil wetlands; for example, North American wetlands contain about 220 PgC, most of which is in peat [8]. Wetlands in the conterminous United States store a total of 11.52 PgC [9]. Another study suggests that the world's wetlands serve as a net sink of 0.83 PgC/year, estimated from 21 wetland simulations that included Russia, Canada, Costa Rica, Botswana and the USA data on carbon sequestration wetland soils [10]. However, information of Mexican carbon pools was not considered in American data. The knowledge regarding carbon sequestration and storage in all the wetland soils is critical for successful pathways to global decarbonization and carbon budgets.

In 2022, the Convention on Wetlands of International Importance, known as theRAM-SAR Convention or RAMSAR treaty, celebrates 51 years of efforts to conserve these crucial ecosystems. The treaty includes 2435 designated wetlands of international importance (Ramsar sites), which involves 254,685,425 hectares of wetland soils around the world (172 contracting parties). Mexico is in second place with wetland Ramsar sites (142) after the United Kingdom (175), covering 8.7 million ha. Many of the priority wetland sites in Mexico are associated with coastal sites on the Gulf of Mexico and the Pacific Ocean [11].

According to this, it is essential to know the Mexican wetlands. Thus, the main objective of this study is to describe the natural wetlands of Mexico, their uses, and soil carbon pool or carbon sequestration function.

2. Materials and Methods

The authors undertook a comprehensive search of the literature on carbon sequestered in Mexican wetland soils (mangroves, flooded palms, swamps, flooded grassland, marshes) based on the most important databases located in Mexican universities, publications of the Mexican carbon program (http://pmcarbono.org/pmc/publicaciones/sintesisn.php (accessed on 5 June 2022)), and the ISI Web of Knowledge (www.isiknowledge.com (accessed on 8 April 2022)) database. The keywords used were: (carbon-pool, -stock, -sinks, -sequestration, soil, wetlands, mangroves, flooded palms, swamps, flooded grassland, marshes using the booleans operators and/or (exclusively in Spanish and English). A total of 482 studies (from the year 2000 to 2022) were identified regarding carbon sinks for Mexican wetland soils; only 32 studies were selected based on studies in situ on carbon pools or carbon sequestration in Mexican wetlands because most were theoretical topics or reviews, of which 56 sites were analyzed because in some papers different types of wetlands were studied. The remaining percentage of studies was used for the introduction section, the justification of the study and the discussion of the data.

To provide context, it is important to mention that, in most cases, the method formeasuring the carbon sequestered in wetland soils consists of sampling points of soil profiles used to analyze carbon and bulk density. Organic carbon is obtained by analyzing the percentage of organic matter (OM) content and calculated as a portion of OM by using the conversion coefficient of 0.58 (Van Bemmelen factor) [3,7,10] or the factor of 0.50 proposed by Mitsch and Gosselink [2].

Statistical analyses to determine differences in carbon pools among wetland soils were performed with IBM SPSS Statistic version 22 for Windows (Armonk, NY, USA: IBM Corp.). A Kolmogorov–Smirnov test was used to check normality; data fitted no normal distributions; thus, the Kruskal–Wallis test at the 5% significant level was used to find differences in the carbon pools between different wetland types.

3. Results and Discussion

3.1. Importance of Natural Wetlands to Ecosystem Services

Natural wetlands are important areas in terms of natural resources and biodiversity. The publication of the Millennium Ecosystem Assessment [12] described a categorization for wetland ecosystem services with four types: (1) Provisioning ecosystem services include products obtained from ecosystems, such as food, water, timber, fiber, or genetic resources. (2) Regulating ecosystem services include air quality and climate regulation, water purification, disease/pest regulation, pollination, and natural hazard regulation. (3) Cultural ecosystem services include benefits that people obtain from ecosystems related to spiritual enrichment, recreation, ecotourism, aesthetics, formal and informal education, inspiration, and cultural heritage, and (4) supporting ecosystem services include basic ecosystem processes of nutrient cycling and primary productivity that may, in turn, lead to the other three services listed above.

Considering the ecosystem services described above, wetlands are known as "the kidneys of the landscape" and "ecological supermarkets", and they provide a potential sink of carbon [2]. Despite the fact that a large percentage of wetlands occur in tropical latitudes, carbon sequestration from coastal tropical Mexican wetlands has not been extensively reported.

3.2. Wetland Types

Natural wetlands include marine, coastal or continental wetland types. The water that flows into the wetland could be freshwater, brackish or saline. According to the vegetation, the wetlands can have herbaceous vegetation or forest plants. However, some specific classifications are described:

Forested wetlands: these are wetlands dominated by trees in temporal or permanent flooded conditions. Wetlands dominated by halophytic trees growing in brackish to saline tidal waters are called Mangroves. Typical mangrove trees are *Avicennia germinans*, *Laguncularia racemosa* and *Rhizophora mangle* L. Wetland trees growing in freshwater conditions are called freshwater swamps. In this study, only the concept "swamp" was used. Typical freshwater swamp trees are *Pachira aquatica*, *Haematoxylum campechianum*, *Ficus insipida*, *Taxodium distichum* and *Nyssa aquatic* [2,13].

Herbaceous wetlands: these are wetlands dominated by emergent herbaceous vegetation adapted to saturated soil conditions. In wetlands with freshwater conditions, common species are *Typha* spp., *Thalia geniculate* L., *Scirpus* spp. and *Pontederia sagittata*. While common species of herbaceous wetlands growing in brackish or saline water are: *Spartina alterniflora*, *Salicornia quinqueflora* and *Galaxias maculates* [3,13]. In this study, these types of wetlands are considered as marshes (wetlands dominated by herbaceous plants, such as grasses, reeds and sedges [2,3]).

On the other hand, other types of herbaceous wetlands are the flooded grasslands: these wetlands are characterized by an abundance of grass (or sedges), as well as periodic flooding with fresh or brackish water or a high-water level during some months of the year, sufficient to influence the vegetation. Typical grasses include species of the *Poaceae* family or sedges with species of the *Cyperaceae* family [14].

3.3. The Carbon Cycle and Dominant Organisms in Wetland Soils

Carbon capture and sequestration is a physical process that involves the capture of atmospheric carbon dioxide (CO_2) and its storage. In wetlands, the major components of the carbon cycle are illustrated in Figure 1. Various reactions utilizing carbon take place within wetlands. The key processes are respiration and photosynthesis in aerobic conditions, fermentation, methanogenesis, methane oxidation and sulfate, iron, and nitrate reduction in the anaerobic areas [15].

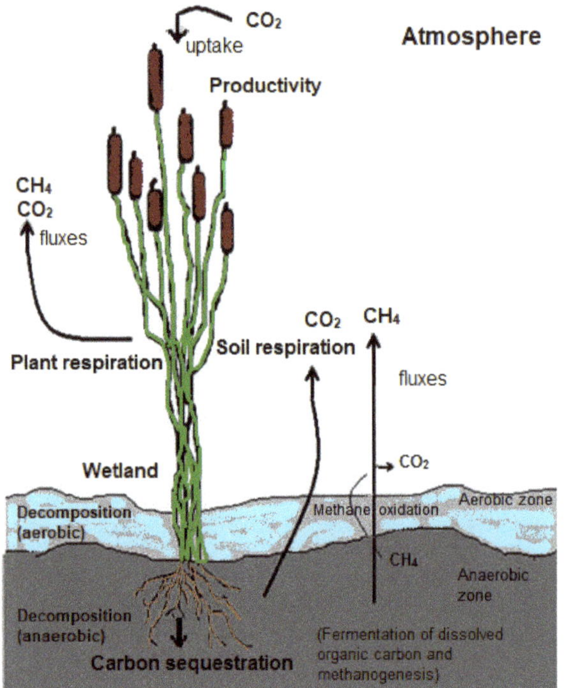

Figure 1. Schematic diagram showing the major components of the carbon budget in a wetland and its carbon exchanges with the atmosphere (adapted from [2,7,10]).

Photosynthesis ($6CO_2 + 12H_2O + \text{light} \rightarrow C_6H_{12}O_6 + 6O_2 + 6H_2O$) and aerobic respiration ($C_6H_{12}O_6 + 6O_2 \rightarrow 6CO_2 + 6H_2O + 12e^- + \text{energy}$) dominate the aerobic areas (aerial and aerobic water and soil), with H_2O as the major electron donor in photosynthesis and oxygen as the terminal electron acceptor in respiration [2].

The fermentation of organic matter or glycolysis for the substrate involved occurs when organic matter is the terminal electron acceptor in the anaerobic respiration of microorganisms. This forms various low-molecular-weight acids and alcohols, as well as carbon dioxide, e.g., lactic acid ($C_6H_{12}O_6 \rightarrow 2CH_3CH_2OCOOH$) and ethanol ($C_6H_{12}O_6 \rightarrow 2CH_3CH_2OH + 2CO_2$). Fermentation represents one of the major ways in which high-molecular-weight carbohydrates are broken down to low-molecular-weight organic compounds, usually as dissolved organic carbon, which is, in turn, available to other microbes [2,16].

The methanogenesis occurs when certain methanogenic bacteria members of the Archaea domain use CO_2 as an electron acceptor for the production of gaseous methane ($CO_2 + 8H+ \rightarrow CH_4 + 2H_2O$). Depending on the wetlands and type of archaea, hydrogenotrophic methanogenesis or acetoclastic processes occur. In non-fertilized soils, it has been observed that acetoclastic methanogenesis represents 51–67% of the produced methane [16]. On the other hand, methane oxidation is carried out by obligate methanotrophic bacteria, which are from a larger group of eubacteria; they convert methane gas in sequence to methanol (CH_3OH), formaldehyde ($HCHO$), and finally CO_2 ($CH_4 \rightarrow CH_3OH \rightarrow HCHO \rightarrow HCOOH \rightarrow CO_2$)

Sulfate reduction: this metabolism is carried out by sulfate-reducing bacteria (SRB) when the redox potential (Eh) decreases to -120 mV. SRB use mainly sulfate as their terminal electron acceptor in the anaerobic oxidation of organic substrates and reduce it to hydrogen sulfide (H_2S) ($2(CH_2O) + SO_4^{-2}$ $2HCO^{3-} + H_2S$). Sulfate reducers are capable of using formate, lactate and H_2; therefore, they compete with methanogens for substrates [2,16]. In coastal sediments, distinct depth distributions have been observed, where SO_4^{2-} reducers are abundant in the first few centimeters, but as SO_4^{2-} is depleted, methanogens become more abundant at greater depths [17].

Iron reduction: this is a process carried out when the Eh descends to –47 mV. Several groups of facultative and anaerobic bacteria participate in it. Manganese-reduction and iron-reduction are relevant processes in those wetlands with high mineral supplies [16].

Denitrification: it is a respiration process in which the electron acceptor is nitrate, and it starts when oxygen concentration is <10 µM. The resulting denitrification products are molecular nitrogen (N_2) and nitrogen oxide (NOx). Anaerobic Gram-negative bacteria performs this process; among them, the genera *Pseudomonas* spp., *Clostridium* spp., *Bacillus* spp. and *Alcaligenes* spp. have been reported [18].

Hydrology and radial O_2 leakage have differences in the oxidation state of metals. Therefore, site mineralogy interacts with hydrology to shape the wetland microbial community. For example, wetland roots are often coated with iron (Fe) (III) and Manganese (Mn) (IV) oxides. Plants supply electron donors in the form of root exudates and oxidize metals through O_2 leakage, supporting metal-reducing bacteria [19].

Soils of organic matter typically contain between 45% and 50% carbon. Organic soils formed from plant debris decompose slowly in very wet settings due to low oxygen conditions, also referred to as anaerobic. Organic soils are very black, porous, and light in weight and are often referred to as "peat" or "musk" [20]. Another process in wetlands is respiration, described as the biological conversion of carbohydrates to carbon dioxide, and fermentation is the conversion of carbohydrates to chemical compounds such as lactic acid, or ethanol and carbon dioxide. In a wetland, organic carbon is converted into compounds including carbon dioxide and methane and/or stored in plants, dead plant matter, microorganisms, or peat [15,16].

Microbial degradation of above-ground plant litter is likely to begin before the material enters the soil, and the role of fungi in wetlands needs more investigation. Some dominant organisms in wetland soils described in a mini review [19] include fungi (*Chaetothyriales*, *Cantharellales*), bacteria (*Bacteroides*, *Planctomycetes*, *Chloroflexi*, *Acidobacteria*, *Actinobacteria*), fermenters (*Chloroflexi*, *Proteobacteria*, *Verrucomicrobia*), iron reducers (*Geobacter* sp., *Desulfovibrio* sp., *Anaeromyxobacter*, *Shewanella*), sulfate reducers (*Desulfarculales*, *Desulfovibrionales*, *Syntrophobacterales*, *Firmicutes*, *Desulfobulbaceae*), methanogens (*Methanoregulaceae*, *Methanosarcinaceae*, *Methanosaetaceae*) and methanotrophs (*Methylobacter*, *Methylocystis*, *Methylobacter*).

3.4. Carbon Sequestration in Mexican Wetland Soils

Considering that the importance of wetlands in carbon sequestration as an environmental service to mitigate global warming was described by the Millennium Ecosystem Assessment [12], it is important to highlight that this regulating ecosystem service is scarcely perceived or recognized by the population.

Generally, provisioning and the cultural ecosystem services are the most identified because they are more visible or palpable [21], so the dissemination of this type of knowledge is important, and this study highlights such function, in which it is important to note that even though there are some global carbon balances, counts or earth system models, these do not include Mexican or tropical data [2,9,12,22]. However, in the last 20 years, measurements of the carbon storage function have been made in Mexico, including mangroves, swamps, marshes, flooded palms and flooded grasslands (Table 1).

Carbon sequestration in Mexican marshes oscillated between 16 and 110 Kg C m^{-2} (Table 1), while data of carbon pool reported for marshes of Old Woman Creek from Ohio were between 9 and 14 Kg C m^{-2}. In Palo Verde, a national park in Costa Rica, only 6 to 7 Kg C m^{-2} were reported [23]. The carbon pool observed in the riverine marsh of

Botswana, Africa, oscillated within 0.8–1.3 Kg C m^{-2} [24], underlining the importance of Mexican tropical wetlands in carbon storage, in addition to highlighting the importance of these data for the generation of regulations or public policies for their protection since these types of wetlands do not have extensive legal protection as in the case of the mangroves.

The flooded grasslands are sites with minimal attention regarding the function of the carbon pool. In Mexican regions, three studies were found between the states of Veracruz, Chiapas and Tabasco (Table 1, Figure 2), with carbon sequestered in the soil within 28 and 31 Kg C m^{-2}. In the same regions, one of the studies mentioned above [25] claims that soil carbon concentration decreases in areas converted from swamps of forested wetlands to flooded grasslands due to decreases in carbon inputs, physical disturbances, and shorter hydroperiods, which enhance higher greenhouse emissions, so the changing land use negatively affects the ecosystem services as a carbon pool.

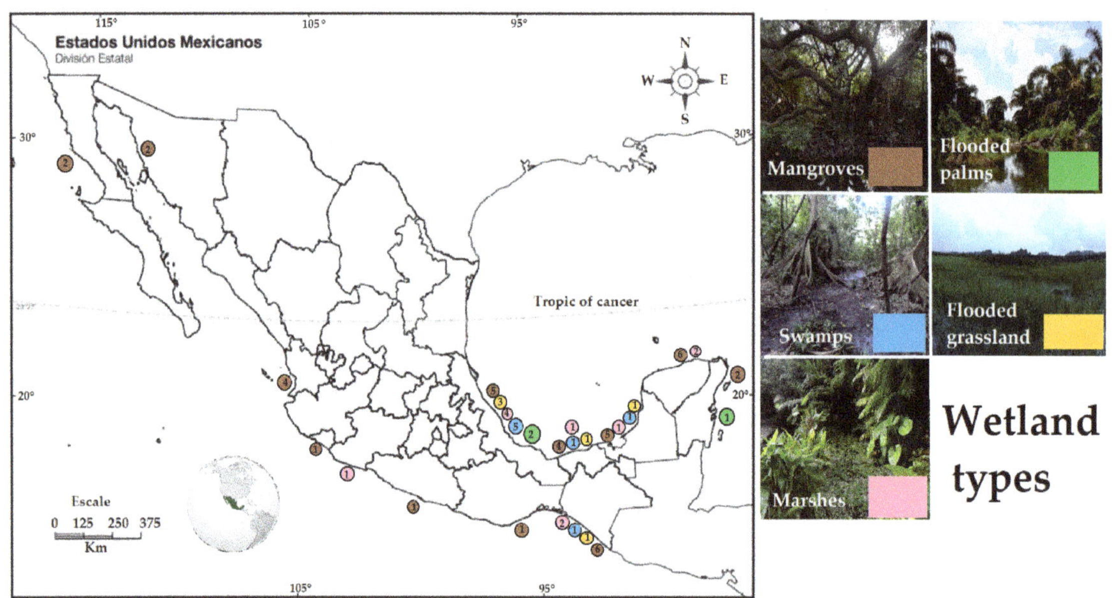

Figure 2. Location of the Mexican wetlands reviewed. Places represented by letters are referenced in Table 1. The number inside the circle is the number of studies in that state/site. The color of the circle represents the wetland type.

Regarding mangroves, in Mexico, they are the wetland type with the best policies and law enforcement for protection and conservation [3]. However, despite this importance, mangroves are being deforested. Throughout the twentieth century, 30–50% of global mangrove cover has been destroyed. Using previously published global models of carbon stocks and Mexico-specific carbon sequestration data and calculating gross deforestation, it was found that the current rate of deforestation will result in a social cost of USD 392.0 ($\pm$7.4) million over the next 25 years [26].

Thus, it is essential to follow up on the policies established in the country and avoid permits for land-use change in these ecosystems, considering that the values of carbon sequestered in mangrove soils oscillate between 7 and 93 Kg C m^2 in almost all the coastal zone of the country (Table 1, Figure 2). The values reported are similar to data in inventories of soil carbon for natural and replanted mangrove forests from tropical and subtropical areas, including Indonesia, Thailand, China and Australia (28–56 Kg C m^{-2}) [27].

Regarding swamps, these ecosystems have been recognized as treasures of the country, and some books have described the importance, history, science and policies of these wetland sites [2,3,28]. In Mexico, for example, some studies have reported their importance as a carbon pool, mainly in the coastal areas of Veracruz and Chiapas with values between 35 and 73 Kg C m^{-2}, while in other tropical zones such as Costa Rica or temperate areas such as Ohio, the values reported for similar wetland types are lower (10–21 Kg C m^{-2}) [10,23].

Flooded palms are a type of wetland that is less common; however, in Mexico, three sites with carbon pool data were identified in Veracruz and Quintana Roo (Table 1, Figure 2), with values of 1.5 to 16 Kg C m^{-2}. These ecosystems are important as the fruits of the palms are widely used for the preparation of traditional recipes, and the stem of the palms is used for house construction [29,30]. At EARTH University, Humedal La Reserva, of Costa Rica, in the middle of the rainforest reserve on the university campus, there is a swamp palm dominated by *Raphia taedigera* with a carbon pool of 15.28 Kg C m^{-2} reported in the soil, similar to the maximum value detected for Mexican flooded palms.

Table 1. Carbon sequestration in Mexican wetland soils.

Forested Wetland Type	Site (Municipality or Area, State)	Carbon Stock (Kg C m^{-2})	Location in the Map (Figure 2)	Reference
Marshes	Tecolutla and Vega de Alatorre, Veracruz	25.9	D	Marín-Muñiz [31]
Marshes	Alto Lucero and Tecolutla, Veracruz	31.0	D	Campos [32]
Marshes	Veracruz, Tabasco/Campeche, Chiapas	110	D, G, E, F	Sjögersten et al. [33]
Marshes	Yucatán Peninsula	17.8	H	Adame et al. [34]
Marshes	Cuitzeo, Michoacán	16.8	K	Paredes-García et al. [35]
Marshes	La Encrucida, Biosphere Reserve, Chiapas	33.7	F	Adame et al. [36]
Marshes	Yucatán Peninsula	21.2	H	Morales-Ojeda et al. [37]
Marshes	Río Blanco, Veracruz	68	D	Hernández et al. [38]
Flooded grassland	Jamapa y Yagual, Veracruz	28	D	Hernández et al. [38]
Flooded grassland	Veracruz, Tabasco/Campeche, Chiapas	27.1	D, E, G, F	Sjögersten et al. [33]
Flooded grassland	Estero Dulce and Boquilla de Oro, Veracruz	30.6	D	Hernández et al. [25]
Mangrove	Yucatán Peninsula	28.0	H	Morales-Ojeda et al. [37]
Mangrove	Veracruz, Tabasco/Campeche, Chiapas	93	D	Sjögersten et al. [33]
Mangrove	Oaxaca and Guerrero	66.3	M, L	Herrera et al. [39]
Mangrove	Huimanguillo and Cárdenas, Tabasco	64.7	E	Moreno et al. [40]
Mangrove	Laguna de Términos, Campeche	25.2	G	Moreno-May et al. [41]
Mangrove	Carmen city, Campeche	11.7	G	Ceron-breton et al. [42]
Mangrove	Yucatán Peninsula	66.4	H	Adame et al. [34]
Mangrove	La Encrucida, Biosphere Reserve, Chiapas	78.5	F	Adame et al. [36]
Mangrove	Pantanos de Centla, Tabasco and Campeche	45.8	E, G	Kauffman et al. [43]
Mangrove	Vega de Alatorre, Veracruz	22	D	Hernández et al. [38]
Mangrove	La Encrucida, Biosphere Reserve, Chiapas	28.4	F	Adame and Fry. [44]
Mangroves	Alvarado, Veracruz	16	D	Moreno-Casasola et al. [45]
Mangrove	Tuxpan, Veracruz	14.7	D	Santiago [46]
Mangrove	Agua Brava Lagooon, Nayarit	4.2	C	Herrera-Silveira et al. [39]
Mangrove	Bahía Tóbari, Sonora	7.9	B	Bautista-Olivas et al. [47]
Mangrove	Cuyutlán, Colima	10.2	J	Herrera-Silveira et al. [39]

Table 1. *Cont.*

Forested Wetland Type	Site (Municipality or Area, State)	Carbon Stock (Kg C m^{-2})	Location in the Map (Figure 2)	Reference
Mangrove	Nayarit	12	C	Valdés et al. [48]
Mangrove	La Paz Bay, Baja California	17.5	A	Ochoa-Gómez et al. [49]
Mangrove	Central coastal plain of Veracruz	37.5	D	Hernández and Junca-Gómez [50]
Mangrove	Paraíso Tabasco	20	E	Arias [51]
Mangrove	Península Yucatán	28.7	H	Gutiérrez-Mendoza and Herrera-Silveira[52]
Mangrove	Celestun, Yucatán	61.6	H	Herrera-Silveira et al. [53]
Mangrove	Nayarit	10	C	Valdés et al. [48]
Mangrove	Magdalena and Malandra bay. Baja California	22.5	A	Ezcurra et al. [54]
Mangrove	Sian Ka'an, Quintana Roo	45	I	Herrera-Silveira et al. [39]
Mangrove	Puerto Morelos, Yucatán	36	H	Herrera-Silveira et al. [39]
Mangrove	Aguiabampo, Sonora	3.5	B	Barreras-Apodaca et al. [55]
Mangrove	El Rabón, Nayarit	30	C	Castillo-Cruz and Rosa-Meza [56]
Mangrove	La Encrucijada, Chiapas	17.9	F	Barreras-Apodaca et al. [55]
Mangrove	Isla Arena, Campeche	30.5	G	Pech-Poot et al. [57]
Mangrove	Celestún, Yucatán	22.4	H	Pech-Poot et al. [57]
Mangrove	Cancún, Quintana Roo	26.4	I	Pech-Poot et al. [57]
Mangrove	La Encrucijada, Chiapas	6.3	F	Velázquez-Pérez et al.[58]
Mangrove	La Encrucijada, Chiapas	140		Sjögersten et al. [33]
Swamp	La Encrucida, Biosphere Reserve, Chiapas	72.2	F	Adame et al. [36]
Swamp	Jamapa, Veracruz	39	D	Hernández et al. [44]
Swamp	Alvarado, Veracruz	60	D	Moreno-Casasola et al. [45]
Swamp	Campeche y Tabasco	300	E, G	Sjögersten et al. [33]
Swamp	Tecolutla, Actopan, and Alto Lucero, Veracruz	45	D	Marín-Muñiz et al. [59]
Swamp	Alto Lucero and Tecolutla, Veracruz	52	D	Campos et al. [32]
Swamp	Tecolutla and Vega de Alatorre, Veracruz	35	D	Marín-Muñiz et al. [31]
Flooded Palm	Sian Ka'an, Quintana Roo	6.5	I	Alamilla, [60]
Flooded Palm	Alvarado, Veracruz	16	D	Moreno-Casasola et al. [45]
Flooded Palm	Jamapa, Veracruz	1.5	D	Sánchez [61]

3.5. Mean Carbon Sequestration or Carbon Pool in Mexican Wetland Soils

The use of natural ecosystems to accumulate carbon in the soil is one of the most cost-effective tools for reducing the net effect of greenhouse gas emissions and abating climate change [62]. Mexican wetlands have been studied mainly in the last 10 years regarding their high productivity in organic matter in the soil; the values reported were averaged according to the wetland type (Figure 3), finding statistical differences of (p = 0.011). The best wetland type for carbon sequestration in the soil was the swamp with 86.17 $\pm$ 35.9 Kg C m^{-2}; this value was significantly higher than flooded grassland (28.57 $\pm$ 1.04 Kg C m^{-2}; p = 0.017), mangroves (34.1 $\pm$ 5.2 Kg C m^{-2}; p = 0.010), flooded palms (8.0 $\pm$ 4.2 Kg C m^{-2}; p = 0.017) or the marshes (40.55 $\pm$ 11.5 Kg C m^{-2}; p = 0.049). These values are very important for climate models of the carbon balance. Marín-Muñiz et al. [63] argued that wetlands should be considered as a sink of carbon in the 100-year time horizon. Thus, the importance of conserving and protecting these ecosystems is worth mentioning.

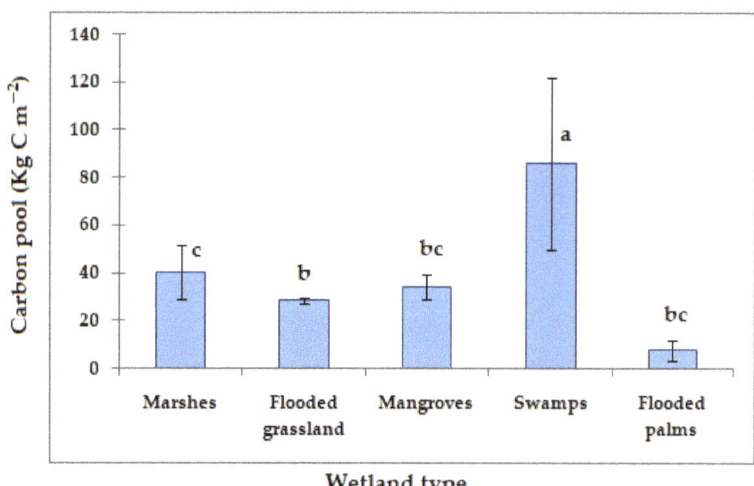

Figure 3. Carbon pool in the different wetland types of Mexico. Lines over the bars are the standard error. Letter over the bars represents statistical analysis (different letters imply values significantly different ($p < 0.05$) form each other).

On the other hand, the carbon pool in mangrove soils revealed significant differences ($p < 0.05$) with respect to the carbon pool of flooded palm and swamp zones. Flooded palm soils sequestered a similar amount of carbon to flooded grassland ($p = 0.100$) but were different compared to the other wetland types. The carbon in marshes was statistically similar to the carbon pool in flooded palms (0.990) and mangrove soils ($p = 0.447$). The importance of the vegetation regarding the quantity of carbon sequestration in the soil has been documented in some studies [31,64]; similarly, other factors such as water level and flooded conditions are also important in the carbon pool in the wetlands [8,31].

Comparing the carbon pool of Mexican wetland soils with other reported values reported of wetlands in other countries or for the Mexican terrestrial ecosystems, it was observed that swamps, marshes, mangroves, and flooded grasslands can store almost 13, 7, 6, and 5 times more carbon in the soil than Mexican terrestrial ecosystems (Table 2), respectively. Only the carbon stored in flooded palm wetlands was similar to the carbon of Mexican terrestrial ecosystems. A similar situation was observed for values of carbon pools in wetlands in the USA and Canada. Comparing the carbon pool function in European and African wetlands with the values observed in Mexican wetlands, both were similar for mangroves, marshes and flooded grasslands. Regarding the carbon pool reported in Africa, this was lesser than that detected in Mexican wetlands (Table 2).

Given the importance of carbon storage in Mexican wetlands, it is necessary to continue promoting the importance of their protection and conservation, their environmental services, and the economic value of these ecosystems. Some authors [29,30,65] in Mexico have established community participation works to rescue traditional uses of wetland resources and festivals on the importance of wetlands as awareness and appreciation strategies. In addition to the climate change threats to wetlands of North and Central America [66,67], it is time to pay attention to conserving the existing wetlands as natural treasures for the well-being of humans.

Table 2. The carbon pool in the different wetland types of Mexico versus the carbon pool in other ecosystems and wetlands in other countries.

Ecosystem	Carbon Pool (Kg Cm^{-2})	Reference
Mexican terrestrial ecosystem	6.26	Vega-López [68].
Everette USA	7.81	Crooks et al. [69].
Clayoquot Sound marsh soils Canada	8.06	Chastain and Kohfeld[70].
African Salt Marshes	10.9	Raw et al. [71].
Okavango Delta, riverine marsh, Botswana, África	1.5	Bernal and Mitsch[24].
Wetlands of Europe	15–30	Abdul et al. [72].
Swamps	86.17	
Flooded grassland	28.57	
Mangroves	34.1	This study
Flooded palms	8.0	
Marshes	40.55	

4. Conclusions

Tropical wetlands are carbon-rich ecosystems. The Mexican carbon pool in the soil was reviewed according to the different wetland types, including swamps, mangroves, flooded grasslands, flooded palms and marshes. In Mexico, the mangrove has been the ecosystem with the most studies on carbon sequestration. This is probably due to the fact that they are the type of wetland that is protected under certain laws. New studies regarding different wetland ecosystems were found in which it was observed that swamps stored more carbon in the soil compared to other wetland types; however, the flooded grasslands and marshes presented a similar carbon pool to mangroves, so new public policies on protection and conservation of this type of wetland are needed. In the case of flooded palms, the average carbon pool of only three sites was 8 Kg C m^{-2}; however, in addition to their importance and function as a carbon pool, such wetlands provide a social benefit due to the fruits of the palms in these ecosystems. Thus, this study claims that Mexican wetlands can be natural and cost-effective tools to store carbon in order to mitigate the effect of greenhouse gas emissions.

Author Contributions: Conceptualization, J.L.M.-M. and S.Z.; methodology, M.S.-H.; software, I.Z.-C.; validation, L.C.S.H., M.V.-Á. and S.Z.; formal analysis, S.Z.; investigation, S.Z. and M.V.-Á.; resources, E.A.B.-T. and S.Z.; data curation, I.Z.-C.; writing—original draft preparation, J.L.M.-M.; writing—review and editing, J.L.M.-M. and S.Z.; visualization, E.A.B.-T.; supervision, M.V.-Á.; project administration, M.S.-H.; funding acquisition, L.C.S.H. All authors have read and agreed to the published version of the manuscript.

Funding: This research received no external funding.

Institutional Review Board Statement: Not applicable.

Informed Consent Statement: Not applicable.

Conflicts of Interest: The authors declare no conflict of interest.

References

1. Pachauri, R.K.; Allen, M.R.; Barros, V.R.; Broome, J.; Cramer, W.; Christ, R.; Church, J.A.; Clarke, L.; Dahe, Q.; Dasgupta, P.; et al. *Climate Change 2014: Synthesis Report. Contribution of Working Groups I, II and III to the Fifth Assessment Report of the Intergovernmental Panel on Climate Change*; Pachauri, R.K., Meyer, L., Eds.; IPCC: Geneva, Switzerland, 2014; p. 151. Available online: https://www.ipcc.ch/site/assets/uploads/2018/02/SYR_AR5_FINAL_full.pdf (accessed on 8 December 2021).
2. Mitsch, W.J.; Gosselink, J.G. *Wetlands*, 5th ed.; John Wiley and Sons Inc.: New York, NY, USA, 2015.
3. Marín-Muñiz, J.L. *Humedales, Riñones del Planeta y Hábitat de MúltiplesEspecies*; Editora de Gobierno del Estado de Veracruz: Xalapa, VER, Mexico, 2018; Available online: https://www.sev.gob.mx/servicios/publicaciones/serie_fueraseries/Humedales_Impresion.pdf (accessed on 5 April 2022). (In Spanish)
4. Salimi, S.; Almuktar, S.A.; Scholz, M. Impact of climate change on wetland ecosystems: A critical review of experimental wetlands. *J. Environ. Manag.* **2021**, *286*, 112160. [CrossRef] [PubMed]

5.   Sood, A.; Vyas, S. Carbon Capture and Sequestration- A Review. *IOP Conf. Ser. Earth Environ. Sci.* **2017**, *83*, 12024. [CrossRef]

6.   Xu, T.; Weng, B.; Yan, D.; Wang, K.; Li, X.; Bi, W.; Li, M.; Cheng, X.; Liu, Y. Wetlands of International Importance: Status, Threats, and Future Protection. *Int. J. Environ. Res. Public Health* **2019**, *16*, 1818. [CrossRef]

7.   Mitsch, W.J.; Nahlik, A.; Wolski, P.; Bernal, B.; Zhang, L.; Ramberg, L. Tropical wetlands: Seasonal hydrologic pulsing, carbon sequestration, and methane emissions. *Wetl. Ecol. Manag.* **2009**, *18*, 573–586. [CrossRef]

8.   Bridgham, S.D.; Megonigal, J.P.; Keller, J.K.; Bliss, N.B.; Trettin, C. The carbon balance of North American wetlands. *Wetlands.* **2006**, *26*, 889–916. [CrossRef]

9.   Nahlik, A.M.; Fennessy, M.S. Carbon storage in US wetlands. *Nat. Commun.* **2016**, *7*, 13835. [CrossRef]

10.   Mitsch, W.J.; Bernal, B.; Nahlik, A.; Mander, Ü.; Zhang, L.; Anderson, C.J.; Jørgensen, S.E.; Brix, H. Wetlands, carbon, and climate change. *Landsc. Ecol.* **2012**, *28*, 583–597. [CrossRef]

11.   Available online: http://www.ramsar.org/ (accessed on 10 December 2021).

12.   Reid, W.V. Millennium Ecosystem Assessment. In *Ecosystems and Human Well-Being: Synthesis*; Island Press: Washington, DC, USA, 2005.

13.   Moreno-Casasola, P.; Infante-Mata, D.M. *Conociendo los Manglares, las Selvas Inundables y los Humedales Herbáceos*; INECOL-OIMT-CONAFOR: Xalapa, Veracruz, Mexico, 2016; 62p, Available online: https://idoc.pub/documents/conociendo-los-manglares-y-selvas-inundables-vnd1j62p9wnx (accessed on 1 May 2022). (In Spanish)

14.   Joyce, C.; Simpson, M.; Casanova, M. Future wet grasslands: Ecological implications of climate change. *Ecosyst. Health Sustain.* **2016**, *2*, e01240. [CrossRef]

15.   Kayranli, B.; Scholz, M.; Mustafa, A.; Hedmark, Å. Carbon storage and fluxes within freshwater wetlands: A critical review. *Wetlands.* **2010**, *30*, 111–124. [CrossRef]

16.   Torres-Alvarado, R.; Ramírez-Vives, F.; Fernández, F.; Barriga-Sosa, I. Methanogenesis and methane oxidation in wetlands. Implications in the global carbon cycle. *Hidrobiológica.* **2005**, *15*, 327–349. Available online: http://www.scielo.org.mx/scielo.php?script=sci_arttext&pid=S0188-88972005000300009 (accessed on 11 June 2022). (In Spanish).

17.   Wilms, R.; Sass, H.; Köpke, B.; Cypionka, H.; Engelen, B. Methane and sulfate profiles within the subsurface of a tidal flat are reflected by the distribution of sulfate-reducing bacteria and methanogenic archaea. *FEMS Microbiol. Ecol.* **2007**, *59*, 611–621. [CrossRef]

18.   Struwe, S.; Kjøller, A. Field determination of denitrification in water-logged forest soils. *FEMS Microbiol. Lett.* **1989**, *62*, 71–78. [CrossRef]

19.   A Yarwood, S. The role of wetland microorganisms in plant-litter decomposition and soil organic matter formation: A critical review. *FEMS Microbiol. Ecol.* **2018**, *94*. [CrossRef] [PubMed]

20.   Vepraskas, M.; Craft, C. *Wetland Soils. Genesis, Hydrology, Landscapes, and Classification*; CRC Press: Boca Ratón, FL, USA, 2016.

21.   Marin-Muñiz, J.L.; Alarcón, M.E.H.; Rivera, E.S.; Moreno-Casasola, P. Percepciones sobre servicios ambientales y pérdida de humedales arbóreos en la comunidad de Monte Gordo, Veracruz. *Madera Bosques.* **2016**, *22*. [CrossRef]

22.   Sjögersten, S.; Black, C.R.; Evers, S.; Hoyos-Santillan, J.; Wright, E.L.; Turner, B. Tropical wetlands: A missing link in the global carbon cycle? *Glob. Biogeochem. Cycles.* **2014**, *28*, 1371–1386. [CrossRef]

23.   Bernal, B.; Mitsch, W.J. A comparison of soil carbon pools and profiles in wetlands in Costa Rica and Ohio. *Ecol. Eng.* **2008**, *34*, 311–323. [CrossRef]

24.   Bernal, B.; Mitsch, W.J. Carbon sequestration in freshwater wetlands in Costa Rica and Botswana. *Biogeochemistry.* **2013**, *115*, 77–93. [CrossRef]

25.   Hernandez, M.E.; Marín-Muñiz, J.L.; Casasola, P.M.; Vázquez, V. Comparing soil carbon pools and carbon gas fluxes in coastal forested wetlands and flooded grasslands in Veracruz, Mexico. *Int. J. Biodivers. Sci. Ecosyst. Serv. Manag.* **2014**, *11*, 5–16. [CrossRef]

26.   Kumagai, J.A.; Costa, M.T.; Ezcurra, E.; Aburto-Oropeza, O. Prioritizing mangrove conservation across Mexico to facilitate 2020 NDC ambition. *Ambio.* **2020**, *49*, 1992–2002. [CrossRef]

27.   Alongi, D.M. Carbon sequestration in mangrove forests. *Carbon Manag.* **2012**, *3*, 313–322. [CrossRef]

28.   Harvey, G.E.; Grünwald, M. The Swamp: The Everglades, Florida, and the Politics of Paradise. *J. South. Hist.* **2007**, *73*, 766. [CrossRef]

29.   Gonzalez-Marín, R.M.; Moreno-Casasola, P.; Orellana, R.G.; Castillo, A. Palm use and social values in rural communities on the coastal plains of Veracruz, Mexico. *Environ. Dev. Sustain.* **2012**, *14*, 541–555. [CrossRef]

30.   González-Marín, R.; Moreno-Casasola, P.; Orellana, R.; Castillo, A. Traditional wetland palm uses in construction and cooking in Veracruz Gulf of Mexico. *Indian J. Tradit. Knowl.* **2012**, *11*, 408–413. Available online: http://nopr.niscair.res.in/handle/123456789/14380 (accessed on 20 April 2022).

31.   Marín-Muñiz, J.L.; Hernández, M.E.; Moreno-Casasola, P. Comparing soil carbon sequestration in coastal freshwater wetlands with various geomorphic features and plant communities in Veracruz, Mexico. *Plant Soil.* **2014**, *378*, 189–203. [CrossRef]

32.   Campos, A.C.; Hernández, M.E.; Moreno-Casasola, P.; Espinosa, E.C.; R., A.R.; Mata, D.I. Soil water retention and carbon pools in tropical forested wetlands and marshes of the Gulf of Mexico. *Hydrol. Sci. J.* **2011**, *56*, 1388–1406. [CrossRef]

33.   Sjögersten, S.; de la Barreda-Bautista, B.; Brown, C.; Boyd, D.; Lopez-Rosas, H.; Hernández, E.; Monroy, R.; Rincón, M.; Vane, C.; Moss-Hayes, V.; et al. Coastal wetland ecosystems deliver large carbon stocks in tropical Mexico. *Geoderma.* **2021**, *403*, 115173. [CrossRef]

34.   Adame, M.F.; Kauffman, J.B.; Medina, I.; Gamboa, J.N.; Torres, O.; Caamal, J.P.; Reza, M.; Herrera-Silveira, J.A. Carbon Stocks of Tropical Coastal Wetlands within the Karstic Landscape of the Mexican Caribbean. *PLoS ONE.* **2013**, *8*, e56569. [CrossRef]

35.   Paredes-García, S.S.; Moreno-Casasola, P.; Barrera, E.d.l.; García-Oliva, F.; Lindig-Cisneros, R. Biomasa y carbonoalmacenadoen un humedal continental enCuitzeo, Michoacán, Mexico. *Tecnol. Cienc. Agua.* **2021**, *12*, 416–441. [CrossRef]

36. Adame, M.F.; Santini, N.S.; Tovilla, C.; Vázquez-Lule, A.; Castro, L.; Guevara, M. Carbon stocks and soil sequestration rates of tropical riverine wetlands. *Biogeosciences.* **2015**, *12*, 3805–3818. [CrossRef]

37. Morales-Ojeda, S.M.; Herrera-Silveira, J.A.; Orellana, R. Almacenes de carbono en un paisaje de humedal cárstico a lo largo de un corredor transversal costero de la Península de Yucatán. *Madera Bosques.* **2021**, *27*, 1–18. [CrossRef]

38. Hernández, M.E.; Campos, A.; Marín-Muñiz, J.L.; Moreno-Casasola, P. Almacenes de carbono en selvas inundables, manglares, humedales herbáceos y potreros inundables. In *Servicios Ecosistémicos de las Selvas y Bosques Costeros de Veracruz*; Moreno Casasola, P., Ed.; Inecol ITTO Conafor INECC: Xalapa, VER, Mexico, 2016; pp. 121–129. Available online: https://www.itto.int/files/itto_project_db_input/3000/Technical/Servicios_Ecosostemicos_de_las_selvas_y_bosques_costeros.pdf (accessed on 6 May 2022). (In Spanish)

39. Herrera, J.; Camacho, A.; Pech, E.; Pech, M.; Ramírez, J.; Teutli, C. Dinámica del carbono (almacenes y flujos) en manglares de Mexico. *Terra Latinoamericana.* **2016**, *34*, 61–72.

40. Cáliz, E.M.; Peña, A.G.; Castorena, M.D.C.G.; Solorio, C.A.O.; López, D.J.P. Los manglares de Tabasco, una reserva natural de carbono. *Madera Bosques.* **2016**, *8*, 115–128. [CrossRef]

41. Moreno-May, G.; Cerón, J.; Cerón, R.; Guerra, J.; Amador, L.; Endañú, E. Evaluation of carbon storage potential in mangrove soils of Isla del Carmen. *Unacar Tecociencia.* **2010**, *4*, 23–39. Available online: https://www.academia.edu/2568197/Estimaci%C3%B3n_del_potencial_de_captura_de_carbono_en_suelos_de_manglar_de_isla_del_Carmen (accessed on 1 November 2021).

42. Cerón-Bretón, J.G.; Cerón-Bretón, R.M.; Rangel-Marrón, M.; Estrella-Cahuich, A. Evaluation of carbon sequestration potential in undisturbed mangrove forest in Términos Lagoon Campeche. Dev. *Energy Environ. Econ.* **2010**, *1*, 295–300. Available online: https://www.researchgate.net/publication/279903142_Evaluation_of_carbon_sequestration_potential_in_undisturbed_mangrove_forest_in_Terminos_Lagoon_Campeche (accessed on 25 November 2019).

43. Kauffman, J.B.; Trejo, H.H.; Del Carmen Jesus Garcia, M.; Heider, C.; Contreras, W.M. Carbon stocks of mangroves and losses arising from their conversion to cattle pastures in the Pantanos de Centla, Mexico. *Wetl. Ecol. Manag.* **2015**, *24*, 203–216. [CrossRef]

44. Adame, M.F.; Fry, B. Source and stability of soil carbon in mangrove and freshwater wetlands of the Mexican Pacific coast. *Wetl. Ecol. Manag.* **2016**, *24*, 129–137. [CrossRef]

45. Moreno-Casasola, P.; Hernández, M.E.; Campos, A.C. Hydrology, Soil Carbon Sequestration and Water Retention along a Coastal Wetland Gradient in the Alvarado Lagoon System, Veracruz, Mexico. *J. Coast. Res.* **2017**, *77*, 104–115. [CrossRef]

46. Santiago, L. Estimación del Potencial de Captura de Carbono (c) del Bosque de Manglar de Tumilco de Tuxpan, Veracruz, México: Tesis Maestría en Manejo de Ecosistemas Marinos y Costeros. Universidad Veracruzana: Tuxpan, Veracruz, México, 2018; Available online: https://www.uv.mx/pozarica/mmemc/files/2020/02/LuisAlbertoSantiagoMolina.pdf (accessed on 3 June 2022).

47. Bautista-Olivas, A.I.; Mendoza-Cariño, M.; Cesar-Rodriguez, J.; Colado-Amador, C.E.; Robles-Zazueta, C.A.; Meling-López, A.E. Above-ground biomass and carbon sequestration in mangroves in the arid area of the northwest of Mexico: Bahía del Tóbari and Estero El Sargento, Sonora. *Rev. Chapingo Ser. Cienc. For. Ambient.* **2018**, *24*, 387–403. [CrossRef]

48. Valdés, V.E.; Valdés, J.I.; Ordaz, V.M.; Gallardo, J.F.; Pérez, J.; Ayala, C. Evaluación del carbono orgánico en los suelos de manglares de Nayarit. *Rev. Mex. Cienc. Forestales.* **2011**, *2*, 807–815. Available online: http://www.scielo.org.mx/scielo.php?script=sci_arttext&pid=S2007-11322011000600005 (accessed on 28 October 2021). (In Spanish).

49. Ochoa-Gómez, J.G.; Lluch-Cota, S.E.; Rivera-Monroy, V.H.; Lluch-Cota, D.B.; Troyo-Diéguez, E.; Oechel, W.; Serviere-Zaragoza, E. Mangrove wetland productivity and carbon stocks in an arid zone of the Gulf of California (La Paz Bay, Mexico). *For. Ecol. Manag.* **2019**, *442*, 135–147. [CrossRef]

50. Hernández, M.E.; Junca-Gómez, D. Carbon stocks and greenhouse gas emissions (CH_4 and N_2O) in mangroves with different vegetation assemblies in the central coastal plain of Veracruz Mexico. *Sci. Total Environ.* **2020**, *741*, 140276. [CrossRef] [PubMed]

51. Arias, X. *Carbono, Nitrógeno y Azufre en Manglares de Paraíso Tabasco*; Tesis Ingeniero en Restauración Forestal; Universidad Au-tónoma: Chapingo, Mexico, 2018; Available online: http://dicifo.chapingo.mx/pdf/tesislic/2018/Arias_Vel%C3%A1zquez_Xochitl_Rosario.pdf (accessed on 9 November 2021).

52. Gutiérrez-Mendoza, J.; Herrera-Silveira, J. Almacenes de Carbono en manglares de tipo Chaparro en un escenario cárstico. In *Estado Actual del Conocimiento del Ciclo del Carbono Y Sus Interacciones en México: Síntesis A 2014*; Paz, F., Wong, J., Eds.; Programa Mexicano del Carbono; Centro de investigación y estudios avanzados del Instituto Politécnico Nacional, unidad Mérida y Centro de investigación y asistencia en tecnología y diseño del estado de Jalisco: Jalisco, México, 2014; pp. 405–414. Available online: http://pmcarbono.org/pmc/publicaciones/Libro_Merida_2014_PMC_ISBN-web.pdf (accessed on 20 October 2021). (In Spanish)

53. Herrera-Silveira, J.; Teutli-Hernández, C.; Caamal-Sosa, J.; Pech-Cardenas, M.; Pech-Poot, E.; Carrillo-Baeza, L.; Zenteno, K.; Erosa, J.; Pérez, O.; Gamboa, S. Almacenes y flujos de carbono en diferentes tipos ecológicos de manglares en Celestun, Yucatán. In *Estado Actual del Conocimiento del Ciclo del Carbono y sus Interacciones en México: Síntesis a 2018*; Paz, F., Torres, R., Velázquez, Eds.; Programa Mexicano del Carbono; Centro de investigación científica y de educación superior de Ensenada, Universidad Autónoma de Baja California: Baja, México, 2018; pp. 219–225.

54. Ezcurra, P.; Ezcurra, E.; Garcillán, P.P.; Costa, M.T.; Aburto-Oropeza, O. Coastal landforms and accumulation of mangrove peat increase carbon sequestration and storage. *Environ. Sci.* **2016**, *113*, 4404–4409. [CrossRef] [PubMed]

55. Barreras-Apodaca, A.; Sánchez-Mejía, Z.; Bejarano, M.; Méndez-Barroso, L.; Borquez-Olguín, R. Carbono almacenado en la capa superficial de suelo de dos manglares geográficamente contrastantes. In *Estado actual del Conocimiento del Ciclo del Carbono y sus Interacciones en México: Síntesis a 2017*; Paz, F., Velázquez, A., Rojo, M., Eds.; Programa Mexicano del Carbono; InstitutoTecnológico de Sonora: Álamos, Mexico, 2017; pp. 258–264. (In Spanish)

56. Castillo-Cruz, I.; De la Rosa-Meza, K. Cuantificación de carbono en manglares en El Rabón, dentro de la RB Marismas Nacionales, Nayarit. In *Estado Actual del Conocimiento del Ciclo del Carbono y sus Interacciones en México: Síntesis a 2017*; Paz, F., Velázquez, A., Rojo, M., Eds.; (In Spanish). Programa Mexicano del Carbono; InstitutoTecnológico de Sonora: Álamos, México, 2017; pp. 252–257.

57. Pech-Poot, E.; Herrera-Silveira, J.; Caamal-Sosa, J.; Cortes-Balan, O.; Carrillo-Baeza, L.; Teutli-Hernández, C. Carbono en sedimentos de manglares de ambientes cársticos: La Península de Yucatán. In *Estado Actual del Conocimiento del Ciclo del Carbono y susInteraccionesen México: Síntesis a 2016*; Paz, F., Torres, M., Eds.; Programa Mexicano del Carbono; Universidad Autónoma del Estado de Hidalgo: Campo de Tiro, México, 2016; pp. 336–343. (In Spanish)

58. Velázquez-Pérez, C.; Tovilla-Hernández, C.; Romero-Berny, E.I.; De Jesús-Navarrete, A. Estructura del manglar y su influencia en el almacén de carbono en la Reserva La Encrucijada, Chiapas, México. *Madera Bosques.* **2019**, *25*. [CrossRef]

59. Marín-Muñiz, J.L.; Hernández, M.E.; Moreno-Casasola, P. Soil carbon sequestration in coastal freshwater wetlands of Veracruz. *Trop. Subtrop. Agroecosyst.* **2011**, *13*, 365–372. Available online: http://www.revista.ccba.uady.mx/ojs/index.php/TSA/article/view/1336 (accessed on 25 May 2022).

60. Alamilla, S. *Gradientes de Carbono por Tipo de Suelo y Vegetación en Quintana Roo*; Tesis Licenciatura en Manejo de Recursos Naturales. Chetumal, Quitana: Roo, México, 2018; Available online: http://risisbi.uqroo.mx/bitstream/handle/20.500.12249/1973/S590.2018-1973.pdf?sequence=1&isAllowed=y (accessed on 1 October 2021). (In Spanish)

61. Sánchez, E. *Caracterización de tres Propiedades del Suelo en Humedales Transformados a Potreros, en el Municipio de Jamapa, Veracruz y su Entorno. Tesis especialista en diagnóstico y gestión Ambiental*; Facultad de Ciencias Químicas; Universidad Veracruzana: Xalapa, Veracruz, México, 2015; Available online: https://cdigital.uv.mx/bitstream/handle/123456789/42319/SanchezGarciaEdgar.pdf?sequence=2&isAllowed=y (accessed on 6 October 2021). (In Spanish)

62. Bernal, B.; Mitsch, W.J. Carbon Sequestration in Two Created Riverine Wetlands in the Midwestern United States. *J. Environ. Qual.* **2013**, *42*, 1236–1244. [CrossRef]

63. Marin-Muñiz, J.L. Carbon balance in tropical freshwater wetland on the coastal plain of the Gulf of Mexico. *Limnetica* **2020**, *39*, 653–665. [CrossRef]

64. Ji, H.; Han, J.; Xue, J.; Hatten, J.A.; Wang, M.; Guo, Y.; Li, P. Soil organic carbon pool and chemical composition under different types of land use in wetland: Implication for carbon sequestration in wetlands. *Sci. Total Environ.* **2020**, *716*, 136996. [CrossRef]

65. Zitácuaro-Contreras, I. Administración de Humedales Artificiales con Perspectiva de Género como Estrategia Sustentable para el Saneamiento de Aguas Residuales Municipales. Tesis de Doctorado, El Colegio de Veracruz, Veracruz, Mexico, 2021.

66. Mitsch, W.J.; Hernandez, M.E. Landscape and climate change threats to wetlands of North and Central America. *Aquat. Sci.* **2012**, *75*, 133–149. [CrossRef]

67. Martínez, A.; Manzanilla, S.; Hidalgo, J. Vulnerability to Climate Change of Marine and Coastal Fisheries in Mexico. *Atmosfera.* **2011**, *24*, 103–123. Available online: http://www.scielo.org.mx/pdf/atm/v24n1/v24n1a8.pdf (accessed on 5 May 2022).

68. Vega-López, E. Valor Económico Potencial de las Áreas Naturales Protegidas Federales de México como Sumideros de Carbono. The Nature Conservancy-Mexico. Available online: https://docplayer.es/67251566-Valor-economico-potencial-de-las-areas-naturales-protegidas-federales-de-mexico-como-sumideros-de-carbono.html (accessed on 11 June 2022). (In Spanish).

69. Crooks, S.; Rybczyk, J.; O'Connell, K.; Devier, D.L.; Poppe, K.; Emmett-Mattox, S. *Coastal Blue Carbon Opportunity Assessment for the Snohomish Estuary: The Climate Benefits of Estuary Restoration*; Report by Environmental Science Associates, Western Washington University, Earth Corps, and Restore America's Estuaries: Bellingham, WA, USA, 2014. [CrossRef]

70. Chastain, S.; Kohfeld, K.E. Blue Carbon in tidal wetlands of the pacific coast of Canada: Commission for Environmental Cooperation's (CEC's) 2015–2016 project, North American Blue Carbon: Next Steps in Science for Policy. 2017. Available online: http://www.cec.org/publications/blue-carbon-in-tidal-wetlands-of-the-pacific-coast-of-canada/ (accessed on 20 May 2022).

71. Raw, J.; Julie, C.; Adams, J. A comparison of soil carbon pools across a mangrove-salt marsh ecotone at the southern African warm-temperate range limit. *S. Afr. J. Bot.* **2019**, *127*, 301–307. [CrossRef]

72. Abdul, D.; Marín, A.; Trombetti, M.; San Roman, S. *Carbon Pools and Sequestration Potential of Wetlands in the European Union*; European Topic Centre on Urban, Land and Soil Systems, Viena and Malaga: Catalonia, Spain, 2021; ISBN 9783200074330. Available online: https://www.eionet.europa.eu/etcs/etc-di/products/etc-uls-report-10-2021-carbon-pools-and-sequestration-potential-of-wetlands-in-the-european-union (accessed on 1 June 2022).

Article

Determining the Effects of Compost Substitution on Carbon Sequestration, Greenhouse Gas Emission, Soil Microbial Community Changes, and Crop Yield in a Wheat Field

Hongzhi Min [1], Xingchen Huang [1], Daoqing Xu [2], Qingqin Shao [1], Qing Li [3], Hong Wang [1,*] and Lantian Ren [1,*]

[1] Engineering Research Center for Smart Crop Planting and Processing Technology, Anhui Science and Technology University, Chuzhou 233100, China
[2] Cotton Research Institute of Anhui Academy of Agricultural Sciences, Hefei 230036, China
[3] College of Plant Science & Technology, Hua Zhong Agricultural University, Wuhan 430070, China
* Correspondence: hongw_21@163.com (H.W.); renlt@ahstu.edu.cn (L.R.)

Abstract: Compost produced by straw and livestock and poultry manure under the action of microorganisms is one of the main forms of organic alternative fertilizers at present. The present study explored the effects of compost substitution on soil greenhouse gas emissions, soil microbial community changes, and wheat yield to determine the best substitution ratio for reducing greenhouse gas emissions and soil microbial community changes and increasing wheat yield. Using the single-factor randomized block trial design, four treatments were employed, the characteristics of greenhouse gas emission, yield and yield components, and the changes of soil microbial community under different compost substitution ratio in the whole wheat growing season were determined by static box-gas chromatography. During the wheat season, both CO_2 and N_2O emissions were reduced, whereas CH_4 emission was increased. That all treatments reduced the Global Warming Potential (GWP) and Greenhouse gas emission intensity (GHGI) in wheat season compared with T0. Compost substitution can alleviate the global warming potential to some extent. Under the condition of compost substitution, the wheat yield under T2 and T3 increased significantly compared with that under the control; however, the spike number and 1000-grain weight did not differ significantly among the treatments. When compost replacement was 30%, the yield was the highest. Under different ratios of compost substitution, the microbial communities mainly comprised Proteobacteria, Actinobacteria, Firmicutes, Patescibacteria, Chloroflexi, Acidobacteria, Bacteroidetes, Gemmatimonadetes, and Verrucomicrobia. The soil microbial community structure differed mainly due to the difference in the compost substitution ratio and was clustered into different groups. In conclusion, to achieve high wheat yield and low greenhouse gas emissions, compost replacement of 30% is the most reasonable means for soil improvement and fertilization.

Keywords: composting; wheat; yield; greenhouse gas; soil micro-organism

Citation: Min, H.; Huang, X.; Xu, D.; Shao, Q.; Li, Q.; Wang, H.; Ren, L. Determining the Effects of Compost Substitution on Carbon Sequestration, Greenhouse Gas Emission, Soil Microbial Community Changes, and Crop Yield in a Wheat Field. *Life* **2022**, *12*, 1382. https://doi.org/10.3390/life12091382

Academic Editor: Tanya Soule and Ling Zhang

Received: 1 August 2022
Accepted: 30 August 2022
Published: 5 September 2022

Publisher's Note: MDPI stays neutral with regard to jurisdictional claims in published maps and institutional affiliations.

1. Introduction

In China, obtaining high crop yield requires the application of a large amount of chemical fertilizers, leading to an increase in the production cost. Moreover, the long-term application of chemical fertilizers leads to soil acidification and consolidation, and the utilization rate of nitrogen fertilizer is low. Increasing use of chemical fertilizers has deteriorated soil fertility and quality of agricultural products, which has not only affected the comprehensive production capacity of soil but also seriously affected the ecological environment [1]. High-temperature composting of crop straw is one of the effective methods to replace chemical fertilizers [2–4]. Straw composting can increase the content of soil organic matter, improve soil physiology and microbial characteristics, promote microbial fixation of nutrients, reduce nutrient loss [5,6], and reduce CH_4 emissions [7]. However, at the same time, it can lead to an increase in N_2O emissions [8,9], and promote the recycling of natural resources. Therefore, transformation from inorganic agriculture to ecological and green agriculture is necessary [10]. Use of compost can promote resource utilization of agricultural waste, reduce the use of chemical fertilizers, and improve crop yield and quality, indicating that it has a high practical application value. The combined application of organic and chemical fertilizers is the most efficient approach for improving soil quality, increasing crop yield, and reducing greenhouse gas emissions. The use of chemical fertilizer alone in farmland can lead to an increase in greenhouse gas emissions [11]. Wang et al. [12] reported that the combined application of bio-organic and chemical fertilizers could increase the number of tillers of wheat by 12.4–18.9% and significantly increase the number of effective spikes, grains per spike, and 1000-grain weight of wheat. Other studies have reported that composting can promote plant growth and development and increase crop yield [13]. According to Chaoui et al. [14], composting can increase the number of micro-organisms in soil and improve the availability of soil nutrients. The effect of straw returning on N_2O emission is uncertain in China and abroad; straw returning has been reported to not only promote N_2O emission [15], but also inhibit N_2O emission [16]. Additionally, a few studies reported that using straw compost instead of some part of chemical fertilizer can increase crop yield. The present study investigated the effect of replacing some part of chemical fertilizer with straw compost in terms of differences in growth indices and yield and quality of wheat with different straw compost substitution ratios. The findings may be useful in promoting resource utilization of agricultural waste and reducing the use of chemical fertilizers, thereby providing a theoretical basis for wheat high-yield cultivation techniques.

2. Materials and Methods

2.1. Experimental Field and Design

This experiment was conducted in the Anhui University of Science and Technology (E117°33′39″, W32°52′49″) from December 2020 to June 2021, during the field experiment average temperature of 17–8 °C, during the field experiment rainfall of 145 mm. The former stubble of the experiment was rice; the soil was yellow cinnamon soil. The organic matter content in the 0–20-cm soil layer is 20.8 g/kg, with the alkali-hydrolysable nitrogen content of 110.9 mg/kg, available phosphorus content of 25.8 mg/kg, and available potassium content of 115.2 mg/kg.

2.2. Test Materials

2.2.1. Production of Compost

The compost used in the experiment comprised cow manure and rice straw harvested after ripening according to the mass ratio of 2:1, and the raw materials were pre-treated before composting. Large chunks of cow dung were simply crushed, and the straw was crushed into small pieces (smaller than 1 cm), which were stacked and set aside. We mixed layers by layers and sprayed water while mixing; the pile was turned repeatedly with a lawn grabber until it was evenly mixed. The overall moisture content was adjusted to 55–65%. The C/N ratio was 25:30, and the oxygen content was 8–18%. The pH was maintained at 6.5–8.0. A small pile with a bottom width of 1.5 m, a height of 1–1.2 m, and a length of 2–3 m was fermented at a high temperature for 50–60 days, and a compost with weak ammonia and manure odor was obtained; the color of rice straw and cow manure changed to granular black–brown after maturity. The nitrogen, P_2O_5, K_2O, and organic matter contents of the compost were 1.03%, 0.87%, 1.35%, and 47.8%, respectively, and the pH was 6.67.

2.2.2. Experimental Wheat

The experimental wheat variety 'Huaimai 44' was selected. The seedling stage of this variety is semi-creeping; the leaves are short and green in color. After attaining maturity, the plants become compact, their sword leaf is erect, and the ripening phase is better. The plants exhibit a strong tillering ability, a large number of panicles, earlier yellowing, and strong ability of cold and lodging resistance.

2.3. Single Factor Experimental Design

The types of straw compost used as a substitute for chemical fertilizer treatment (T) were as follows: 1. conventional fertilizer control (T0 treatment; compound fertilizer ($N18\%$-P_2O_5-18%-$K_2O18\%$): 600 kg/hm^2, urea: 300 kg/hm^2); 2. composting to replace 10% chemical fertilizer (T1 treatment; compound fertilizer ($N18\%$-P_2O_5-18%-$K_2O18\%$): 540 kg/hm^2, urea: 270 kg/hm^2, straw compost: 3 t/hm^2); 3. composting to replace 20% chemical fertilizer (T2 treatment; compound fertilizer ($N18\%$-P_2O_5-18%-$K_2O18\%$): 480 kg/hm^2, urea: 240 kg/hm^2, compost: 6 t/hm^2); and 4. composting to replace 30% chemical fertilizer (T3 treatment; compound fertilizer ($N18\%$-P_2O_5-18%-$K_2O18\%$): 420 kg/hm^2, urea: 210 kg/hm^2, compost: 9 t/hm^2). Each treatment was performed in triplicate. On 5 December 2020, the seeds were sown with an equal row spacing of 15 cm. The sowing rate of 14 m^2 was 52.50 g/row and that of 15 m^2 was 56.25 g/row. Except for different fertilization treatments, the other field-management measures adopted the unified management mode of local high-yield wheat field in each district. Wheat was harvested on 9 June 2021.

2.4. Observation Indicators and Methods

2.4.1. Collection and Determination of Greenhouse Gases

The closed static chamber method was used to measure the emission fluxes of CH_4, CO_2, and N_2O in wheat field. The sampling device consists of three parts made of opaque organic plastic, namely top box, middle box, and base. In the inner surface of the box (50 cm × 50 cm × 50 cm), a small fan is placed that ensures the uniform distribution of gas. The outer surface consists of a sponge aluminum foil, which is used for reflection and heat insulation, and a thermometer, which is inserted in the upper part of the box and measures the temperature of the box. The upper surface of the base (50 cm × 50 cm × 25 cm) has grooves. During gas production, the grooves are sealed with water, and the box is covered to prevent air leakage between the box and base. The sample was collected through a syringe from the sampling port. The base was buried between the crop rows, exposing only grooves, and it did not move throughout the wheat growing season.

Gas samples were collected during the jointing stage (13 March 2021), booting stage (10 April 2021), flowering stage (24 April 2021), filling stage (9 May 2021), and mature stage (6 May 2021). The sampling time was fixed from 8:00 am to 11:00 am. Water injection in the base tank was closed during gas production; the small fan box at the top of the top box was opened and placed on the base, and the samples were collected every 5 min. Air was extracted from the gas inlet by using a 60 mL syringe at the time intervals of 5, 10, and 15 min, and 60 mL gas was injected into the vacuum air bag for preservation. A total of 3 gas samples were collected. Simultaneously, the temperatures of air and the box were observed and recorded.

The collected samples were analyzed using the Agilent7890B (Agilent, Waldbronn, Germany) gas chromatograph in the laboratory. The analysis column was Porpak.Q packed column. The temperature of the column box was 40 °C, and the carrier gas was high purity N_2. N_2O electron capture detector (ECD) was used. The working temperature was 300 °C, and the lowest detection limit was 32 $\mu g \cdot kg^{-1}$. CO_2 and CH_4 were determined using hydrogen detector (FID). The working temperature was 300 °C. The lowest detection limit of CO_2 and CH_4 was 4 and 0.2 $mg \cdot kg^{-1}$, respectively. The standard gas of the national standard metrology center was used to calibrate the gas chromatograph, and the external standard working curve was obtained for all 60 samples. The unobserved daily emission fluxes were calculated using the interpolation method; further, the measured values were summed up with the calculated values, and the respective emissions of CO_2, CH_4, and N_2O were obtained. The greenhouse gas emission fluxes were calculated using the following formula:

$$F = \frac{dc}{dt} \times \frac{M}{V_0} \times \frac{P}{P_0} \times \frac{T_0}{T} \times H \tag{1}$$

where F is the greenhouse gas emission flux (mg/(m^2·h)); dc/dt is the slope of the regression curve of the gas volume fraction with time during sampling; M is the molar mass of the gas (g/mol); V_0 is the molar volume of the gas under the standard gas (22.41 L/mol); P and P_0 are the air pressure at the sampling point (Pa) and the air pressure at the standard state (101,325 Pa), respectively; T and T_0 are the absolute temperatures at the time of sampling (K) and in the standard state (273.15 K), respectively; and H is the height of the sampling box (m).

The cumulative greenhouse gas emissions during the wheat growth period were obtained by multiplying the average greenhouse gas emission flux during the two adjacent sampling periods with the time interval between two sampling periods.

2.4.2. Calculation of Comprehensive Warming Potential and Greenhouse Gas Emission Intensity

Using comprehensive warming potential (*GWP*), other greenhouse gas emissions can be converted into equivalent CO_2, and their climates can be compared [17]. In this study, the *GWP* was used to express the potential effects of different greenhouse gases on global warming. On the 100-year warming scale, the warming potential of CH_4 and N_2O is 28 and 265 times of that of CO_2. The *GWP* can be calculated from the cumulative emission of each gas and its corresponding temperature increasing coefficient.

$$GWP = CO_2 + 25 \times CH_4 + 298 \times N_2O \tag{2}$$

GHGI was used to evaluate the comprehensive greenhouse effect of each treatment. The algorithm of *GHGI* is:

$$GHHI = GWP/Y \tag{3}$$

where *GHGI* is the greenhouse gas emission intensity of the treatment, and *Y* is the crop yield of each treatment (kg·hm^{-2}).

2.4.3. Determination of Yield

Representative 1 m, 2-row samples were selected from each plot during the wheat maturity stage. The average spike number was investigated, and the spike number per unit area was calculated. Overall, 20 plants were selected from each plot for indoor seed testing, and the average grain number per spike and 1000-grain weight were investigated.

2.4.4. Soil sample Collection and Treatment

Soil samples were collected in June 2021 (after wheat harvest). After removing the surface floating soil, according to the 'Z'-shaped sampling route, 0–20 cm soil cores were randomly selected using a sampler (diameter 2.5 cm). Then, the samples were mixed evenly to remove impurities such as gravel and plant residual roots. The fresh soil samples were divided into two parts: one part was air-dried for the determination of soil physical and chemical properties, and the other part was preserved at $-80\,^\circ$C for the extraction of soil genomic DNA.

2.4.5. Extraction and Sequencing of Soil DNA

In total, 0.5 g of soil stored in a refrigerator at $-80\,^\circ$C was taken, and total DNA of the soil was extracted using the PowerSoil$^\circledR$ DNA extraction kit (MoBio Laboratories, Inc., Carlsbad, CA, USA), according to the manufacturer's instructions. DNA content was quantified using a NanoDrop spectrophotometer (ND-2000, Thermo Scientific, Waltham, MA, USA), diluted to 10 ng·μL^{-1}, and stored in a refrigerator at $-20\,^\circ$C util use. Each DNA sample had 9 PCR repeats and included 2 negative controls without DNA template. The PCR reaction system and reaction procedure were based on the method described by Zhao et al. [18]. The PCR product was purified through gel cutting, and the concentration was determined using PicoGreen$^\circledR$ (Promega, Madison, WI, USA). After equimolar mixing, the product was sent to Guangdong MAGIGENE Technology limited company for sequencing.

2.4.6. Statistical Analysis

The test data were processed and analyzed using MS Excel 2010 and DPS7.05 software, and the significance was tested using the LSD method. The relationship between soil physical and chemical properties and greenhouse gases and their functional bacteria was analyzed using the 'psych' package in R, and the principal coordinate analysis (pCoA) and redundancy analysis (RDA) of community structure differences and environmental factors on community structure were completed using 'vegan' in R. The 'pheatmap' and 'ggplot2' packages in Excel and R were used for drawing.

3. Results

3.1. Effects of Different Compost Substitution Ratios on Soil N_2O, CH_4, and CO_2 Emissions

Figure 1 shows the dynamic changes in N_2O, CH_4, and CO_2 emission fluxes during the wheat growth period under different compost substitution ratios. As shown in Figure 1, under different compost substitution ratios, the flux emission of each treatment displayed a unimodal pattern, and the emission flux demonstrated first an increasing trend and then a decreasing trend with the seasonal change; the maximum emission flux was obtained at the jointing stage after sowing. The emission of N_2O after T1 treatment was higher than that after T0 treatment, and the amount of N_2O uptake after T2 treatment was higher than that after T3 treatment.

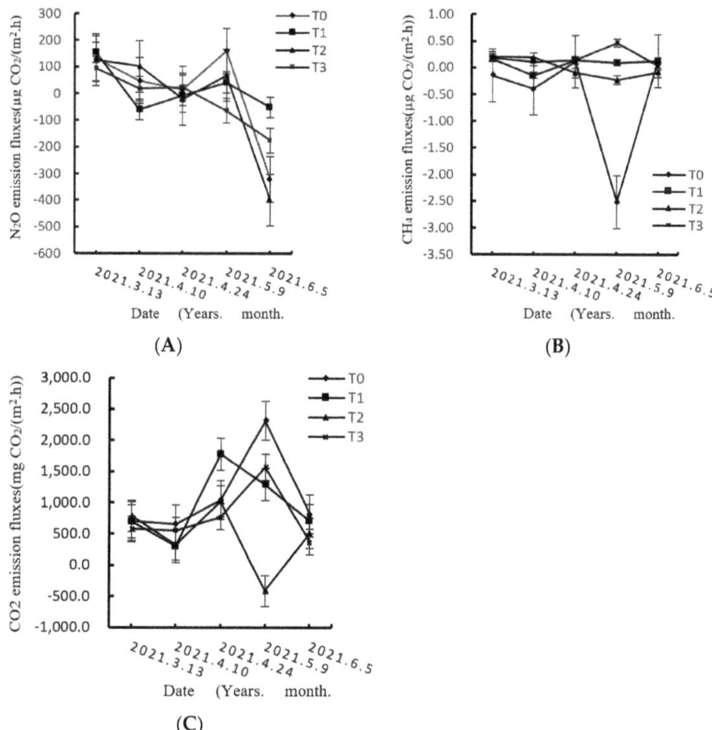

Figure 1. Dynamics of (**A**) N_2O flux, (**B**) CH4 flux, (**C**) CO_2 flux under different treatments.

The flux of CH_4 was both positive and negative, and it fluctuated rapidly. As shown in Table 1, during the wheat growth period, the cumulative emission of CH_4 in each treatment was negative, suggesting absorption. The amount of absorption under T0 treatment was higher than that under T2 treatment, and there was a positive value. The emission under

T3 treatment was higher than that under T1 treatment. Outcomes under each treatment differed significantly from those under T0 ($p < 0.05$).

Table 1. Accumulation of greenhouse gas emissions, comprehensive greenhouse gas warming potential, and greenhouse gas emission intensity under different compost substitution ratios.

Treatment	N_2O Cumulative Emission (kg N·hm^{-2})	CH_4 Cumulative Emission (kgC·hm^{-2})	CO_2 Cumulative Emission (kg C·hm^{-2})	GWP (kgCO$_2$-eq·hm^{-2})	GHGI (g CO$_2$-eq·kg^{-1})
T0	0.32 ± 0.01 b	-28.21 ± 0.71 d	$55,731.98 \pm 1393.30$ a	$55,121.75$ a	6.93 a
T1	0.74 ± 0.02 a	3.74 ± 0.09 b	$48,128.64 \pm 1203.22$ b	$48,442.61$ b	6.56 a
T2	-1.26 ± 0.03 c	-0.22 ± 0.01 c	$22,451.49 \pm 561.29$ d	$22,070.57$ d	2.63 c
T3	-1.07 ± 0.03 c	9.39 ± 0.23 a	$38,729.71 \pm 968.24$ c	$38,645.56$ c	4.01 b

Note: Values followed by different lowercase letters in a column indicate significant difference among treatments at the 0.05 level.

The emission flux of CO_2 from each treatment clearly changed during the growth period of wheat, showing a single peak. The emission flux of CO_2 increased gradually after low performance at the booting stage, reached the peak at the filling stage, and then decreased. According to Table 1, the cumulative CO_2 emission of each treatment was in the order T0 > T1 > T3 > T2 during the wheat growth period. The cumulative emission of 55,731.98 kgC·hm^{-2} under T1, T2, and T3 decreased significantly by 13.6%, 59.7%, and 30.5%, respectively, compared with that under T0 treatment ($p < 0.05$). Outcomes under each treatment differed significantly from those under T0 ($p < 0.05$).

3.2. GWP and GHGI under Different Compost Substitution Ratios

The estimated results of GWP and GHGI under different compost substitution ratios during the wheat growth period are presented in Table 1. T0 treatment made the greatest contribution to the comprehensive warming potential of farmland. The comprehensive warming potential of TI, T2, and T3 decreased by 13.8%, 149.8%, and 42.6%, respectively, compared with that of T0. With significant variations in the yield under each treatment, both GHGI and GWP exhibited a different trend. Under different compost substitution ratios, the GHGI under different treatments was in the order T1 > T3 > T2 > T0. The GHGI under T1 was the highest (42.53 g·CO$_2$-eq kg^{-1}), although T1 exhibited the lowest yield among all treatments.

3.3. Effects of Different Compost Substitution Ratios on Wheat Yield and Yield Components

The effects of different compost substitution ratios on wheat yield, quality, and yield components are shown in Table 2. The number of grains per spike in T2 was significantly higher than those in T0, T1, and T3; however, the difference in the number of grains per spike between T0 and T3 was nonsignificant. The number of panicles in T1 was significantly lower than those in T0, T2, and T3. In terms of the 1000-grain weight, T0, T1, and T2 exhibited no significant difference; however, 1000-grain weight after T3 was significantly higher than that after other treatments. Overall, 30% substitution of compost significantly increased the wheat yield, and the yield after T3 (9647.01 kg·hm^{-2}) increased by 21.3% compared with that under T0. The yield under T0 and T2 was significantly higher than that under T1; however, it did not differ significantly between T0 and T2.

Table 2. Effects of different compost substitution ratios on wheat yield and yield components.

Treatments	Spike Number (Ear/each)	Kernels Per Spike (Per Spike)	1000-Grainweight (g)	Yield (kg/hm^2)
T0	620.00 ± 2.65 a	32.38 ± 1.00 c	39.61 ± 4.03 b	7950.35 ± 33.93 b
T1	566.00 ± 4.93 b	35.90 ± 0.29 b	36.32 ± 1.61 b	7380.70 ± 64.32 c
T2	624.67 ± 16.34 a	38.92 ± 0.67 a	34.41 ± 2.55 b	8363.91 ± 218.83 b
T3	630.67 ± 11.84 a	31.18 ± 0.74 c	49.07 ± 0.59 a	9647.01 ± 181.06 a

Note: Values followed by different lowercase letters in a column indicate significant difference among treatments at the 0.05 level.

3.4. Effect of Microbial Community Composition under Different Compost Substitution Ratios

Figure 2A shows the relative abundance of microbial community structure under different compost substitution ratios. The microbial community under different treatments comprised mainly Proteobacteria, Actinobacteria, Firmicutes, Patescibacteria, Chloroflexi, Acidobacteria, Bacteroidetes, Gemmatimonadetes, and Verrucomicrobia. With an increase in the compost substitution ratio, the relative abundance of Actinomycetes under T0 increased significantly compared with those under T1, T2, and T3 ($p < 0.05$), but it decreased when the substitution ratio increased to 30%. The relative abundance of Proteus was significantly lower under T2 than those under T0, T1, and T3 ($p < 0.01$). The relative abundance of thick-walled bacteria was significantly higher under T2 than those under T0, T1, and T3 ($p < 0.01$). The relative abundance of Patescibacteria was significantly lower under T3 than those under T0, T1, and T2 ($p < 0.01$). The relative abundance of Campylobacter was significantly higher under T0 and T3 than that under T1 and T2 ($p < 0.01$). Likewise, the relative abundance of acid bacilli was significantly lower under T1 and T2 than that under T0 and T3 ($p < 0.01$).

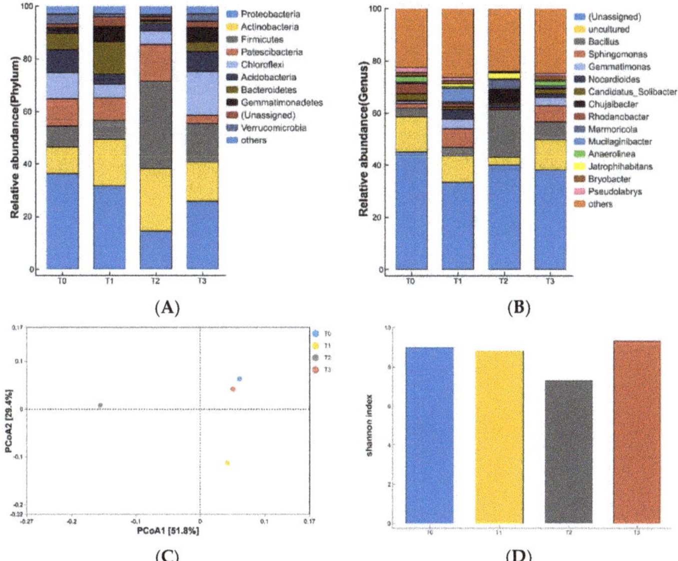

Figure 2. The relative abundance (**A**), the relative abundance of dominant functional bacteria (>1%) (**B**), the PCoA (**C**), and fragrance index (**D**) of the differences in community structure under different compost substitution ratios.

Figure 2B shows the relative abundance of microbial communities under different compost substitution ratios at the genus level, including 13 dominant functional bacteria (relative abundance more than 1%) and various α-, β-, and γ-type amoeba (Proteobacteria). The relative abundance of *Bacillus* was significantly higher under T2 than those in T0, T1, and T3. For *Sphingomonas*, the relative abundance was significantly higher under T1 and T3 than that under T0 and T2. The relative abundance of *Gemmatimonas* was significantly lower under T0 and T2 than that under T1 and T3. The relative abundance of *Nocardioides* was significantly higher under T1 than that under T0, T2, and T3 ($p < 0.01$).

Figure 2C presents the OTU data based on 97% similarity, and the weighted (Weighted Unifrac) algorithm was used to stack the results of pCoA of microbial community structure under different compost substitution ratios. Among them, the variance contribution rates of the first and second principal components were 51.8% and 29.4%, respectively, and the cumulative variance contribution rate was 81.2%. On the first axis, T1 and T2 could be well separated on the first axis, and the amount of variance explained was 51.8%.

Figure 2D shows the results of Shannon index diversity analysis of the microbial community structure. The higher the Shannon index, the richer is the microbial community diversity, and vice-a-versa. The Shannon index under T2 was significantly lower than that under other three treatments ($p < 0.05$); however, no significant difference was observed in terms of diversity among the other three treatments.

3.5. Redundancy Analysis of the Microbial Community Structure, Environmental Factors, and Soil Nutrient Content

The correlation between the microbial community structure, environmental factors, and soil nutrient content was further examined, and the results are shown in Figure 3. A total of 80% of the relationship between microbial community structure, environmental factors, and soil nutrient content could be explained by the two axes, reflecting the effects of environmental factors and soil nutrient content on the soil microbial community structure. Further analysis indicated that the amount of variance explained by RDA1 and RDA2 was 67.1% and 12.9%, respectively, and the responses of soil microbial communities to environmental factors and soil nutrients were different under different compost treatments. Among them, T0, T1, and T3 positively correlated with AN, N_2O, SOM, and CO_2, whereas T2 positively correlated with TN and CH_4.

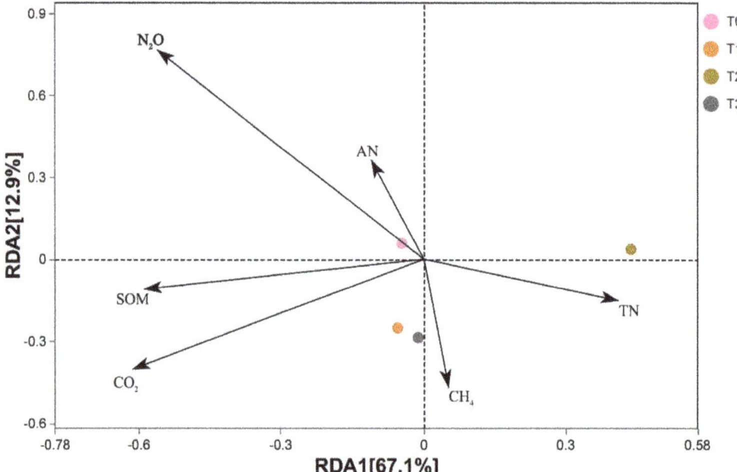

Figure 3. Redundancy analysis of environmental factors, soil nutrient content, and soil microbial community structure.

4. Discussion

Many studies have reported that the application of compost can improve the physical and chemical properties of soil [19–22], increase the content of soil nutrients [19,20,23,24], and increase the diversity and abundance of soil micro-organisms [14,25–27]. Moreover, compost application is beneficial to the transformation and release of soil nutrients [14]. Greenhouse gas emissions from farmland are affected by many factors such as climatic conditions, soil characteristics, fertilizer types, and agricultural management measures [27–30].

N_2O is emitted from soil by the production of ammonium salts during the process of oxidation involving nitrifying bacteria [31]. Some studies have suggested that compost substitution can change the soil characteristics [32], and stimulate the soil microbial activity to increase soil microbial biomass, thereby promoting denitrification to increase N_2O

emission [33–35]. A study by Guo et al. [36] indicated that compost substitution can significantly reduce soil N_2O emission, consistent with the findings of the present study.

The emission flux of N_2O is relatively high at the jointing stage, which may be related to fertilization. Fertilization can provide a large amount of available nitrogen to soil micro-organisms and accelerate the processes of microbial nitrification and denitrification, thus promoting N_2O emission [37]. Under the compost substitution ratio of 20%, the cumulative N_2O emission was the lowest, which may be attributed to the low Shannon index and poor microbial community diversity. Studies have reported that micro-organisms, particularly Proteus, thick-walled bacteria, and Actinomycetes, are involved in straw decomposition [38]. Nitrogen-fixing micro-organisms mainly include various α-, β-, and γ-type Proteobacteria [39]. Among these groups, Firmicutes can participate in straw decomposition and are the most abundant bacteria that play a leading role in nitrogen utilization and compete with nitrifying bacteria for nitrogen sources; thus, they promote the fixation of plant available nitrogen, reduce the substrate for N_2O production [40,41], and inhibit N_2O emissions.

Previous studies have reported the involvement of micro-organisms in various carbon cycle metabolic processes such as the conversion of inorganic carbon to organic carbon, production of methane and methane oxidation, and decomposition of organic matter [42]. Some studies have reported that the amount of CH_4 emitted from dryland soil is relatively low, which is mostly shown as absorption. This may be due to soil dryness, good aeration conditions, and abundant oxygen in the soil, which make CH_4 easy to be oxidized [43]. It may also be due to the rapid decomposition of organic matter and slow accumulation of organic carbon in dryland soil, thus affecting the production and emission of CH_4 [44]. In the present study, the cumulative emission of CH_4 increased under different compost substitution ratios, and the soil emission of CH_4 was increased by the compost substitution treatment. This may be attributed to the fact that the relative abundance of *Bacillus*, Actinobacteria, and *Sphingomonas* increased after compost substitution, which reduced the decomposition of soil organic matter and promoted its rapid accumulation, thereby increasing the production and emission of CH_4. The absorption under T0 was higher than that under T1, T2, and T3, which may be due to the presence of fewer *Proteus* in the soil without composting and increased fixation of organic matter in the soil [39], leading to a decrease in the methanogen matrix and CH_4 emission.

In this study, CO_2 emissions reduced under different compost substitution ratios, and the relative abundance of *Bacillus* increased after compost substitution. This resulted in soil carbon sequestration, leading to a decrease in the decomposition of organic matter by micro-organisms and the transformation of mineral nutrients. Different compost substitution ratios lead to differences in the amount of soil organic carbon fixed. The stability of soil organic carbon increases gradually, and the deep organic carbon is not easy to be used by biology. Therefore, soil respiration is relatively weak, and CO_2 emissions decrease.

In this study, all treatments reduced the comprehensive warming potential of greenhouse gases in the wheat season, and the comprehensive warming potential under T3 treatment was relatively low, which may be due to the lack of carbon-fixing micro-organisms in T3, rendering the soil to fix a limited amount of organic carbon. The comprehensive warming potential under T2 decreased by 60% compared with that under T0, which indicated that soil organic carbon was fixed due to compost substitution. The GHGI under T2 was the weakest among all compost substitution treatments, indicating that compost substitution could not only reduce the GHGI but also enhance the carbon sequestration of soil micro-organisms and improve soil productivity.

Additionally, compost substitution increased wheat yield, which is consistent with the findings of a study by Wang Jiabao [12]. The yield increased the most under compost replacement of 30%. The compost replacement of 30% increased the abundance of micro-organisms causing soil carbon sequestration and improved soil productivity, in addition to significantly increasing the effective panicles and 1000-grain weight of wheat. Thus, compost substitution is an ideal cultivation measure for soil improvement and fertilization.

5. Conclusions

Throughout the wheat season, compost substitution significantly reduced CO_2 and N_2O emissions, and the accumulation of CO_2 under all treatments was lower than that under T0. 20% of compost substitution was the smallest and the largest without composting. N_2O accumulation was also the lowest under compost replacement of 20%; whereas for CH_4, it was emission. In terms of GWP and GHGI, compost substitution of 30% could reduce GHGI and significantly increase wheat yield by 21.3% on a 100-year scale. Therefore, compost replacement of 30% could relatively reduce greenhouse gas emissions and significantly increase wheat production, indicating that compost replacement of 30% could be beneficial to both the economy and environment.

Author Contributions: Conceptualization, H.W. and L.R.; methodology, H.W.; software, H.M.; validation, H.M., X.H. and Q.S.; formal analysis, Q.L.; investigation, H.M.; resources, L.R.; data curation, H.M.; writing—original draft preparation, H.M.; writing—review and editing, H.W.; visualization, H.M.; supervision, L.R.; project administration, D.X.; funding acquisition, D.X. All authors have read and agreed to the published version of the manuscript.

Funding: This research was funded by the Natural Science Foundation of Anhui Province, China, grant number 2008085QD181; and the Major Provincial Science and Technology Projects, grant numbers 201903a06020001, 201903A06020023, 202003a06020003.

Institutional Review Board Statement: Not applicable.

Informed Consent Statement: Not applicable.

Data Availability Statement: Not applicable.

Conflicts of Interest: The authors declare no conflict of interest.

References

1. Rawluk, C.D.L.; Grant, C.A.; Razc, G.J. Ammonia volatilization from soils fertilizer with urea and varying rates of urease inhibitor NBPT. *Can. J. Soil Sci.* **2001**, *81*, 239–246. [CrossRef]
2. Wakase, S.; Sasaki, H.; Itoh, K.; Otawa, K.; Kitazume, O.; Nonaka, J.; Satoh, M.; Sasaki, T.; Nakai, Y. Investigation of the microbial community in a microbiological additive used in a manure composting process. *Bioresour. Technol.* **2008**, *99*, 2687–2693. [CrossRef]
3. Mahmoodabadi, M.; Heydarpour, E. Sequestration of organic carbon influenced by the application of straw residue and farmyard manure in two different soils. *Int. Agrophys.* **2014**, *28*, 169–176. [CrossRef]
4. Turmel, M.S.; Speratti, A.; Baudron, F. Crop residue management and soil health: A systems analysis. *J. Agric. Syst.* **2015**, *134*, 6–16. [CrossRef]
5. Ngoran, k.; Zapata, F.; Sanginga, N. Availability of N from casuarina residues and inorganic N to maize using15N reisotope technique. *Biol. Fert. Soils* **1998**, *8*, 95–100. [CrossRef]
6. Naher, U.A.; Biswas, J.C.; Maniruzzaman, M. Bioorganic fertilizer: A green technology to reduce synthetic N and P fertilizer for rice production. *J. Front. Plant Sci.* **2021**, *12*, 602052. [CrossRef]
7. Farmaha, B.S. Evaluating Animo model for predicting nitrogen leaching in rice and wheat. *J. Arid. Land. Res. Manag.* **2014**, *28*, 25–35. [CrossRef]
8. Lenaerts, S.; Van Der Borght, M.; Callens, A. Suitability of microwave drying for mealworms(Tenebriomolitor) as alternative tofreeze drying: Impact on nutritional quality and colour. *J. Food Chem.* **2018**, *254*, 129–136. [CrossRef]
9. Azam, F.; Benckiser, G.; Müller, C. Release, movement and recovery of 3, 4-dimethylpyrazole phosphate (DMPP) ammonium and nitrate from stabilized nitrogen fertilizer granules in a silty clay soil under laboratory conditions. *J. Biol. Fertil. Soils* **2001**, *34*, 118–125. [CrossRef]
10. Halvorson, A.D.; Wienhold, B.J.; Black, A.L. Tillage and nitrogen fertilization influence grain and soil nitrogen in an annual cropping system. *Agron. J.* **2001**, *93*, 836–841. [CrossRef]
11. Liu, H.; Li, J.; Li, X.; Zheng, Y.; Feng, S.; Jiang, G. Mitigating greenhouse gas emissions through replacement of chemical fertilizer with organic manure in a temperate farmland. *Sci. Bull.* **2015**, *60*, 598–606. [CrossRef]
12. Cong, T.; Frank, J.L.; Nancy, G.C. Responses of soil microbial biomass and N availability to transitionstrategies from conventional to organic farming systems. *J. Agric. Ecosyst. Environ.* **2005**, *113*, 206–215. [CrossRef]
13. Bastida, F.; Kandeler, E.; Hernandez, T. Longtern effect of municipal solid waste amendment on microbial abundance and humus associated enzymed activities under semiarid conditions. *J. Microb. Ecol.* **2008**, *55*, 651–661. [CrossRef]
14. Chaoui, H.I.; Zibilske, L.M.; Ohno, T. Effects of earthworm casts and compost on soil microbial activity and plant nutrient availability. *Soil Biol. Biochem.* **2003**, *35*, 295–302. [CrossRef]

15. Chotte, J.L.; Ladd, J.L. Measurement of biomass C N and 14C of soil at different water contents using a fumigation extraction assay. *Soil Biol. Biochem.* **1998**, *30*, 1221–1224. [CrossRef]

16. Mahmood, T.; Ali, R.; Malik, K.A.; Shamsi, S.R.A. Nitrous oxide emissions from an irrigated sandyclay loam cropped to maize and wheat. *J. Biol. Fertil. Soils* **1998**, *27*, 189–196. [CrossRef]

17. Wang, C.; Luo, X.; Zhang, H. Differences between the shares of greenhouse gas emissions calculated with GTP and GWP for major countries. *Clim. Change Res.* **2013**, *9*, 49–54. [CrossRef]

18. Zhao, J.; Ni, T.; Li, Y.; Xiong, W.; Ran, W.; Shen, B.; Shen, Q.; Zhang, R. Responses of bacterial communities in arable soils in a rice-wheat cropping system to different fertilizer regimes and sampling times. *PLoS ONE* **2014**, *9*, e85301. [CrossRef]

19. Miron, J.; Yosef, E.; Nikbachat, M.; Zenou, A.; Zuckerman, E.; Solomon, R.; Nadler, A. Fresh dairy manure as a substitute for chemical fertilization in growing wheat forage; effects on soil properties, forage yield and composition, weed contamination, and hay intake and digestibility by sheep. *Anim. Feed Sci. Technol.* **2011**, *168*, 179–187. [CrossRef]

20. Haynes, R.J.; Naidu, R. Influence of lime, fertilizer and manure applications on soil organic matter content and soil physical conditions: A review. *Nutr. Cycl. Agroecosyst.* **1998**, *51*, 123–137. [CrossRef]

21. Asada, K.; Toyota, K.; Nishimura, T.; Ikeda, J.I.; Hori, K. Accumulation and mobility of zinc in soil amended with different levels of pig-manure compost. *J. Environ. Sci. Health Part B* **2010**, *45*, 285–292. [CrossRef]

22. Quast, C.; Pruesse, E.; Yilmaz, P.; Gerken, J.; Schweer, T.; Yarza, P.; Peplies, J.; Glockner, F.O. The SILVA ribosomal RNA gene database project: Improved data processing and web-based tools. *J. Nucl. Acids Res.* **2013**, *41*, D590–D596. [CrossRef]

23. Eghball, B.; Ginting, D.; Gilley, J.E. Residual effects of manure and compost applications on corn production and soil properties. *Agron. J.* **2004**, *96*, 442–447. [CrossRef]

24. Erhart, E.; Hartl, W.; Putz, B. Biowaste compost affects yield, nitrogen supply during the vegetation period and crop quality of agricultural crops. *Eur. J. Agron.* **2005**, *23*, 305–314. [CrossRef]

25. Pérez-Piqueres, A.; Edel-Hermann, V.; Alabouvette, C.; Steinberg, C. Response of soil microbial communities to compost amendments. *Soil Biol. Biochem.* **2006**, *38*, 460–470. [CrossRef]

26. Patrick, D.S.; Anthony, G.H.; David, B.W.; Larry, P.W. Tracking temporal changes of bacterial community fingerprints during the initial stages of composting. *FEMS Microbiol. Ecol.* **2003**, *46*, 1–9. [CrossRef]

27. Bronson, K.F.; Neue, H.U.; Singh, U. Automated chamber measurements of methane and nitrous oxide flux in a flooded rice soil. I. Residue, nitrogen, and water management. *J. Soil Sci. Soc. Am.* **1997**, *61*, 981–987. [CrossRef]

28. Bichel, A.; Oelbermann, M.; Echarte, L. Impact of residue addition on soil nitrogen dynamics in intercrop and sole crop agroecosystems. *J. Geoderma* **2017**, *304*, 12–18. [CrossRef]

29. Lal, R. Soil carbon sequestration impacts on global climate change and food security. *J. Sci.* **2004**, *304*, 1623–1627. [CrossRef]

30. Ewel, K.C.; Cropper, W.P.; Gholz, H.L. Soil CO2 evolution in Florida slash pine plantations I. Changes through time. *J. Can. J. For. Res.* **1987**, *17*, 325–329. [CrossRef]

31. Ugalde, D.; Brungs, A.; Kaebernick, M.; McGregor, A.; Slattery, B. Implications of climate change for tillage practice in Australia. *Soil Till. Res.* **2007**, *97*, 318–330. [CrossRef]

32. Sodhi, G.P.S.; Beri, V.; Benbi, D.K. Soil aggregation and distribution of carbon and nitrogen in different fractions under long-term application of compost in rice wheat system. *J. Soil Tillage Res.* **2009**, *103*, 412–418. [CrossRef]

33. Skiba, U.; Smith, K.A. Nitrification and denitrification as sources of nitric and nitrous oxide in a sandy loam soil. *Soil Biol. Biochem.* **1993**, *25*, 1527–1536. [CrossRef]

34. Sey, B.; Whalen, J.; Gregorich, E.; Rochette, P.; Cue, R. Carbon dioxide and nitrous oxide content in soils under corn and soybean. *Soil Sci. Soc. Am. J.* **2008**, *72*, 931–938. [CrossRef]

35. Owen, L.B.; Edwards, W.M.; van Keuren, R.W. Groundwaer nitrate levels under fertilized grass and grass-legume pastures. *J. Environ. Qual.* **1994**, *23*, 752–758. [CrossRef]

36. Frolking, S.; Qiu, J.J.; Boles, S. Combing remote sensing and ground census data to develop new maps of the distribution of rice agriculture in China. *J. Glob. Biogeochem. Cycle* **2002**, *16*, 1091. [CrossRef]

37. Mutegi, J.K.; Munkholm, L.J.; Petersen, B.M. Nitrous oxide emissions and controls as influenced by tillage and crop residue management strategy. *J. Soil Biol. Biochem.* **2010**, *42*, 1701–1711. [CrossRef]

38. Fan, F.; Yin, C.; Tang, Y.; Li, Z.; Song, A.; Wakelin, S.A.; Zou, J.; Liang, Y. Probing potential microbial coupling of carbon and nitrogen cycling during decomposition of maize residue by 13C-DNA-SIP. *Soil Biol. Biochem.* **2014**, *70*, 12–21. [CrossRef]

39. Beatty, P.H.; Good, A.G. Plant science. Future prospects for cereals that fix nitrogen. *Science* **2011**, *333*, 416–417.

40. Das, S.; Adhya, T.K. Effect of combine application of organic manure and inorganic fertilizer on methane and nitrous oxide emissions from a tropical flooded soil planted to rice. *Geoderma* **2014**, *213*, 185–192. [CrossRef]

41. Yao, Z.; Zheng, X.; Wang, R.; Xie, B.; Butterbach-Bahl, K.; Zhu, J. Nitrous oxide and methane fluxes from a rice–wheat crop rotation under wheat residue incorporation and no-tillage practices. *Atmos. Environ.* **2013**, *79*, 641–649. [CrossRef]

42. Tseng, C.H.; Tang, S.L. Marine microbial metagenomics: From individual to the environment. *Int. J. Mol. Sci.* **2014**, *15*, 8878–8892. [CrossRef] [PubMed]

43. Okuda, H.; Noda, K.; Sawamoto, T. Emission of N2O and CO2 and uptake of CH4 in soil from a Satsuma mandarin orchard under mulching cultivation in central Japan. *J. Jpn. Soc. Hortic. Sci.* **2007**, *76*, 279–287. [CrossRef]

44. Brix, H.; Sorrell, B.K.; Lorenzen, B. Are phragmites-dominated wet-lands a net source or net sink of greenhouse gases. *J. Aquat. Bot.* **2001**, *69*, 313–324. [CrossRef]

life

Article

Soil Organic Carbon Mineralization and Its Temperature Sensitivity under Different Substrate Levels in the Mollisols of Northeast China

He Yu [1], Yueyu Sui [2,3], Yimin Chen [2,3], Tianli Bao [1,*] and Xiaoguang Jiao [1,*]

[1]   Heilongjiang Provincial Key Laboratory of Ecological Restoration and Resource Utilization for Cold Region, College of Modern Agriculture and Ecological Environment, Heilongjiang University, Harbin 150080, China; yuhehlju@163.com

[2]   Key Laboratory of Mollisols Agroecology, Northeast Institute of Geography and Agroecology, Chinese Academy of Sciences, Harbin 150081, China; suiyy@iga.ac.cn (Y.S.); chenyimin@iga.ac.cn (Y.C.)

[3]   University of Chinese Academy of Sciences, Beijing 100049, China

*   Correspondence: 2020048@hlju.edu.cn (T.B.); 2004086@hlju.edu.cn (X.J.); Tel.: +86-18246106025 (T.B.); +86-18845637170 (X.J.)

Abstract: Soil organic carbon (SOC) mineralization plays an important role in global climate change. Temperature affects SOC mineralization, and its effect can be limited by the substrate available. However, knowledge of the effects of temperature and substrate quality on SOC mineralization in the Mollisols of Northeast China is still lacking. In this study, based on a spatial transplant experiment, we conducted a 73-day incubation to examine the effects of temperature on SOC mineralization and its temperature sensitivity under different carbon levels. We found that the SOC content, incubation temperature and their interaction had significant effects on SOC mineralization. A higher SOC content and higher incubation temperature resulted in higher SOC mineralization. The temperature sensitivity of SOC mineralization was affected by the substrate quality. The temperature sensitivity of SOC mineralization, showed a downward trend during the incubation period, and the range of variation in the Q_{10} declined with the increment in the SOC content. The study suggested that there was a higher SOC mineralization in high levels of substrate carbon when the temperature increased. Further, SOC mineralization under higher SOC contents was more sensitive to temperature changes. Our study provides vital information for SOC turnover and the CO_2 sequestration capacity under global warming in the Mollisols of Northeast China and other black soil regions of the world.

Keywords: substrate quality; carbon dioxide; Kinetic theory; Mollisols; temperature sensitivity

Citation: Yu, H.; Sui, Y.; Chen, Y.; Bao, T.; Jiao, X. Soil Organic Carbon Mineralization and Its Temperature Sensitivity under Different Substrate Levels in the Mollisols of Northeast China. *Life* **2022**, *12*, 712. https://doi.org/10.3390/life12050712

Academic Editor: Dmitry L. Musolin and Ling Zhang

Received: 8 April 2022
Accepted: 5 May 2022
Published: 10 May 2022

Publisher's Note: MDPI stays neutral with regard to jurisdictional claims in published maps and institutional affiliations.

1. Introduction

Soil organic carbon (SOC) is one of the most important carbon pools in terrestrial ecosystems, accounting for 60~80% of the global terrestrial carbon [1]. SOC decomposition in soils affects atmospheric CO_2 concentrations [2] and subsequently influences global climate change [3]. The study of SOC decomposition has been the focus of attention worldwide. Additionally, terrestrial ecosystems and their carbon dynamics significantly impact the global carbon budget [4]. Most studies have suggested that the changes in soil carbon content are associated with different types of land use [5,6] and different soil moisture and temperature conditions [7,8]. Better understanding of the dynamics of terrestrial SOC in different ecosystems is essential to determine SOC decomposition and turnover. The quantitative dynamics of SOC are important for predicting ecological processes and assessing soil fertility.

SOC mineralization, which mediates critical ecosystem processes important for the **decomposition and turnover** of organic matter, is an important monitor of the soil carbon cycle and can provide an estimate of soil carbon decomposition [9]. Carbon mineralization can be affected by many factors, such as **soil organic matter (SOM) quality and quantity, oxygen (O_2)**, water availability, soil biota and soil temperature [10]. One of the most important drivers affecting SOC mineralization is soil temperature, which might be considered a driver of organic matter turnover in soils [1]. The temperature dependence of SOC mineralization, which is the key source of soil heterotrophic respiration, has been the subject of intense scientific debate [8]. Liu et al. [11] suggested that elevated temperature resulted in an exponential reduction in dissolved organic carbon (DOC). Xiao et al. [12] found that rising temperatures tended to result in a higher portion of stable C in soils. The observed contradictory views indicate that the effects of soil temperature on SOC are complex. The determination of SOC mineralization under different soil temperatures may provide more insights into the causal processes of global warming.

Kinetic theory indicates that the temperature sensitivity of SOM decomposition should increase with substrate recalcitrance [13]. Carney et al. [14] have suggested that carbon inputs stimulate microbial activity and result in higher SOC turnover. Giardina and Ryan [15] found that SOC mineralization in mineral soils was controlled more by substrate quality than temperature. A study on Mollisols demonstrated that when incubated at 25 °C, soils with a higher SOC content had higher CO_2 production, but when incubated at 15 °C, no link between the SOC content and CO_2 production was detected [16]. Substrate quality and availability affect the decomposition of SOM [17], and soil temperature influences SOC mineralization by its effects on microbial metabolic activity. The combination of the above two drivers has a interactive effect on SOC turnover rates compared with either factor alone [18]. Thus, the warming effect may be limited by the amount of substrate available and water for decomposition [19]. A better understanding of SOC mineralization is needed to assess the effects of soil temperature and substrate inputs on soil carbon storage.

Northeast China is one of the major regions of Mollisols worldwide. Mollisols, which have a large C pool, are characterized by high SOC and nutrient contents [20]. High amounts of CO_2 can be released from the carbon in black soil. Differences in the SOC density and structure among different soil types result in the characteristics of SOC mineralization and its temperature sensitivity [21]. Therefore, increased recognition of regional SOC mineralization and its temperature sensitivity are important for understanding the C cycle as well as for improving C management strategies in the Mollisols of Northeast China [6,7].

Based on a spatial transplant experiment, we conducted an incubation experiment to examine how substrate C contents affect SOC mineralization under different temperatures and investigate its temperature sensitivity (Q_{10}). We hypothesized that higher SOC content would induce larger CO_2 production than those low levels of SOC content with the increment of temperature. In order to test the hypothesis, our objectives were to (1) examine how temperature affects SOC mineralization; (2) evaluate the effect of different SOC contents on SOC stability; and (3) determine the temperature sensitivity of SOC mineralization under substrates with different C contents. This study is necessary for determining SOC stocks and the CO_2 sequestration potential in response to global warming in Mollisols of Northeast China and other black soil regions of the world.

2. Materials and Methods

2.1. Study Region

In October 2004, five sampling sites in Northeast China were selected for this study based on the SOC content. These sites were located in Lishu County and Dehui city in Jilin Province and in Hailun city, Bei'an city, and Nenjiang County in Heilongjiang Province in China. These regions have a typical temperate continental monsoon climate. The soil is classified as typical black soil, i.e., a Mollisol according to USDA soil taxonomy. Soil inorganic C could be ignored due to the lack of carbonates in these soils. Maize (*Zea mays*) is a common cultivated crop in these regions.

2.2. Experimental Design

We collected soil samples from a size of 1.4 m (length) × 1.2 m (width) × 1.0 m (depth) by wooden boxes from Lishu County, Dehui city, Hailun city, Bei'an city and Nenjiang County. These samples were transported to the Hailun Agroecology Experiment Station of the Chinese Academy of Sciences. The mean annual temperature is 1.5 °C in this region, and the mean annual precipitation ranges from 500 mm to 600 mm. With the use of a spatial transplant method (The crops were planted in a unified mode and management that eliminated the complexity of different climate conditions and land management ways), same standard (1.4 m × 1.2 m × 1.0 m) plots were established. The plots were separated by a 20-cm-thick brick wall, which was covered with cement and pasted with tarpaulin inside to minimize the risk of sampling nonindependent areas. In the study, treatments were divided into SOC10, SOC19, SOC29, SOC34, and SOC63 according to the C contents in the soils from the above sampling regions. Each treatment consisted of three replicates. Site descriptions are shown in Table 1.

Table 1. Sampling site descriptions.

Site	Coordinates	Mean Annual Temperature (°C)	Mean Annual Precipitation (mm)	Crop
Lishu	N 43°20′, E 124°28′	5.4	556.2	Maize
Dehui	N 44°12′, E 125°33′	4.5	457.6	Maize
Hailun	N 47°27′, E 126°56′	1.5	549.3	Maize
Bei'an	N 48°09′, E 126°44′	1.1	523.4	Maize
Nenjiang	N 49°08′, E 125°37′	−0.2	532.1	Maize

2.3. Soil Sampling and Preparation

The selected samples from different regions were used to determine the effect of carbon levels on SOC mineralization. Clearly, under the premise of satisfying different C contents, the other physical-chemical properties of the soil samples were not consistent. In fact, the situation in the field was complex and changed. The samples, obtained from different regions, can better reflect the actual soil conditions in the field. A single-variable experiment is necessary to explore the complicated effects of the soil C level on SOC mineralization in the future.

After maize harvest in October 2016, twenty 10 cm × 5 cm × 20 cm (length × width × depth) samples were randomly collected to a depth of 0–20 cm at each sampling site. The twenty subsamples were thoroughly mixed to generate one composite sample, and fine roots and other residues were removed from the samples. The composite sample was the total amount of the twenty samples. Soon after collection, the samples were transported to the Key Laboratory of Ecological Restoration and Resource Utilization for Cold Region in Harbin, Heilongjiang Province. All samples were air-dried and sieved through a 1 mm mesh for the incubation experiment, the measurement of soil physical and chemical properties. The rest of the samples were sieved through a 0.25 mm mesh to determine SOC, total nitrogen (TN), total phosphorus (TP) and total potassium (TK). Soil pH was determined with an automatic acid-base titrator using a 1:2.5 soil:water suspension. The SOC content was measured by the Walkley Black method [22]. TN was determined by the kjeldahl method. TP was determined via the H_2SO_4-$HClO_4$ digestion method. TK was analyzed according to NaOH melting with flame photometry. Available nitrogen (AN) was measured by alkali hydrolysis diffusion method. Available phosphorus (AP) was determined using the $NaHCO_3$ leaching molybdenum antimony colorimetric technique, and available potassium (AK) was tested by NH_4OAc extraction with flame photometry [23]. Soil properties are presented in Table 2.

Table 2. Soil properties of selected sites.

Site	SOC (g kg^{-1})	TN (g kg^{-1})	TP (g kg^{-1})	TK (g kg^{-1})	AN (mg kg^{-1})	AP (mg kg^{-1})	AK (mg kg^{-1})	pH	C/N
Lishu	9.63 ± 2.44 a	0.79 ± 0.03 a	0.64 ± 0.02 a	12.58 ± 1.69	97.39 ± 9.88 a	64.10 ± 8.98 a	145.11 ± 10.99	6.63 ± 0.12 a	12.21 ± 0.89
Dehui	18.56 ± 2.67 b	1.68 ± 0.04 b	0.84 ± 0.01 a	12.97 ± 1.27	120.73 ± 14.87 a	26.29 ± 7.31 b	151.24 ± 9.54	5.95 ± 0.06 b	11.05 ± 1.19
Hailun	29.35 ± 3.56 c	2.55 ± 0.11 c	1.60 ± 0.03 b	13.84 ± 2.76	218.00 ± 25.98 b	57.81 ± 4.77 c	156.77 ± 15.33	6.10 ± 0.02 b	11.48 ± 2.18
Bei'an	34.11 ± 2.98 c	2.86 ± 0.08 c	1.93 ± 0.07 bc	14.37 ± 2.38	338.74 ± 19.87 c	58.34 ± 3.99 c	167.88 ± 14.86	5.42 ± 0.11 c	11.87 ± 0.99
Nenjiang	63.17 ± 4.76 d	4.87 ± 0.13 d	2.43 ± 0.05 c	15.56 ± 3.04	366.72 ± 26.56 c	57.15 ± 4.68 c	163.44 ± 16.21	6.34 ± 0.05 a	12.99 ± 1.21

Note: Different lowercase letters indicate significant differences among sampled sites. Unmarked means that there are no significant differences between treatments. C/N = SOC: TN. Values are means ± standard error (n = 9).

2.4. Laboratory Incubation Experiment

We incubated soils in the laboratory using a factorial design of substrate (5 levels: SOC10, SOC19, SOC29, SOC34, and SOC63) and temperature (4 levels: 5 °C, 15 °C, 25 °C, and 35 °C) with three replicates. Air-dried soils (100 g) were placed in a 1000 mL jar and preincubated at 60% water-holding capacity (WHC) at 25 °C in the dark for 7 days. The WHC was determined by the method described by Alef and Nannipieri (1995) [24]. After preincubation, the samples were incubated at selected temperatures (5, 15, 25 or 35 °C) in the dark for 73 days. It is worth noting that the changes in the indices in the study tended to be stable after 73 days of incubation. During the incubation, samples were sealed in the incubator, and soil moisture was adjusted to 60% of WHC with deionized water.

2.5. CO$_2$ Production

In this study, the accumulation of CO$_2$ production was used to represent SOC mineralization in the soils. A beaker containing 25 mL 1.0 mol L^{-1} NaOH solution was placed in each jar to capture evolved CO$_2$. The NaOH solution was exchanged on days 1, 3, 5, 7, 14, 21, 28, 35, 42, 49, 56, 63, and 73 of the incubation. Then, the CO$_2$ was titrated with 0.5 mol L^{-1} HCl in excessive amounts of BaCl$_2$ [25]. We calculated the cumulative CO$_2$ production, which characterized SOC mineralization, during two successive sampling intervals. The cumulative CO$_2$ production was expressed as mg CO$_2$ kg^{-1} soil, representing the amount of CO$_2$ released from the soil.

2.6. Evaluations and Calculations

In our study, the SOC mineralization rate and the temperature sensitivity of SOC mineralization were expressed as Rs and Q_{10} [26], respectively. The following equations were used.

$$R_s = a \cdot e^{bT} \tag{1}$$

$$Q_{10} = e^{10b} \tag{2}$$

where R_s is the SOC mineralization rate (mg SOC kg^{-1} soil d^{-1}), T represents the incubation temperature, a is the SOC mineralization rate at a temperature of 0 °C, and b is the temperature coefficient, which is related to the Q_{10} (increase in the rate of respiration over a 10 °C increase in temperature).

2.7. Data Analysis

The Kolmogorov-Smirnov test and Levene's test were used to determine the normality and equality of the data, respectively. One-way analysis of variance (ANOVA) and least significant difference (LSD) multiple comparisons ($p < 0.05$) were used to assess the soil physical-chemical properties, soil cumulative CO$_2$ production and Q_{10}. Two-way ANOVA was performed to evaluate the effects of SOC level and soil temperature on cumulative CO$_2$ production. The relationship between cumulative CO$_2$ production and selected soils was evaluated with Pearson correlation analysis. The effects of the interaction of SOC contents and incubation temperatures on cumulative CO$_2$ production were determined by a univariate general linear model with Duncan's test. Curve estimation with an exponential model was used to evaluate parameter b in Equation (1). All statistical analyses were performed using SPSS statistical software ver. 20 (SPSS Inc., Chicago, IL, USA). Graphs were generated using SigmaPlot 12.5.

3. Results

3.1. Changes in Cumulative CO_2 Production under Different SOC Contents at 5, 15, 25 and 35 °C

The cumulative CO_2 production from soils with different SOC contents under different incubation temperatures can be seen as Figure 1, and the cumulative CO_2 production ranged from 567.10 to 1003.91 mg CO_2 kg^{-1} soil at the end of the incubation. During the incubation period, we found the cumulative CO_2 production increased with the increase in the SOC content. After 7 days of incubation, the cumulative CO_2 production under different SOC treatments began to show obvious differences as follows: SOC63 > SOC34 > SOC29 > SOC19 > SOC10. The cumulative CO_2 production stabilized across the treatments at different temperatures after 49 days of incubation. We also found that cumulative CO_2 production increased as the incubation temperature increased, with significantly higher cumulative CO_2 production at higher temperatures with increasing SOC content (Supplementary Tables S1–S9).

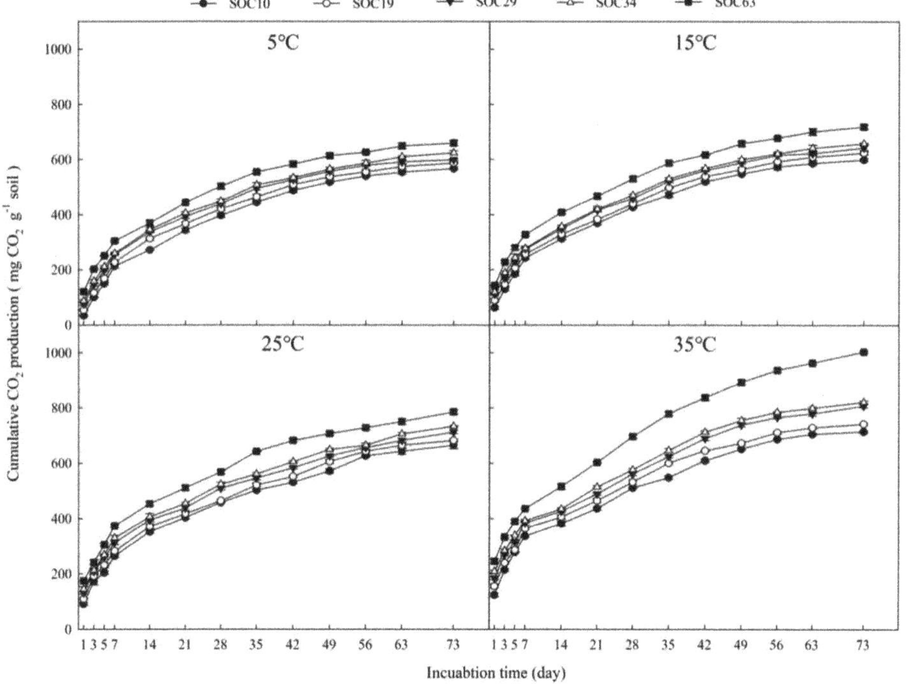

Figure 1. Changes in cumulative CO_2 production throughout the incubation at different soil temperatures (5 °C, 15 °C, 25 °C and 35 °C). Data scatters are means $\pm$ standard error ($n = 9$).

3.2. Characteristics of the Q_{10} Value under Different SOC Contents

We averaged the Q_{10} values from the different incubation temperatures to reflect the temperature sensitivity of SOC mineralization in the study, and we found that the Q_{10} value showed a descending trend with the extension of the incubation time (Figure 2). During the incubation, the Q_{10} value varied in response to different SOC contents, ranging from 1.07 to 1.53, 1.08~1.40, 1.07~1.32, 1.08~1.32, and 1.11~1.26 in SOC10, SOC19, SOC29, SOC34 and SOC63, respectively.

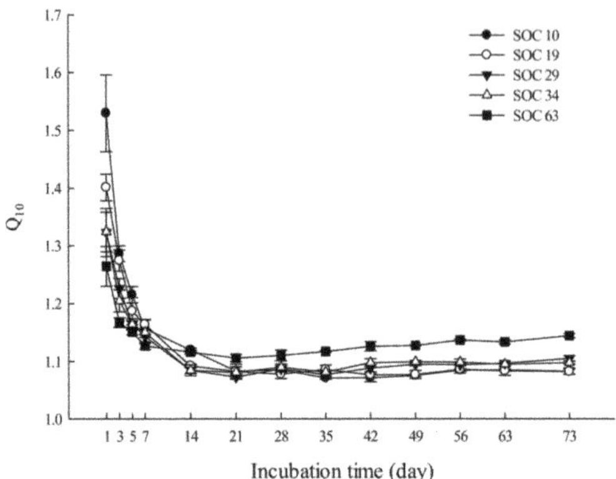

Figure 2. Effects of the soil carbon content on the Q_{10} value. Data scatters are means $\pm$ standard error ($n = 9$).

Both the highest and lowest Q_{10} values appeared in SOC10 on day 1 and day 35, respectively. The Q_{10} value showed a downward trend with little fluctuation and roughly leveled off by the end of the incubation in SOC10, SOC19 and SOC34. In SOC29 and SOC63, the Q_{10} value continued to decrease until day 21 and showed a modest upward trend with fluctuation thereafter. During the first five days of the incubation, soils with lower SOC contents had higher Q_{10} values. At the end of the incubation, the Q_{10} value in descending order was SOC63, SOC29, SOC34, SOC10 and SOC19 (Supplementary Table S10). Additionally, we also examined the changes in the Q_{10} value between the first day and the end of the incubation (the 73rd day of the incubation), and we found that the range of variation in the Q_{10} value decreased with the increase in the SOC content. There were significant differences of the Q_{10} between SOC63 and other treatments (Figure 3 and Supplementary Table S10).

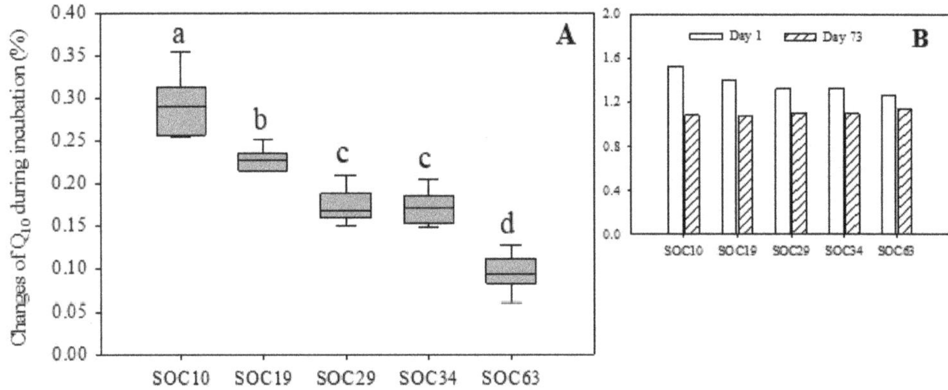

Figure 3. Changes in the Q_{10} value under different carbon levels during the incubation. Vertical bars of (**A**) = (Value of Q_{10} on the first day of incubation−value of Q_{10} on the 73rd day of incubation)/value of Q_{10} on the first day of incubation. (**B**) Day 1 = Value of Q_{10} on the first day of incubation; Day 73 = Value of Q_{10} on the 73rd day of incubation. Different lowercase letters indicate significant differences between SOC levels.

3.3. Key Factors That Drive Changes in SOC Mineralization

According to the two-way ANOVA, we found that the SOC contents, incubation temperature and their interactions significantly affected SOC mineralization, and the temperature had a larger effects on SOC mineralization than the SOC contents (Table 3).

Additionally, Pearson correlation analysis showed that soil organic carbon, total N, total P, total K, available N and available K were significantly positively correlated with cumulative CO_2 production, indicating that soil nutrients were the main factor that determined the changes in SOC mineralization (Table 4).

Table 3. Effects of SOC level and soil temperature on cumulative CO_2 production based on analysis of variance.

Index	Factors	F Value	df	*p*
	C level	380.25	22	$p < 0.01$
CO_2 production	Soil temperature	1126.20	28	$p < 0.01$
	C level $\times$ Soil temperature	35.30	19	$p < 0.01$

Note: Data for statistical analysis are based on a n = 12 samples per treatment.

Table 4. Correlations between soil physical-chemical properties and cumulative CO_2 production.

	SOC	TN	TP	TK	AN	AP	AK	pH	C/N
Pearson Correlation	0.936 **	0.927 **	0.847 **	0.915 **	0.896 **	0.194	0.761 *	−0.106	0.331
Sig. (2-tailed)	<0.001	<0.001	<0.001	<0.001	<0.001	0.241	0.021	0.389	0.281

Note: Data for statistical analysis are based on a n = 12 samples per treatment. * and ** represent significance differences at $p < 0.05$ and $p < 0.01$, respectively.

4. Discussion

4.1. Effects of SOC Contents and Temperature on SOC Mineralization

The temperature effect may be limited by the substrate available for decomposition [19]. In this study, we found that the temperature and substrate C contents had significant effects on cumulative CO_2 production (Figure 1; Table 3).

A temperature increase resulted in a higher cumulative CO_2 production (Figure 1). Many incubation experiments have shown that increasing temperature can promote SOC mineralization [17,27,28]. Temperature affects SOC mineralization by its effects on microbial metabolic activity. An increase in temperature is favorable to microbial activity and therefore increased the turnover rate of SOC [29,30]. Leifeld and Fuhrer [28] also suggested that temperature is a major controlling factor for SOC turnover. The responses of SOC to temperature are important for the evaluation of possible atmospheric feedbacks from the SOM reservoir [31]. In the current study, the incubation environment was not suitable for the collection of microbial data. Additional research needs to be conducted to explore how changes in the microbial community drive variations in SOC mineralization.

As expected, we found that the cumulative CO_2 production increased with increasing substrate carbon contents (Figure 1), and the alterations in cumulative CO_2 production were significantly correlated with soil nutrients (Table 4). Substrate quality affected cumulative CO_2 production, with the magnitudes varying across soils and substrate SOC [16]. There are higher C mineralization rates in the soils have been found in high contents of SOC [32]. Additionally, these results may primarily be attributed to microbial effects. Many studies have indicated that changes in soil nutrients modify microbial growth [33–35]. Greater C content lead to an increase in the abundance of microorganisms. A larger microbial community may be more efficient at SOC decomposition [14,15]. Competition is less crucial to limiting soil microbes because more niches may appear on account of improvements in soil nutritional resources [36]. Thus, the high levels of carbon may stimulate microbial activity and result in higher SOC turnover. As a result, the increase in substrate carbon might enhance SOC mineralization.

We found the increased SOC contents and warming had strong effects on SOC mineralization, and these effects were pronounced when warming and substrate treatments were applied together (Figure 1, Table 3). These two drivers has a different effect on SOC mineralization has a different effect on than either treatment alone [37]. Increased rates of SOC cycling caused by increased C inputs were exacerbated by warming [17]. It may be that warming affected the SOC turnover rate, the amount of CO_2 production is determined by the amount of substrate C in soils [38]. Steinweg et al. [39] demonstrated that there are distinct mechanisms by which the temperature and substrate quality affect microbial respiration. The increased temperature promoted microorganisms to take up and metabolize substrates more quickly, and higher substrate C made greater amounts of C available to soil microorganisms in general. Furthermore, many other factors can influence the SOC mineralization, such as oxygen (O_2), soil moisture and the fraction of substrate C and so on. Further research is necessary to explore the mechanism of these factors on SOC mineralization.

4.2. Influences of SOC Contents on the Temperature Sensitivity of SOC Mineralization

Temperature sensitivity of SOC mineralization (Q_{10}) in all carbon treatments showed a downward trend as incubation times increased, and a higher carbon content resulted in a smaller drop in Q_{10} in this study (Figure 2). The reason for the decline in Q_{10} may be the reduction in the SOC mineralization rate over time. Reichstein et al. [40] suggested that Q_{10} changed with incubation time, which may be caused by the alteration of labile C and recalcitrant C during incubation. Generally, the stability of SOC play a key role on the temperature sensitivity of the SOC decomposition, additional research needs to be conducted to explore whether changes in labile C and recalcitrant C drive changes in the SOC decomposition.

To further determine the changes in Q_{10}, we examined the variation amplitude of Q_{10} between the first day and the end of the incubation (the 73rd day of the incubation), and we found that Q_{10} was higher under a high carbon content than under low levels of carbon at the end of the incubation (Figure 3). The availability of SOC affects the response of soil heterotrophic respiration to temperature change, and Q_{10} decreases with the reduction in labile SOC content in soils [41]. The availability of soil nutrients was relatively high under the high soil carbon content compared with the low levels of soil carbon when the external temperature increased.

We also demonstrated that SOC mineralization under higher SOC contents was more sensitive to temperature changes in the middle and later period of SOC mineralization, and SOC mineralization under lower SOC contents was more sensitive to temperature changes at the early stage of SOC mineralization. To be degree, high levels of substrate carbon can stimulate SOC mineralization and result in greater SOC turnover. These findings can provide theoretical support for the impacts of different substrate gradients on SOC mineralization. Determining the temperature sensitivity of the decomposition of the different levels of SOC pools is critical for predicting the long-term impacts of climate change on SOC storage of Mollisols in Northeast China in the context of global warming.

5. Conclusions

In the study, the cumulative CO_2 production was relatively high under the high levels of soil carbon content. It suggested that high contents of SOC in soils resulted in a higher C mineralization. Changes in temperature can also affect the SOC mineralization. Larger cumulative CO_2 production were found with the increasing of temperature. We also found the effects of increased SOC contents and warming on SOC mineralization were significant when these two drivers were applied together. High levels of substrate carbon can stimulate SOC mineralization and result in greater SOC turnover when the temperature increased. Q_{10}, which represents the temperature sensitivity of SOC mineralization, varied between different SOC contents. The results suggested that the SOC mineralization under higher SOC contents was more sensitive to temperature changes. These fndings are important for achieving a better understanding of SOC turnover and the CO_2 sequestration capacity under global warming in the Mollisols of Northeast China and other black soil regions of the world.

Supplementary Materials: The following supporting information can be downloaded at: https://www.mdpi.com/article/10.3390/life12050712/s1. Table S1. Changes in cumulative CO_2 production with varied SOC contents when incubated at 5 °C. Table S2. Variations of cumulative CO_2 production among different SOC contents when incubated at 15 °C. Table S3. Alterations in cumulative CO_2 production between SOC contents when incubated at 25 °C. Table S4. Changes in cumulative CO_2 production with different SOC contents when incubated at 35 °C. Table S5. Alterations of cumulative CO_2 production under SOC10 among incubation temperatures. Table S6. Changes in cumulative CO_2 production in the SOC19 with different incubation temperatures. Table S7. Variations of cumulative CO_2 production under SOC29 between incubation temperatures. Table S8. Changes in cumulative CO_2 production under SOC34 with incubation temperatures. Table S9. Variations of cumulative CO_2 production in the SOC63 among incubation temperatures. Table S10. Changes in the Q_{10} values with SOC contents in soils.

Author Contributions: Conceptualization, X.J.; methodology, Y.S.; formal analysis, T.B. and Y.C.; data curation, T.B. and Y.C.; writing-original draft preparation, H.Y. and T.B.; writing-review and editing, T.B.; visualization, T.B.; supervision, X.J.; project administration, X.J.; funding acquisition, X.J.; All authors have read and agreed to the published version of the manuscript.

Funding: This research was funded by National Nature Science Foundation Program of China (grant NOs. 42077081) and National Science and Technology Basic Resources Investigation Special Project (grant NOs. 2021FY100400).

Institutional Review Board Statement: Not applicable.

Informed Consent Statement: Not applicable.

Data Availability Statement: The datasets generated and analyzed during the current study are available from the corresponding author on reasonable request.

Conflicts of Interest: The authors declare no conflict of interest.

References

1. Doetter, S.; Stevens, A.; Six, J.; Merckx, R.; Van, O.K.; Pinto, M.C.; Casanova-Katny, A.; Muñoz, C.; Boudin, M.; Venegas, E.Z.; et al. Soil carbon storage controlled by interactions between geochemistry and climate. *Nat. Geosci.* **2015**, *8*, 780–784. [CrossRef]

2. Lal, R. Soil carbon sequestration impacts on global climate change and food security. *Science* **2004**, *304*, 1623–1627. [CrossRef] [PubMed]

3. Meinshausen, M.; Meinshausen, N.; Hare, W.; Raper, S.C.B.; Frieler, K.; Knutti, R.; Allen, M.R. Greenhouse-gas emission targets for limiting global warming to 2 °C. *Nature* **2009**, *458*, 1158–1162. [CrossRef] [PubMed]

4. Schlesinger, W.H.; Andrews, J.A. Soil respiration and the global carbon cycle. *Biogeochemistry* **2000**, *48*, 7–20. [CrossRef]

5. Cambule, A.H.; Rossiter, D.G.; Stoorvogel, J.J.; Smaling, E.M.A. Soil organic carbon stocks in the Limpopo National Park, Mozambique: Amount, spatial distribution and uncertainty. *Geoderma* **2014**, *213*, 46–56. [CrossRef]

6. Wang, S.; Xu, L.; Zhuang, Q.L.; He, N.P. Investigating the spatio-temporal variability of soil organic carbon stocks in different ecosystems of China. *Sci. Total. Environ.* **2021**, *758*, 143644. [CrossRef]

7. Deng, B.L.; Yuan, X.; Siemann, E.; Wang, S.L.; Fang, H.F.; Wang, B.H.; Gao, Y.; Shad, N.; Liu, X.J.; Zhang, W.Y.; et al. Feedstock particle size and pyrolysis temperature regulate effects of biochar on soil nitrous oxide and carbon dioxide emissions. *Waste Manag.* **2021**, *120*, 33–40. [CrossRef]

8. Moonis, M.; Lee, J.; Jin, H.; Kim, D.G.; Park, J.H. Effects of warming, wetting and nitrogen addition on substrate-induced respiration and temperature sensitivity of heterotrophic respiration in a temperate forest soil. *Pedosphere* **2021**, *31*, 363–372. [CrossRef]

9. Bahn, M.; Rodeghiero, M.; Anderson-Dunn, M.; Dore, S.; Gimeno, C.; Drösler, M.; Jones, S. Soil respiration in European grasslands in relation to climate and assimilate supply. *Ecosystems* **2008**, *11*, 1352–1367. [CrossRef]

10. Mande, H.K.; Abdullah, A.M.; Aris, A.Z.; Nuruddin, A.A. A comparison of soil CO_2 efflux rate in young rubber plantation, oil palm plantation, recovering and primary forest ecosystems of Malaysia. *Pol. J. Environ. Stud.* **2014**, *23*, 1649–1657.

11. Liu, C.; Chu, W.; Li, H.; Boyd, S.A.; Teppen, B.J.; Mao, J.; Lehmann, J.; Zhang, W. Quantification and characterization of dissolved organic carbon from biochars. *Geoderma* **2019**, *335*, 161–169. [CrossRef]

12. Xiao, X.; Chen, B.; Chen, Z.; Zhu, L.; Schnoor, J.L. Insight into multiple and multilevel structures of biochars and their potential environmental applications: A critical review. *Environ. Sci. Technol.* **2018**, *52*, 5027–5047. [CrossRef] [PubMed]

13. Hartley, I.P.; Ineson, P. Substrate quality and the temperature sensitivity of soil organic matter decomposition. *Soil. Biol. Biochem.* **2008**, *40*, 1567–1574. [CrossRef]

14. Carney, K.M.; Hungate, B.A.; Drake, B.G.; Megonigal, J.P. Altered soil microbial community at elevated CO_2 leads to loss of soil carbon. *Proc. Natl. Acad. Sci. USA* **2007**, *104*, 4990–4995. [CrossRef]

15. Giardina, C.P.; Ryan, M.G. Evidence that decomposition rates of organic carbon in mineral soil do not vary with temperature. *Nature* **2000**, *404*, 858–861. [CrossRef]

16. Dai, S.S.; Li, L.J.; Ye, R.; Zhu-Barker, X.; Horwath, W.R. The temperature sensitivity of organic carbon mineralization is affected by exogenous carbon inputs and soil organic carbon content. *Eur. J. Soil. Biol.* **2017**, *81*, 69–75. [CrossRef]

17. Hopkins, F.M.; Filley, T.R.; Gleixner, G.; Lange, M.; Top, S.M.; Trumbore, S.E. Increased belowground carbon inputs and warming promote loss of soil organic carbon through complementary microbial responses. *Soil. Biol. Biochem.* **2014**, *76*, 57–69. [CrossRef]

18. Nie, M.; Pendall, E.; Bell, C.; Gasch, C.K.; Raut, S.; Tamang, S.; Wallenstein, M.D. Positive climate feedbacks of soil microbial communities in a semi-arid grassland. *Ecol. Lett.* **2012**, *16*, 234–241. [CrossRef]

19. Melillo, J.M.; Steudler, P.A.; Aber, J.D.; Newkirk, K.; Lux, H.; Bowles, F.P.; Catricala, C.; Magill, A.; Ahrens, T.; Morrisseau, S. Soil warming and carbon-cycle feedbacks to the climate system. *Science* **2002**, *298*, 2173–2176. [CrossRef]

20. Huang, Y.; Sun, W.J. Changes in topsoil organic carbon of croplands in mainland China over the last two decades. *Chin. Sci. Bull.* **2006**, *51*, 1785–1803. [CrossRef]

21. Song, G.; Li, L.; Pan, G.; Zhang, Q. Topsoil organic carbon storage of China and its loss by cultivation. *Biogeochemistry* **2005**, *74*, 47–62. [CrossRef]

22. Nelson, D.; Sommers, L. *Total Carbon, Organic Carbon and Organic Matter*; ASA Publication: Madison, WI, USA, 1982; pp. 539–577.

23. Institute of Soil Science, Chinese Academy Science (ISSCAS). *Physical and Chemical Analysis Methods of Soils*; Shanghai Science Technology Press: Shanghai, China, 1978; pp. 2–145.

24. Alef, K.; Nannipieri, P. *Methods in Applied Soil Microbiology and Biochemistry*; Academic Press: London, UK, 1995.

25. Cotrufo, M.; Ineson, P. Effects of enhanced atmospheric CO_2 and nutrient supply on the quality and subsequent decomposition of fine roots of *Betula pendula* Roth. and *Picea Sitchensis* (Bong.) Carr. *Plant Soil* **1995**, *170*, 267–277. [CrossRef]

26. Rey, A.; Pegoraro, E.; Tedeschi, V.; Parri, I.D.; Jarvis, P.G.; Valentini, R. Annual variation in soil respiration and its components in a coppice oak forest in Central Italy. *Global. Change. Biol.* **2002**, *8*, 851–866. [CrossRef]

27. Xu, X.; Inubushi, K.; Sakamoto, K. Effect of vegetations and temperature on microbial biomass carbon and metabolic quotients of temperate volcanic forest soils. *Geoderma* **2006**, *136*, 310–319. [CrossRef]

28. Leifeld, J.; Fuhrer, J. The Temperature Response of CO_2 production from bulk soils and soil fractions is related to soil organic matter quality. *Biogeochemistry* **2005**, *75*, 433–453. [CrossRef]

29. Sofi, J.A.; Lone, A.H.; Canie, M.A.; Dar, N.A.; Bhat, S.A.; Mukhtar, M.; Dar, M.A.; Ramzan, S. Soil Microbiological Activity and Carbon Dynamics in the Current Climate Change Scenarios: A Review. *Pedosphere* **2016**, *26*, 577–591. [CrossRef]

30. Wixon, D.L.; Balser, T.C. Toward conceptual clarity: PLFA in warmed soils. *Soil. Biol. Biochem.* **2013**, *57*, 769–774. [CrossRef]

31. Sarmiento, J.L.; Gruber, N. Sinks for anthropogenic carbon. *Phys. Today* **2002**, *55*, 30–36. [CrossRef]

32. Kimetu, J.M.; Lehmann, J.; Kinyangi, J.M.; Cheng, C.H.; Thies, J.; Mugendi, D.N.; Pell, A. Soil organic C stabilization and thresholds in C saturation. *Soil. Biol. Biochem.* **2009**, *41*, 2100–2104. [CrossRef]

33. Moyano, F.E.; Manzoni, S.; Chenu, C. Responses of soil heterotrophic respiration to moisture availability: An exploration of processes and models. *Soil. Biol. Biochem.* **2013**, *59*, 72–85. [CrossRef]

34. Degens, B.P.; Schipper, L.A.; Sparling, G.P.; Vojvodic-Vukovic, M. Decreases in organic C reserves in soils can reduce the catabolic diversity of soil microbial communities. *Soil. Biol. Biochem.* **2000**, *32*, 189–196. [CrossRef]

35. Bao, T.L.; Gao, L.Q.; Wang, S.S.; Yang, X.Q.; Ren, W.; Zhao, Y.G. Moderate disturbance increases the PLFA diversity and biomass of the microbial community in biocrusts in the Loess Plateau region of China. *Plant Soil* **2020**, *451*, 499–513. [CrossRef]

36. Li, L.J.; Han, X.Z.; You, M.Y.; Yuan, Y.R.; Ding, X.L.; Qiao, Y.F. Carbon and nitrogen mineralization patterns of two contrasting crop residues in a Mollisol: Effects of residue type and placement in soils. *Eur. J. Soil Biol.* **2013**, *54*, 1–6. [CrossRef]

37. Gray, S.B.; Classen, A.T.; Kardol, P.; Yermakov, Z.; Miller, R.M. Multiple climate change factors interact to alter soil microbial community structure in an oldfield ecosystem. *Soil. Sci. Soc. Am. J.* **2011**, *75*, 2217–2226. [CrossRef]

38. Dilly, O.; Zyakun, A. Priming effect and respiratory quotient in a Forest soil amended with glucose. *Geomicrobiol. J.* **2008**, *25*, 425–431. [CrossRef]

39. Steinweg, J.M.; Plante, A.F.; Conant, R.T.; Paul, E.A.; Tanaka, D.L. Patterns of substrate utilization during long-term incubations at different temperatures. *Soil. Biol. Biochem.* **2008**, *40*, 2722–2728. [CrossRef]

40. Reichstein, M.; Bednorz, F.; Broll, G.; Kätterer, T. Temperature dependence of carbon mineralisation. *Soil. Biol. Biochem.* **2000**, *32*, 947–958. [CrossRef]

41. Davidson, E.A.; Janssens, I.A. Temperature sensitivity of soil carbon decomposition and feedbacks to climate change. *Nature* **2006**, *440*, 165–172. [CrossRef]

Article

Effects of Partial Blackwater Substitution on Soil Potential NI-Trogen Leaching in a Summer Maize Field on the North China Plain

Tao Zhang [1,2,3], Hao Peng [2,3], Bo Yang [2,*], Haoyu Cao [2], Bo Liu [2] and Xiangqun Zheng [2]

[1] Aerospace Environmental Engineering Co., Ltd., Tianjin 300301, China; 2516209097@tju.edu.cn
[2] Agro-Environmental Protection Institute, Ministry of Agriculture and Rural Affairs, Tianjin 300191, China; penghao@caas.cn (H.P.); 82101205238@caas.cn (H.C.); 82101185126@caas.cn (B.L.); zhengxiangqun@caas.cn (X.Z.)
[3] School of Environmental Science and Engineering, Tianjin University, Tianjin 300072, China
* Correspondence: yangbo@caas.cn

Abstract: In China, promoting harmless blackwater treatment and resource utilization in rural areas is a priority of the "toilet revolution". Exploring the effects of blackwater application in arid areas on soil nitrogen losses can provide a basis for more effective water and fertilizer management. This study analyzed nitrogen leaching and maize yield under blackwater application in the summer maize season of 2020. A total of 5 treatments were used: no fertilizer, single chemical fertilizer application (CF), single blackwater application (HH), and combined chemical fertilizer and blackwater application ratios of 1:1 (CH1) and 2:1 (CH2). The total nitrogen leached from the fertilization treatments was 53.14–60.95 kg·ha^{-1} and the leached nitrate nitrogen was 34.10–40.62 kg·ha^{-1}. Nitrate nitrogen accounted for 50–62% of the total leached nitrogen. Compared with blackwater treatments, nitrate nitrogen moved into deeper soil layers (80–100 cm depth) during the CF treatment. Compared with CF, HH significantly reduced the maize yield by 24.39%. The nitrogen surplus of HH was higher than that of other fertilizer treatments. Considering nitrogen leaching, maize yield, and economic benefits, the CH2 treatment presented the optimal results. These findings address knowledge gaps and assist in guiding policy-makers to effectively promote China's "toilet revolution".

Keywords: nitrogen leaching; blackwater; wastewater reuse; maize fertilization; N surplus

Citation: Zhang, T.; Peng, H.; Yang, B.; Cao, H.; Liu, B.; Zheng, X. Effects of Partial Blackwater Substitution on Soil Potential NI-Trogen Leaching in a Summer Maize Field on the North China Plain. *Life* **2022**, *12*, 53. https://doi.org/10.3390/life12010053

Academic Editor: Ling Zhang

Received: 9 December 2021
Accepted: 29 December 2021
Published: 31 December 2021

Publisher's Note: MDPI stays neutral with regard to jurisdictional claims in published maps and institutional affiliations.

1. Introduction

Farmers worldwide have long used human excrement as a quick-acting fertilizer, owing to its high nitrogen content [1]; this is a traditional practice which has been followed over generations [2]. In China, the "toilet + septic tank + blackwater utilization" model is widely used to prevent pollution and promote recycling of human excrement [3]. In this model, after toilet sewage enters the septic tank, the decomposed manure liquid (i.e., blackwater) is used as a fertilizer for crops [4]. A previous study showed that blackwater use can improve soil structure and porosity while increasing soil organic carbon, and that reusing blackwater as a fertilizer for agriculture can help address soil productivity issues [5]. However, toilet flushing water dilutes the nutrient content of blackwater; therefore, considerably more blackwater is required to ensure normal crop growth, which increases the risk of nutrient loss.

After nitrogen fertilizers are applied to farmland soils, their fate can be roughly divided into three parts: some nitrogen is transformed into effective nutrients and is absorbed and utilized by the crops [6]; some is fixed in the crystal lattice of soil minerals, and thus remains in the soil [7]; the remainder is lost through leaching, nitrification, and denitrification [8,9]. In China, the overall utilization efficiency of nitrogen fertilizers in agriculture is only 30–40% [10], and the data of the first national pollution census showed that total nitrogen loss from agricultural sources accounted for 57.2% of total emissions in China [11]; therefore, significant economic losses are caused by this inefficiency.

Gradual nitrogen leaching below the root zone (i.e., nitrogen mineralization is not synchronized with nitrogen absorption by plants) is an important N-loss pathway which easily occurs in the presence of rainfall and irrigation events [12]. Soil is mostly composed of negatively charged colloids; therefore, it easily adsorbs a large amount NH_4^+-N, whereas the adsorption of NO_3^--N is weak [13]. Therefore, the nitrogen element in the soil easily moves vertically downward with water in the form of NO_3^--N, which characterizes the nitrogen leaching and represents approximately 60% of total dissolved nitrogen loss [14]. Nutrients that leach out of the active layer of plant roots are not easily absorbed by plants, which greatly reduces the nutrient use efficiency of the soil. If the leached nitrogen flows into groundwater, it can lead to exceedingly high levels of nitrate, which can endanger human health.

It has been previously reported that nitrification is stronger in alkaline than acid and neutral soils [15]; therefore, higher concentrations of NO_3^--N increase the risk of nitrogen migration in soil. Moreover, nitrogen movement in soil is not only controlled by the soil environment and hydrological processes, but also by crops and management measures [16–18]. The North China Plain is China's main dryland food production area, with alkaline soil, low water-holding capacity, low organic-matter content, and weak fertilizer-retention capacity [19]. Large amounts of fertilizer and irrigation water are needed to achieve a relatively high yield in these areas [20,21]. However, the use of blackwater as fertilizer may lead to a lower nitrogen utilization rate and higher nitrogen leaching due to the high moisture content [4]. In Beijing-Tianjin-Hebei and other intensive cultivation areas, >40% of the groundwater has a nitrogen content higher than the country's standards for drinking water (the Standards for Drinking Water Quality of China for NO_3^--N (GB5749-2006) is 20 mg·L^{-1}) [22]. However, to the best of our knowledge, there has been no research on nitrogen leaching from blackwater that is returned to fields in these areas.

In this study, we aimed to identify the potential environmental risk of blackwater for agriculture utilization and offer an available strategy to recover the energy and nutrients provided by blackwater. The main objectives of this experiment were to: (i) evaluate the effects of blackwater application levels on nutrient loss and crop yield in the alkaline soils of North China and (ii) explore the threshold of blackwater input under natural rainfall conditions in North China. The findings provide a reference to guide the "toilet revolution", reduce nitrogen loss, and decrease non-point source pollution from farmland.

2. Materials and Methods

2.1. Research Area Overview

The study area was located in a maize field (39°33′ N, 117°82′ E) in Dongjiituo Township, Ninghe County, Tianjin. The area has a continental monsoon climate and is in a warm temperate climatic and semi-arid, semi-humid wind zone, with relatively high summer temperatures and concentrated precipitation, as well as relatively cold and dry winters. The annual mean temperature is 11.2 °C; the minimum and maximum temperatures occur in January and July, respectively; the annual frost-free period is 240 d. Annual mean precipitation is approximately 642 mm, which mainly occurs between June and August, accounting for 70% of the annual precipitation. The cultivated soil in the study site was fluvo-aquic. The basic physical and chemical properties of the soil were as follows: pH = 8.38; organic matter = 9.70 g·kg^{-1}; total nitrogen (TN) = 1.19 g·kg^{-1}; total phosphorus (TP) = 0.64 g·kg^{-1}; alkali hydrolyzable nitrogen = 81.30 mg·kg^{-1}; available phosphorus = 23.05 mg·kg^{-1}; cation exchange capacity = 16.3 cmol·L^{-1}.

2.2. Experimental Setup

A randomized block design was used in the experiments. The chemical fertilizers used were urea (N = 46%), superphosphate (P_2O_5 = 16%), and potassium oxide (K_2O = 60%). The blackwater utilized was the effluent from a 3-grid septic tank (N: 5.1 $g\cdot kg^{-1}$; P: 3.1 $g\cdot kg^{-1}$; K: 3.7 $g\cdot kg^{-1}$). A total of 5 fertilizer treatments were explored: no fertilizer (CK); single application of chemical fertilizer (CF); single application of blackwater (HH); combined application of chemical fertilizer and blackwater at ratios of 1:1 (CH1) and 2:1 (CH2). The experiments for each treatment were repeated four times. The area of a single test plot was approximately 24 m^2 (4 m × 6 m), and a completely randomized block arrangement design was used.

The amount of fertilizer applied to the crops grown in the test site was determined based on the local fertilizing habits. In the summer maize season, the nitrogen application rate was 200 $kg\cdot ha^{-1}$, the phosphorus application rate was 150 $kg\cdot P\cdot ha^{-1}$, and the potassium application rate was 150 $kg\cdot K\cdot ha^{-1}$. The chemical fertilizers P_2O_5 and K_2O were used to remediate insufficient blackwater phosphorus and potassium contents, and were applied as a base fertilizer on a single occasion (Table 1). Blackwater and chemical nitrogen fertilizers were both applied to the surface at a ratio of 4:6.

Table 1. Fertilizer application rates of experimental treatments at different growth stages of maize ($kg\cdot ha^{-1}$).

Treatments	Applied Fertilizers Rate ($kg\cdot ha^{-1}$)			Fertilizer Form	Application Date
	N	P	K		
CK	-	-	-	-	-
	-	-	-	-	
CF	80	150	150	urea + superphosphate + potassium oxide	
	120	-	-	urea	
HH	80	29 + 48	5 + 58	blackwater + superphosphate + potassium oxide	21 June 2020
	120	73	87	blackwater	
CH1	40 + 40	40 + 24	77 + 29	urea + blackwater + superphosphate + potassium oxide	
	60 + 60	36	44	urea + blackwater	
CH2	53 + 27	110 + 16	101 + 20	urea + blackwater + superphosphate + potassium oxide	3 August 2020
	80 + 40	24	29	urea + blackwater	

CK: no fertilizer, CF: chemical fertilizer, HH: blackwater, CH1: combined application of chemical fertilizer and blackwater at 1:1, CH2: combined application of chemical fertilizer and blackwater at 2:1.

2.3. Sample Collection

2.3.1. Leachate Samples

In this experiment, infiltration tanks were used for in-situ monitoring of soil leachate (Figure 1). The leachate collection device was buried in each treatment plot in October 2019, and the leaching tube was planted and domesticated after a crop of winter wheat. The upper part of the collection device was composed of a sampling bottle, a buffer bottle, a vacuum pump, and a connecting pipe, and the underground part was composed of a filter sand layer, a liquid-collecting film, and a leachate collection barrel. The filter sand layer was composed of quartz sand with particle size of 2–3 mm, which was repeatedly cleaned with diluted acid and water. The liquid-collecting film included 2 pieces of polyethylene film with a 0.1-mm thickness. The leachate collection barrel was a cylindrical water barrel composed of a polyethylene material, with a volume of approximately 69 L (50 cm in diameter, 35 cm in height), and was buried at a depth of 80 cm.

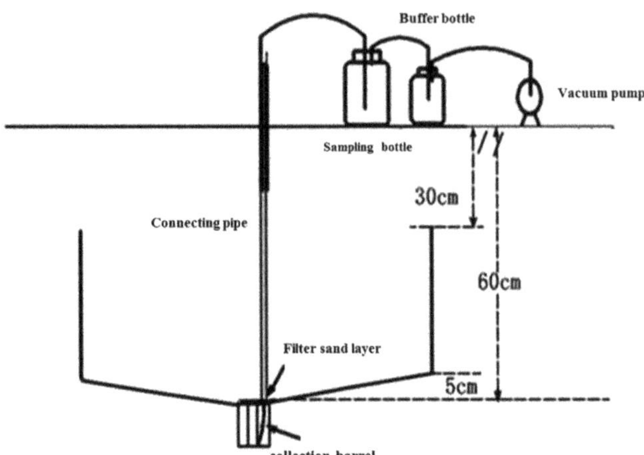

Figure 1. Leaching sample collection device and sampling scene.

We used a vacuum pump to generate negative pressure and extract all of the leachate for analysis. After evenly mixing the leachate samples, 500 mL of the sample was placed in a washed and dried polyethylene bottle and stored at 4 °C. A continuous flow injection analyzer (AA3 HR Auto Analyzer, SEAL Analytical, Germany) was used to determine the TN, NH_4^+, and NO_3^- contents in the eluent within 24 h of the collection. The sampling time was determined according to rainfall events, and the leachate samples of all sampling points were collected within 1 d. A total of 7 sampling campaigns were conducted throughout the experiment, on the day of 15 July, 30 July, 4 August, 15 August, 24 August, 19 September, and 19 October in 2020. Temperature and rainfall information were obtained from a small weather station.

2.3.2. Plant Sampling

The test crop was a summer maize variety, Jingdan 58, the main local variety. The crop was sown on 27 June 2020, with row spacing of 60 cm and plant spacing of 25 cm. The crop was harvested on 15 October 2020 after a 110-d growth period. The maize yield of each plot was determined. In addition, 3 representative plants were randomly selected from each plot. The dried samples were ground into powder and passed through a 100-mesh sieve. After digestion with concentrated H_2SO_4-H_2O_2, the TN content was determined by the semi-micro Kjeldahl method according to the maize yield. Subsequently, the nutrient absorption of the maize was calculated based on its nutrient content.

2.4. Data Analysis

The cumulative nitrogen (TN, NH_4^+, and NO_3^-) leaching amount was calculated according to Equation (1) [23].

$$N_L = \sum_{i=1}^{n} \frac{\left(C_i \times V_i \times 10^{-3}\right)}{1 \times 10^{-2} \times 10^{-4}} \tag{1}$$

where N_L represents the N loss loadings via surface runoff or leaching (kg·ha^{-1}), C_i represents the N concentration of the water sample of each leaching sampling (mg·L^{-1}), V_i represents the water volume of each leaching sampling (L), and 1×10^{-2} is the monitoring area (m^2).

The nitrogen surplus was estimated from total harvested nitrogen and all of the nitrogen inputs based on nitrogen balance in the summer maize cropping system [24]. The nitrogen surplus was calculated according to Equation (2):

$$N\ surplus = input\ N - output\ N \tag{2}$$

where the main external nitrogen inputs in our experiment were nitrogen brought by chemical fertilizer and blackwater. Other inputs, such as nitrogen from atmospheric deposition and irrigation water, were ignored. Nitrogen output included the nitrogen harvested in aboveground biomass (shoots and grains).

All of the statistical analyses were carried out in JMP version 9.0 (SAS Institute Inc., Cary, NC, USA, 2010). All of the data were checked for homogeneity of variances (Levene's test) and normality (Shapiro–Wilk test), and were normally distributed and had homogeneous variances. Differences among treatments in crop yields, nitrogen uptake, nitrogen surplus, nitrate distribute and cumulate nitrogen leaching were further examined with Student's multiple range tests. The effects of soil profiles and fertilization treatments on nitrate content distribution were examined by two-way analysis of variance (ANOVA). Origin 2019 was used to draw the soil nitrate distribution in the soil profile.

3. Results

3.1. Rainfall Characteristics

The maize growing season is the wet season in the basin, with a total rainfall of 350.4 mm; this accounts for 72.3% of the annual rainfall. There were 5 rainfall events with precipitation of >20 mm (Figure 2). The largest rainfall event during the study period occurred on 29 July 2020 and reached 48.7 mm, accounting for 10.1% of the annual rainfall.

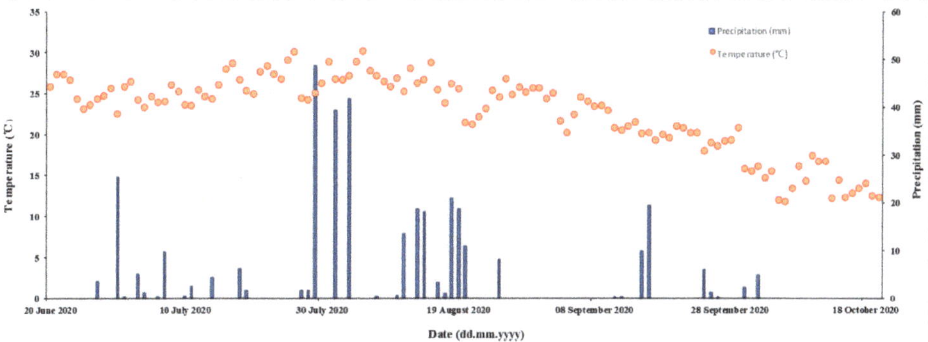

Figure 2. Daily precipitation (mm) and temperature (°C) from 20 June 2020 to 20 October 2020.

3.2. Maize Production and Nitrogen Surplus

The results of the experiments showed that the yield of maize kernels under different treatments was 5.2–8.2 t·ha^{-1}, with an average yield of 7.1 t·ha^{-1} (Figure 3a). Compared with the CF treatment, the combined treatments presented no significant effects on the maize yield, whereas the HH treatment significantly reduced the maize yield by 24.4% ($p < 0.05$). For the treatments using blackwater, a higher proportion of chemical fertilizer led to a higher maize yield; therefore, the maize yield under HF2 was significantly higher than that under HH, with an increase of 23.5% ($p < 0.05$), although there was no significant difference between the 2 combined treatments.

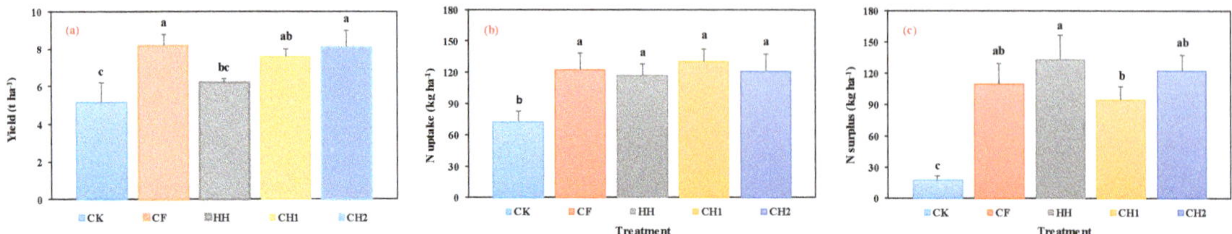

Figure 3. (**a**) Grain yield, (**b**) nitrogen uptake, and (**c**) nitrogen surplus of maize under different treatments. CK: no fertilizer, CF: chemical fertilizer, HH: blackwater, CH1: combined application of chemical fertilizer and blackwater at ratio of 1:1, CH2: combined application of chemical fertilizer and blackwater at ratio of 2:1. Bars indicate the standard error of the mean (+SE) for three replicates of each treatment. Letters above columns indicate significant differences according to the Tukey's multiple range test ($p < 0.05$) among all treatments.

The nitrogen uptake of maize under different treatments ranged from 72.4–130.0 kg·ha^{-1} (Figure 3b). Compared with the CK treatment, the 4 fertilization treatments significantly increased the nitrogen uptake in the aboveground part of maize; however, there was no statistical significance among the 4 fertilizer treatments ($p > 0.05$). An analysis of nitrogen surplus showed that the nitrogen surplus of each fertilization treatment was significantly higher than that of the CK treatment (Figure 3c). Among the fertilization treatments, the HH treatment had the largest nitrogen surplus, reaching 132.4 kg·ha^{-1}, which was significantly higher than that of the CH1 treatment (51.2%; $p < 0.05$).

3.3. Nitrate Nitrogen Migration in Soil Profile

Figure 4 shows the soil nitrate nitrogen profile at a depth of 0–120 cm under different fertilization strategies. The 3 treatments with blackwater application (HH, HF1, and HF2) reached the highest nitrate nitrogen content at a 40–80 cm depth. The highest nitrate nitrogen content of CF was observed at a depth of 80–100 cm, which indicates that the nitrogen from chemical fertilizers leached more easily downward. This result was supported by the one-way ANOVA.

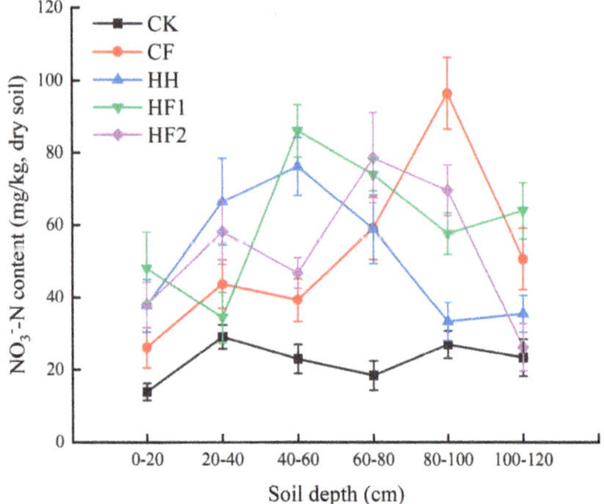

Figure 4. Nitrate nitrogen distribution in different soil profiles. Bars indicate the standard error of the mean (+SE) for three replicates of each treatment. CK: no fertilizer, CF: chemical fertilizer, HH: blackwater, CH1: combined application of chemical fertilizer and blackwater at a ratio of 1:1, CH2: combined application of chemical fertilizer and blackwater at a ratio of 2:1.

The nitrate nitrogen content in the soil profile and its spatial distribution characteristics are important indicators to characterize the leaching risk (Table 2). At a depth of 80–100 cm, the nitrate nitrogen content of the CF treatment was significantly higher than those of the blackwater application treatments, which were 63.4% (HH), 40.1% (HF1), and 27.7% (HF2) ($p < 0.05$). However, there was no statistical difference between the accumulation of nitrate nitrogen in the 4 fertilized soils at a depth of 60–80 cm ($p > 0.05$). The two-factor ANOVA showed that the fertilization strategy and soil depth significantly affected nitrate nitrogen leaching (Table 2).

Table 2. Nitrate nitrogen content in different soil profiles. CK: no fertilizer, CF: chemical fertilizer, HH: blackwater, CH1: combined application of chemical fertilizer and blackwater at a ratio of 1:1, CH2: combined application of chemical fertilizer and blackwater at a ratio of 2:1.

Soil Profile (cm)	CK	CF	HH	CH1	CH2	Two-Way ANOVA
0–20	13.9 ± 2.4bC	26.1 ± 5.6cBC	37.8 ± 7.4bcAB	48.3 ± 9.9cdA	37.9 ± 6.4bcAB	Treatment (T)
20–40	29.1 ± 3.3aC	43.8 ± 6.8bcBC	66.6 ± 12.0aA	34.5 ± 7.1dC	58.2 ± 8.8abAB	$p < 0.001$
40–60	23.0 ± 3.9abC	39.4 ± 6.0bcB	76.3 ± 7.9aA	86.2 ± 7.2aA	46.9 ± 4.3bcB	Soil profile (S)
60–80	18.5 ± 4.0abB	59.3 ± 8.6bA	59.1 ± 9.4abA	74.1 ± 4.5abA	78.7 ± 12.4aA	$p < 0.001$
80–100	26.9 ± 3.8aC	96.5 ± 9.8aA	33.4 ± 5.3cC	57.8 ± 5.7bcB	69.7 ± 7.1aB	T × S
100–120	23.4 ± 5.1abC	50.8 ± 8.5bAB	35.6 ± 5.1cBC	64.1 ± 7.8bcA	26.4 ± 6.5cC	$p < 0.001$

Data are mean values ± standard error (SE). Different small letters within the same column and different capital letters within the same row for each treatment indicate a significant difference at $p < 0.05$, determined by Tukey's multiple range tests.

3.4. Nitrogen Leaching

Table 3 shows the different forms of nitrogen in the leachate of each treatment. The total nitrogen leached from the different treatments was 34.91–60.95 kg·ha^{-1}. Compared with CK, all of the fertilization treatments significantly increased the amount of leached nitrate nitrogen and TN, with increases of 57.80–87.97% and 52.22–74.59%, respectively ($p < 0.05$). However, there was no significant difference among the four fertilization treatments. Compared with the other treatments, the amount of leached ammonium nitrogen was significantly higher in the HH treatment. The amount of leached nitrate nitrogen accounted for 17.05–20.31% of the total leached nitrogen, and the leaching rate of ammonium nitrogen was relatively small, accounting for 0.38–1.41% of the total nitrogen leached from all the different treatments.

Table 3. Cumulative leaching amount of nitrogen (NH_4^+-N, NO_3^--N, total nitrogen [TN]) from different treatments in the summer maize growing season. CK: no fertilizer, CF: chemical fertilizer, HH: blackwater, CH1: combined application of chemical fertilizer and blackwater at ration of 1:1, CH2: combined application of chemical fertilizer and blackwater at ratio of 2:1.

Nitrogen Form	Cumulative Leaching Amount (kg·ha^{-1})				
	CK	CF	HH	CH1	CH2
NH_4^+-N	0.75 ± 0.12b	0.76 ± 0.13b	2.82 ± 0.73a	1.31 ± 0.53b	1.08 ± 0.32b
NO_3^--N	21.61 ± 5.88b	40.62 ± 7.87a	37.63 ± 2.53a	36.52 ± 6.47a	34.10 ± 5.64a
TN	34.91 ± 4.81b	60.95 ± 11.00a	56.31 ± 16.47a	56.21 ± 7.57a	53.14 ± 10.50a

Data are mean values ± SE. Different small letters within the same row for each treatment indicate a significant difference at $p < 0.05$, determined by Tukey's multiple range tests.

4. Discussion

4.1. Effect of Fertilization Strategies on Nitrogen Leaching

Leaching is an important mechanism of nitrogen fertilizer loss, and the form of fertilizer used is an important farmland management measure affecting nitrogen leaching. In this study, compared with CF, nitrogen leaching decreased by 7.6–12.8% in the 3 treatments based on blackwater, although there were no statistically significant differences. Previous studies have shown that nitrogen leaching may vary with fertilizer type [25]. Under the same nitrogen application levels, the combined application of organic and inorganic fertilizers can significantly reduce nitrogen leaching (the content of organic matter in blackwater is higher than that in chemical fertilizer) compared to single application of chemical fertilizers [26]. This is due to the inherent ability of organic matter to improve the soil quality, increase the soil water retention capacity, and promote crop nitrogen uptake [27,28]. Studies have also shown that a high C/N ratio helps to promote the conversion of mineral nitrogen to organic forms (i.e., nitrogen immobilization.) [29,30], thereby reducing nitrogen leaching and runoff.

Interestingly, although no fertilizers were used in the CK treatment, nitrogen leaching of 34.91 $kg \cdot ha^{-1}$ occurred, which we presumed was due to: (i) nitrogen fertilizer remaining from previous crops, since residual nitrate can move continuously downwards and be lost even if it is not leached during the season of application [11]; (ii) nitrogen deposition, for example, a 3-year study investigated atmospheric deposition of different nitrogen species at 10 sites in Northern China and the results indicated that nitrogen deposition levels in Northern China were high, with an average of 59.8 $kg \cdot N \cdot ha^{-1} \cdot yr^{-1}$ [31].

4.2. Effect of Fertilizer with Blackwater on Nitrate-Nitrogen Migration

In our study, the NO_3^- leaching of the 4 fertilization treatments accounted for approximately 17.05–20.31% of the fertilizer input, which was slightly higher than some other studies, such as a meta-analysis conducted by Zhou & Butterbach-Bahl [32], who collected 32 published studies reporting NO_3^- leaching losses in maize and determined that 15% of applied fertilizer nitrogen to maize systems worldwide are leached in the form of NO_3^-. However, our obtained results are within the value estimated by Cui et al. [33], which conducted a meta-analysis of 17 published studies from 19 study sites, including 94 observations from maize system in China, and found that with typical farming practices, an average of 20.8% of the applied nitrogen was either leached or lost as runoff from the maize systems. The difference between the results may be attributed to soil physico-chemical properties such as texture, pH [34], soil organic carbon [35], crop type [25], or annual precipitation. It is worth mentioning that the residual NO_3^- in the soil profile showed that the NO_3^- leaching depth was deeper in the CF treatment, with a peak value at the 80–100-cm depth. The root system of the maize plats was mainly concentrated in the soil layer above 90 cm, which indicated that CF was more likely to cause NO_3^- leaching to groundwater.

The total nitrate accumulations in the 0–4 m soil layer of maize fields was as high as 749 $\pm$ 75 $kg \cdot N \cdot ha^{-1}$ in China [36]. However, the average accumulation of nitrate in the 0–120 cm soil layer was 51.5–60.8 $kg \cdot nitrogen \cdot ha^{-1}$ in this study, considerably lower than the national average value. There are 3 possible reasons that may explain this finding: (1) the nitrogen application rate in our experiment (200 $kg \cdot nitrogen \cdot ha^{-1}$) was lower than the typical rates (263 $kg \cdot nitrogen \cdot ha^{-1}$) for wheat in the North China Plain [37]; (2) we conducted the experiment in the rainy season of the North China Plain, which facilitated the rapid transport of nitrate deeper into the underground water [38]; (3) the soils have high permeability and low cation exchange capacity [20].

4.3. Effects of Blackwater Application on Soil Nitrogen Surplus and Maize Yield

Nitrogen surplus is an effective indicator for measuring nitrogen input productivity, environmental impact, and soil fertility changes [39]. Maintaining the nitrogen balance of the soil-crop system can achieve higher target yields without consuming soil nitrogen. In our study, the highest nitrogen surplus was obtained in the HH treatment, which may have been due to this treatment yielding the lowest maize yield of the four fertilizer treatments. The lowest nitrogen surplus in CH1 treatment may have been related to its high nitrogen uptake.

Chemical fertilizers (especially nitrogen fertilizers) are applied at high rates for food production in China, which leads to decreases in crop yield and quality, and an increase in fertilization costs [40]. This study showed that the application of pure blackwater significantly reduced maize yield compared to the application of conventional fertilizer, which may have been attributed to the higher content of base ions in blackwater [41]. Studies have shown that maize is susceptible to soil salinity, which significantly decreases seed germination, causes harmful effects in growth, and leads to low yield [42]. However, in this study, the yields under the combined treatments were not significantly different from that under CF, which indicated that an appropriate amount of blackwater can maintain the maize yield. Nitrogen fertilizer is a costly component of crop production [43], and at present, the average price of chemical fertilizers in China is 3 yuan/kg. Therefore, according to our results, the use of a combined fertilization treatment can lead to savings of approximately 190–300 yuan/ha, which would reduce the economic burden of fertilization to farmers.

4.4. Feasibility and Prospect of Returning Blackwater to the Field

A large-scale survey revealed that the proportion of pathogenic bacteria in the effluent from septic tanks is very low [44], and ensured environmental health and agricultural application safety. While approximately 86% of the households stated that they would prefer their excreta to be used in agriculture as fertilizer [45], there is no instructional document to teach farmers how to use blackwater to fertilize, and the usual practice of farmers is to return all the collected blackwater to the field. However, according to our study, excessive blackwater may lead to reduced crop production, which has a huge impact on farmers. A balance is needed between increasing farmers' income and decreasing environmental impact, and the fertilizer strategy involving the CH2 treatment appeared to meet both requirements.

In addition, our previous study shows that the application of a reasonable proportion of blackwater and chemical fertilizers did not significantly increase reactive nitrogen emissions [46]. However, application of blackwater-based fertilizers in agriculture will alleviate the environmental impacts of phosphorus mining and synthetic ammonia production [47]. In summary, exploration of the means by which to recycle blackwater or using excreta-derived fertilizers in agriculture is urgently for decision makers.

5. Conclusions

In summary, this study showed that compared to chemical fertilizer, blackwater application could prevent nitrate nitrogen from moving to deeper soils (below 80 cm), and that there was no statistical difference in soil nitrogen surplus and crop nitrogen uptake. Furthermore, the blackwater fertilizer strategy decreased the nitrate nitrogen and total nitrogen leaching by 7.4–16.1% and 7.6–12.8%, respectively. However, the application of blackwater at $200 \ kg \cdot nitrogen \cdot ha^{-1}$ reduced the maize yield by approximately 24.4% compared to application of chemical fertilizer, which may have been due to the high salt content of blackwater. The combined application of blackwater and chemical fertilizers maintained the maize yield without increasing the risk of nitrogen leaching, especially when the ratio of chemical fertilizer to blackwater was 2:1 (i.e., chemical fertilizer provided $133 \ kg \cdot nitrogen \cdot ha^{-1}$, blackwater provided $67 \ kg \cdot nitrogen \cdot ha^{-1}$). Our study shows that a potential reduction in nitrogen leaching and obtainable high maize yield can be achieved by the appropriate blackwater substitution of chemical fertilizers. We suggest that promoting the return of blackwater to fields not only involves allowing farmers to utilize it as fertilizer, but also includes introducing, demonstrating, and teaching them how it can be optimally carried out.

Author Contributions: Conceptualization, T.Z. and X.Z.; methodology, H.P.; software, B.Y.; validation, H.C.; formal analysis, B.L.; investigation, H.P. and H.C.; resources, B.Y.; data curation, B.L.; writing—original draft preparation, B.Y.; writing—review and editing, B.L.; visualization, H.C. All authors have read and agreed to the published version of the manuscript.

Funding: This research was funded by the Nature Science Foundation of Tianjin (No. 19JCQNJC13400) and the Central Public-interest Scientific Institution Basal Research Fund (No. Y2021LM01).

Institutional Review Board Statement: Not applicable.

Informed Consent Statement: Not applicable.

Acknowledgments: We acknowledge the anonymous reviewers and editor for critical and valuable comments which assisted in improving this manuscript.

Conflicts of Interest: The authors declare no conflict of interest.

References

1. Kawa, N.C.; Ding, Y.; Kingsbury, J.; Goldberg, K.; Lipschitz, F.; Scherer, M.; Bonkiye, F. Night Soil: Origins, Discontinuities, and Opportunities for Bridging the Metabolic Rift. *Ethnobiol. Lett.* **2019**, *10*, 40–49. [CrossRef]
2. Nakasaki, K.; Ohtaki, A.; Takemoto, M.; Fujiwara, S. Production of well-matured compost from night-soil sludge by an ex-tremely short period of thermophilic composting. *Waste. Manag.* **2011**, *31*, 495–501. [CrossRef]
3. Fan, B.; Hu, M.; Wang, H.; Xu, M.; Qu, B.; Zhu, S. Get in sanitation 2.0 by opportunity of rural China: Scheme, simulating application and life cycle assessment. *J. Clean. Prod.* **2017**, *147*, 86–95. [CrossRef]
4. Tan, L.; Zhang, C.; Liu, F.; Chen, P.; Wei, X.; Li, H.; Yi, G.; Xu, Y.; Zheng, X. Three-compartment septic tanks as sustainable on-site treatment facilities? Watch out for the potential dissemination of human-associated pathogens and antibiotic resistance. *J. Environ. Manag.* **2021**, *300*, 113709. [CrossRef]
5. Liu, B.; Yang, B.; Zhang, C.; Wei, X.; Cao, H.; Zheng, X. Human Waste Substitute Strategies Enhanced Crop Yield, Crop Quality, and Soil Fertility in Vegetable Cultivation Soils in North China. *Agronomy* **2021**, *11*, 2232. [CrossRef]
6. Zhao, J.; Tao, H.-B.; Liao, S.-H.; Wang, P. Establishment of ANEDr model for evaluating absorbed-nitrogen effects on wheat dry matter production. *J. Integr. Agric.* **2016**, *15*, 2257–2265. [CrossRef]
7. Cao, Y.; He, Z.; Zhu, T.; Zhao, F. Organic-C quality as a key driver of microbial nitrogen immobilization in soil: A meta-analysis. *Geoderma* **2021**, *383*, 114784. [CrossRef]
8. Azad, N.; Behmanesh, J.; Rezaverdinejad, V.; Abbasi, F.; Navabian, M. An analysis of optimal fertigation implications in dif-ferent soils on reducing environmental impacts of agricultural nitrate leaching. *Sci. Rep.* **2020**, *10*, 7797. [CrossRef]
9. Yan, J.; Wang, S.; Wu, L.; Li, S.; Li, H.; Wang, Y.; Wu, J.; Zhang, H.; Hong, Y. Long-term ammonia gas biofiltration through simultaneous nitrification, anammox and denitrification process with limited N_2O emission and negligible leachate production. *J. Clean. Prod.* **2020**, *270*, 122406. [CrossRef]
10. Zhang, W.F.; Zhang, F.S. *Research Report of Chinese Fertilizer Development*; China Agricultural University Press: Beijing, China, 2013. (In Chinese)
11. Ju, X.; Liu, X.; Zhang, F.; Roelcke, M. Nitrogen fertilization, soil nitrate accumulation, and policy recommendations in several agricultural regions of China. *Ambio* **2004**, *33*, 300–305. [CrossRef] [PubMed]
12. Ju, X.; Lu, X.; Gao, Z.; Chen, X.; Su, F.; Kogge, M.; Römheld, V.; Christie, P.; Zhang, F. Processes and factors controlling N_2O production in an intensively managed low carbon calcareous soil under sub-humid monsoon conditions. *Environ. Pollut.* **2011**, *159*, 1007–1016. [CrossRef]
13. Wang, Z.-H.; Li, S.-X. Nitrate N loss by leaching and surface runoff in agricultural land: A global issue (a review). *Adv. Agron.* **2019**, *156*, 159–217. [CrossRef]
14. Ramos, L.; Bettin, A.; Herrada, B.M.P.; Arenas, T.L.; Becker, S.J. Effects of Nitrogen Form and Application Rates on the Growth of Petunia and Nitrogen Content in the Substrate. *Commun. Soil Sci. Plant Anal.* **2013**, *44*, 473–479. [CrossRef]
15. Meinhardt, K.A.; Stopnisek, N.; Pannu, M.W.; Strand, S.E.; Fransen, S.C.; Casciotti, K.L.; Stahl, D.A. Ammonia-oxidizing bacteria are the primary N_2O producers in an ammonia-oxidizing archaea dominated alkaline agricultural soil. *Environ. Microbiol.* **2018**, *20*, 2195–2206. [CrossRef]
16. Wu, D.; Xu, X.; Chen, Y.; Shao, H.; Sokolowski, E.; Mi, G. Effect of different drip fertigation methods on maize yield, nutrient and water productivity in two-soils in Northeast China. *Agric. Water Manag.* **2018**, *213*, 200–211. [CrossRef]
17. Hondebrink, M.; Cammeraat, L.; Cerdà, A. The impact of agricultural management on selected soil properties in citrus orchards in Eastern Spain: A comparison between conventional and organic citrus orchards with drip and flood irrigation. *Sci. Total Environ.* **2017**, *581–582*, 153–160. [CrossRef] [PubMed]
18. Chen, Y.; Peng, J.; Wang, J.; Fu, P.; Hou, Y.; Zhang, C.; Fahad, S.; Peng, S.; Cui, K.; Nie, L.; et al. Crop management based on multi-split topdressing enhances grain yield and nitrogen use efficiency in irrigated rice in China. *Field Crop. Res.* **2015**, *184*, 50–57. [CrossRef]
19. Liu, M.; Wang, C.; Wang, F.; Xie, Y. Maize (*Zea mays*) growth and nutrient uptake following integrated improvement of ver-micompost and humic acid fertilizer on coastal saline soil. *Appl. Soil Ecol.* **2019**, *142*, 147–154. [CrossRef]
20. Ha, N.; Feike, T.; Back, H.; Xiao, H.; Bahrs, E. The effect of simple nitrogen fertilizer recommendation strategies on product carbon footprint and gross margin of wheat and maize production in the North China Plain. *J. Environ. Manag.* **2015**, *163*, 146–154. [CrossRef] [PubMed]
21. Cui, Z.; Zhang, F.; Chen, X.; Miao, Y.; Li, J.; Shi, L.; Xu, J.; Ye, Y.; Liu, C.; Yang, Z.; et al. On-farm evaluation of an in-season nitrogen management strategy based on soil Nmin test. *Field Crop. Res.* **2008**, *105*, 48–55. [CrossRef]

22. Zhang, W.L.; Wu, S.X.; Ji, H.J.; Kolbe. The current situation and controlling manures of non-point source pollutions in China. *Sci. Agric. Sinica* **2004**, *37*, 1008–1017, (In Chinese with English abstract).

23. Chen, Y.-M.; Zhang, J.-Y.; Xu, X.; Qu, H.-Y.; Hou, M.; Zhou, K.; Jiao, X.-G.; Sui, Y.-Y. Effects of different irrigation and fertilization practices on nitrogen leaching in facility vegetable production in northeastern China. *Agric. Water Manag.* **2018**, *39*, 165–170. [CrossRef]

24. Wang, B.; Guo, C.; Wan, Y.; Li, J.; Ju, X.; Cai, W.; You, S.; Qin, X.; Wilkes, A.; Li, Y. Air warming and CO2 enrichment increase N use efficiency and decrease N surplus in a Chinese double rice cropping system. *Sci. Total. Environ.* **2020**, *706*, 136063. [CrossRef] [PubMed]

25. Wang, X.; Zou, C.; Gao, X.; Guan, X.; Zhang, Y.; Shi, X.; Chen, X. Nitrate leaching from open-field and greenhouse vegetable systems in China: A meta-analysis. *Environ. Sci. Pollut. Res.* **2018**, *25*, 31007–31016. [CrossRef]

26. Kramer, S.B.; Reganold, J.P.; Glover, J.D.; Bohannan, B.J.M.; Mooney, H.A. Reduced nitrate leaching and enhanced denitrifier activity and efficiency in organically fertilized soils. *Proc. Natl. Acad. Sci. USA* **2006**, *103*, 4522–4527. [CrossRef]

27. Oldfield, E.E.; Bradford, M.A.; Wood, S.A. Global meta-analysis of the relationship between soil organic matter and crop yields. *Soil* **2019**, *5*, 15–32. [CrossRef]

28. Wei, Z.; Hoffland, E.; Zhuang, M.; Hellegers, P.; Cui, Z. Organic inputs to reduce nitrogen export via leaching and runoff: A global meta-analysis. *Environ. Pollut.* **2021**, *291*, 118176. [CrossRef]

29. Malcolm, B.; Cameron, K.; Curtin, D.; Di, H.; Beare, M.; Johnstone, P.; Edwards, G. Organic matter amendments to soil can reduce nitrate leaching losses from livestock urine under simulated fodder beet grazing. *Agric. Ecosyst. Environ.* **2018**, *272*, 10–18. [CrossRef]

30. Shepherd, M.; Menneer, J.; Ledgard, S.; Sarathchandra, U. Application of carbon additives to reduce nitrogen leaching from cattle urine patches on pasture. *N. Z. J. Agric. Res.* **2010**, *53*, 263–280. [CrossRef]

31. Fu, Y.; Wang, W.; Han, M.; Kuerban, M.; Wang, C.; Liu, X. Atmospheric dry and bulk nitrogen deposition to forest environment in the North China Plain. *Atmos. Pollut. Res.* **2019**, *10*, 1636–1642. [CrossRef]

32. Zhou, M.; Butterbach-Bahl, K. Assessment of nitrate leaching loss on a yield-scaled basis from maize and wheat cropping systems. *Plant Soil* **2013**, *374*, 977–991. [CrossRef]

33. Cui, Z.; Wang, G.; Yue, S.; Wu, L.; Zhang, W.; Zhang, F.; Chen, X. Closing the N-Use Efficiency Gap to Achieve Food and Environmental Security. *Environ. Sci. Technol.* **2014**, *48*, 5780–5787. [CrossRef]

34. Zhou, M.; Zhu, B.; Butterbach-Bahl, K.; Wang, T.; Bergmann, J.; Brüggemann, N.; Wang, Z.; Li, T.; Kuang, F. Nitrate leaching, direct and indirect nitrous oxide fluxes from sloping cropland in the purple soil area, southwestern China. *Environ. Pollut.* **2012**, *162*, 361–368. [CrossRef] [PubMed]

35. Benoit, M.; Garnier, J.; Anglade, J.; Billen, G. Nitrate leaching from organic and conventional arable crop farms in the Seine Basin (France). *Nutr. Cycl. Agroecosystems* **2014**, *100*, 285–299. [CrossRef]

36. Zhou, J.; Gu, B.; Schlesinger, W.H.; Ju, X. Significant accumulation of nitrate in Chinese semi-humid croplands. *Sci. Rep.* **2016**, *6*, 25088. [CrossRef] [PubMed]

37. Ju, X.-T.; Xing, G.-X.; Chen, X.-P.; Zhang, S.-L.; Zhang, L.-J.; Liu, X.-J.; Cui, Z.-L.; Yin, B.; Christie, P.; Zhu, Z.-L.; et al. Reducing environmental risk by improving N man-agement in intensive Chinese agricultural systems. *Proc. Natl. Acad. Sci. USA* **2009**, *106*, 3041–3046. [CrossRef]

38. Lee, K.-H.; Jose, S. Nitrate leaching in cottonwood and loblolly pine biomass plantations along a nitrogen fertilization gradient. *Agric. Ecosyst. Environ.* **2005**, *105*, 615–623. [CrossRef]

39. De Notaris, C.; Rasmussen, J.; Sørensen, P.; Olesen, J.E. Nitrogen leaching: A crop rotation perspective on the effect of N surplus, field management and use of catch crops. *Agric. Ecosyst. Environ.* **2018**, *255*, 188–197. [CrossRef]

40. Yang, B.; Ma, Y.; Zhang, C.; Jia, Y.; Li, B.; Zheng, X. Cleaner Production Technologies Increased Economic Benefits and Greenhouse Gas Intensity in an Eco-Rice System in China. *Sustainability* **2019**, *11*, 7090. [CrossRef]

41. Paulo, P.L.; Galbiati, A.F.; Filho, F.J.C.M.; Bernardes, F.S.; Carvalho, G.A.; Boncz, M. Evapotranspiration tank for the treatment, disposal and resource recovery of blackwater. *Resour. Conserv. Recycl.* **2019**, *147*, 61–66. [CrossRef]

42. Costa-Gutierrez, S.B.; Raimondo, E.E.; Lami, M.J.; Vincent, P.A.; Red, C. Inoculation of pseudomonas mutant strains can im-prove growth of soybean and corn plants in soils under salt stress. *Rhizosphere* **2020**, *16*, 100255. [CrossRef]

43. Yin, L.; Dai, X.; He, M. Delayed sowing improves nitrogen utilization efficiency in winter wheat without impacting yield. *Field Crop. Res.* **2018**, *221*, 90–97. [CrossRef]

44. Gao, Y.; Li, H.; Yang, B.; Wei, X.; Zhang, C.; Xu, Y.; Zheng, X. The preliminary evaluation of differential characteristics and factor evaluation of the microbial structure of rural household toilet excrement in China. *Environ. Sci. Pollut. Res.* **2021**, *28*, 43842–43852. [CrossRef] [PubMed]

45. Zhou, X.; Prithvi, P.S.; Perez-Mercado, L.F.; Barton, M.A.; Lyu, Y.; Guo, S.; Nie, X.; Wu, F.; Li, Z. China should focus beyond access to toilets to tap into the full potential of its Rural Toilet Revolution. *Resour. Conserv. Recycl.* **2022**, *178*, 106100. [CrossRef]

46. Yang, B.; Zhang, T.; Zhang, M.; Li, B. Reactive nitrogen releases and nitrogen footprint during intensive vegetable production affected by partial human manure substitution. *Environ. Sci. Pollut. Res.* **2021**, 1–11. [CrossRef]

47. Ishii, S.K.L.; Boyer, T.H. Life cycle comparison of centralized wastewater treatment and urine source separation with struvite precipitation: Focus on urine nutrient management. *Water Res.* **2015**, *79*, 88–103. [CrossRef] [PubMed]

Review

Effects of Biofuel Crop Switchgrass (*Panicum virgatum*) Cultivation on Soil Carbon Sequestration and Greenhouse Gas Emissions: A Review

Jian Bai [1], Laicong Luo [1], Aixin Li [1], Xiaoqin Lai [1], Xi Zhang [1], Yadi Yu [1], Hao Wang [1], Nansheng Wu [2,*] and Ling Zhang [1,*]

[1] Jiangxi Provincial Key Laboratory of Silviculture, College of Forestry, Jiangxi Agricultural University, Nanchang 330045, China

[2] National Innovation Alliance of *Choerospondias axillaris*, Nanchang 330045, China

* Correspondence: rensh111@126.com (N.W.); lingzhang09@126.com (L.Z.)

Abstract: Under the macroenvironmental background of global warming, all countries are working to limit climate change. Internationally, biofuel plants are considered to have great potential in carbon neutralization. Several countries have begun using biofuel crops as energy sources to neutralize carbon emissions. Switchgrass (*Panicum virgatum*) is considered a resource-efficient low-input crop that produces bioenergy. In this paper, we reviewed the effects of switchgrass cultivation on carbon sequestration and greenhouse gas (GHG) emissions. Moreover, the future application and research of switchgrass are discussed and prospected. Switchgrass has huge aboveground and underground biomass, manifesting its huge carbon sequestration potential. The net change of soil surface 30 cm soil organic carbon in 15 years is predicted to be 6.49 Mg ha^{-1}, significantly higher than that of other crops. In addition, its net ecosystem CO_2 exchange is about -485 to -118 g C m^{-2} yr^{-1}, which greatly affects the annual CO_2 flux of the cultivation environment. Nitrogen (N) fertilizer is the main source of N_2O emission in the switchgrass field. Nitrogen addition increases the yield of switchgrass and also increases the N_2O flux of switchgrass soil. It is necessary to formulate the most appropriate N fertilizer application strategy. CH_4 emissions are also an important indicator of carbon debt. The effects of switchgrass cultivation on CH_4 emissions may be significant but are often ignored. Future studies on GHG emissions by switchgrass should also focus on CH_4. In conclusion, as a biofuel crop, switchgrass can well balance the effects of climate change. It is necessary to conduct studies of switchgrass globally with the long-term dimension of climate change effects.

Keywords: biofuel crops; carbon sequestration; greenhouse gas emissions; net ecosystem CO_2 exchange; phytoremediation

Citation: Bai, J.; Luo, L.; Li, A.; Lai, X.; Zhang, X.; Yu, Y.; Wang, H.; Wu, N.; Zhang, L. Effects of Biofuel Crop Switchgrass (*Panicum virgatum*) Cultivation on Soil Carbon Sequestration and Greenhouse Gas Emissions: A Review. *Life* **2022**, *12*, 2105. https://doi.org/10.3390/life12122105

Academic Editor: Dmitry L. Musolin

Received: 17 November 2022
Accepted: 13 December 2022
Published: 14 December 2022

1. Introduction

In recent years, global climate change, especially global warming, has attracted widespread attention from all walks of life worldwide. Internationally, the United Nations Framework Convention on Climate Change (UNFCCC) reached The Paris Agreement at the Paris Climate Change Conference. The Paris Agreement aims to limit the increase in global average temperatures to 2 °C from pre-industrial periods and to limit temperature increases to 1.5 °C to constrain global temperature rise as soon as possible. The leading cause of global warming is the increase in greenhouse gases produced by human activities [1,2]; the main greenhouse gases are carbon dioxide (CO_2), nitrous oxide (N_2O), and methane (CH_4) [3].

To limit temperature growth to 2 °C, the remaining global cumulative CO_2 emissions should not exceed 400–1000 Gt by the end of the century. Therefore, how to effectively control carbon emissions, especially human-induced carbon emissions, has attracted more attention from the international community. For non-CO_2 greenhouse gases, CH_4 and N_2O

are of concern. According to the global warming potential (GWP) calculation, the GWP of CH_4 is about 23–25 times that of CO_2 and the GWP of N_2O is about 296 times that of CO_2 [4].

Carbon emitted from fossil fuels since the industrial revolution is about 420 Gt C [5]. Globally, CH_4 and N_2O emissions from agriculture exceed 610 million tons per year, accounting for 12% of total emissions [6]. Therefore, reducing agriculture's carbon emissions is a crucial issue. Biofuel crops are mainly perennial (herbaceous or woody) that improve soil quality, promote nutrient cycling and carbon fixation, and can produce large quantities of high-carbon biomass. Compared with fossil fuels, biofuel crops have greater advantages in energy utilization [7] (Figure 1). Furthermore, biofuel crops require less maintenance and input and can be adapted to marginal soils. Eggelston et al. [8] showed that 300–1300 Mt C fossil fuels can be replaced if 10–15% of agricultural land is used to grow biofuel crops. Moreover, under the circumstances, CH_4 emissions from agriculture can be reduced by 15–56% and N_2O emissions can be reduced by 9–26%.

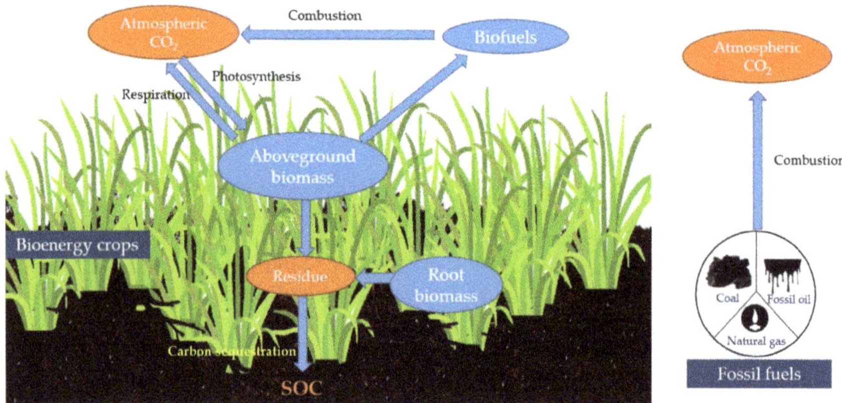

Figure 1. Carbon turnover process of biofuel crops vs. fossil fuels.

Switchgrass (*Panicum virgatum*), a species of grass in the family Poaceae, is an adaptable perennial herbaceous C4 plant native to North America. It is mainly distributed in several countries south of 55° north latitude. There are two ecotypes, including upland and lowland. In general, lowland types, which can grow up to more than 3 m, have larger biomass than upland types [9]. The tillers of the upland ecotype are usually shorter and better adapted to cold and dry habitats [10]. Since the mid-1980s, switchgrass has been mainly used as a renewable biofuel source for research. So far, switchgrass has been used in various forms of biofuel conversion processes, including cellulosic ethanol production, biogas, and direct combustion [11,12]. As a biofuel source, switchgrass has a lower demand for fertilizers and pesticides, which allows switchgrass to produce good yields on the land of the best part of soil types [13]. The climate benefits of biofuels are mainly manifested in (1) the use of alternative fossil fuels; (2) reducing greenhouse gas emissions during biofuel production, mainly through soil C accumulation and avoidance of greenhouse gas emissions. This paper discusses the potential contribution of switchgrass in carbon sequestration and greenhouse gas emission reduction. The future application and study of switchgrass are discussed and prospected.

2. Carbon Sequestration by Switchgrass

Soil and plant carbon sequestration is a practical way to mitigate CO_2 emissions [14,15]. As early as the 1990s, Ma et al. [16] studied the effects of soil management measures, including nitrogen (N) application, row spacing, and harvest frequency, on carbon sequestration in switchgrass fields established for 2–3 years. The results found that the soil management measures of switchgrass did not change the soil carbon concentration. Interestingly, they compared the soils of the switchgrass and their adjacent fallow soils that had been established for some time (10 years). The results showed that the soil organic carbon (SOC) of the switchgrass was significantly higher than that of the fallow land; the SOC of the 0–15 cm soil increased by 44.8% and in the 15–30 cm soil it increased by 28.2% [16]. Therefore, switchgrass soil can store more soil carbon, although detecting it may take several years. Carbon sequestration in the switchgrass field does not occur only in the topsoil. Liebig et al. [17] show that switchgrass soils below 30 cm can also effectively sequester SOC. C stored in deep soils is not prone to mineralization and erosion. According to a four-year measurement, after four growing seasons, the SOC produced by switchgrass is 9.45 Mg ha^{-1} [18]. Different ages of switchgrass have different changes in the underground 30 cm SOC. A prediction from Anderson et al. [19] of net changes in SOC indicated that the change of switchgrass to the underground SOC increases with time and the switchgrass cultivated for 15 years increases by about 6.49 Mg ha^{-1} (Table 1). Hong et al. [20] found that the biomass of switchgrass fields across locations in the USA increased significantly in the first three years after the establishment (Figure 2). The total yield in the third and fourth years was similar (Figure 2). At a soil depth of 1 m, the SOC of switchgrass soil was 9.4% higher than that of farmland and 8.1% higher than that of *Andropogon gerardi*, while the quality of soil N is basically the same as that of farmland [21].

Table 1. Projected net changes in SOC (Mg C ha^{-1}) in the top 30 cm of soil under biofuel crops of various ages. Adapted from Anderson et al. [19].

Ages (Year)	Net Change in SOC (Mg ha^{-1} per 30 cm)		
	Switchgrass	Sugarcane	*Miscanthus*
5	2.66	−34.21	2.31
10	4.64	−31.57	2.97
15	6.49	−28.93	3.63

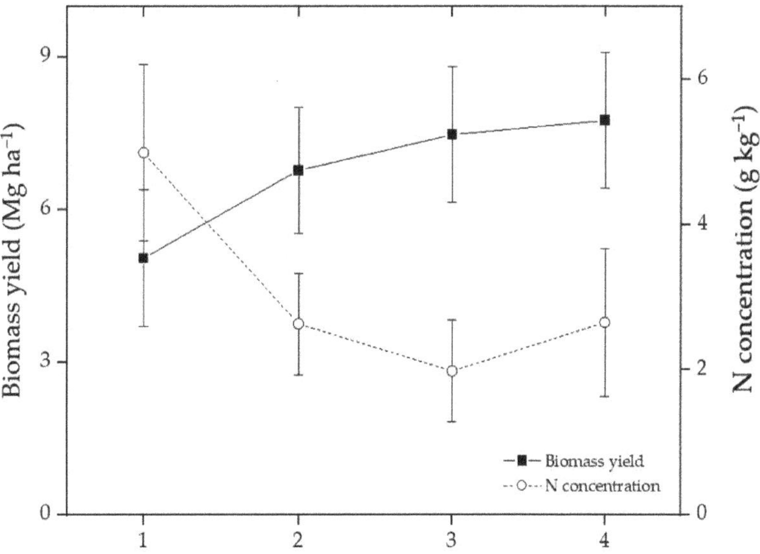

Figure 2. Average biomass yield and N concentrations in biomass of switchgrass across locations in the USA. Data were replotted from Hong et al. [20].

Although the effects of switchgrass soil management measures on soil carbon sequestration did not have a significant effect in the study of Ma et al. [16], some studies have shown that fertilizer management measures and harvesting methods have essential effects on switchgrass carbon sequestration [22–25]. On the Conservation Reserve Program (CRP) land dominated by switchgrass in South Dakota, there is no benefit if the N applied exceeds 56 kg ha^{-1} [24]. The application of NH_4NO_3 and manure can effectively increase switchgrass's soil carbon sequestration, especially at soil depths of 30–90 cm [23]. Switchgrass is a perennial herb whose roots can grow deep in the soil. It has considerable root biomass, which is more than the aboveground biomass [16]. The root biomass of switchgrass in different soil types at different depths is shown as follows (Table 2). Zan et al. [26] showed that switchgrass has a biomass 4–5 times that of maize and can store 2.2 Mg C ha^{-1} yr^{-1}. Liebig et al. [17] found that the cumulative rate of C was 1.1 Mg C ha^{-1} yr^{-1}, most of which occurred at depths of 30 cm underground. Tulbure et al. [27] used RF (Random Forest packet in R) to analyze the effects of multiple factors such as fertilizer, genetics, and precipitation on yield. The results showed that the total variance of RF interpretation was 75%, with N fertilizer being the most important explanatory variable, followed by genetics, precipitation, and management measures.

Table 2. The root biomass (kg m^{-2}) of switchgrass in different soil types [28].

Depth (cm)	Clay Loam	Sandy Loam
	Root Biomass (kg m^{-2})	
0–20	7.28 ± 0.44	7.44 ± 0.39
20–40	2.66 ± 0.10	1.97 ± 0.43
40–60	1.75 ± 0.07	1.84 ± 0.33
60–80	1.25 ± 0.08	3.23 ± 0.31
80–100	1.16 ± 0.07	2.26 ± 0.25

3. Net Ecosystem CO_2 Exchange of Switchgrass

Net ecosystem CO_2 exchange (NEE) is the result of imbalances between total primary production (GPP) and ecosystem respiration (Re), which can affect carbon dynamics and budgets [29]. A better understanding of switchgrass's NEE changes will help assess switchgrass's potential for climate change mitigation. Some NEE of biofuel crops are shown below (Table 3). Zeri et al. [30] found that switchgrass has a stronger carbon sink capacity at the initial establishment stage than *Miscanthus* × *giganteus* (giant miscanthus, a sterile hybrid of *Miscanthus sinensis* and *Miscanthus sacchariflorus*). Compared with corn, switchgrass absorbs more carbon. The NEE of switchgrass is -336 ± 40 g C m^{-2} and that of corn is 64 ± 41 g C m^{-2} [31]. From 2012 to 2013, the analysis of the NEE of switchgrass [32–34] showed that it had a stronger carbon sink capability than sorghum land. This may be because that switchgrass has a net carbon sink of about 4–5 months (April/May–August) and sorghum has only 3 months of net carbon sink (June–August).

Table 3. Four energy crops' net ecosystem CO_2 exchange (NEE) of biofuel crops since 2005.

Location	Year	Crop	NEE (g C m^{-2} yr^{-1})	Citation
Urbana, IL, USA	2009	Switchgrass	-453 ± 20	[30]
		Miscanthus	-281 ± 30	
		Corn	-307 ± 40	
	2010	Switchgrass	-485 ± 20	
Guelph, ON, Canada	2014	Switchgrass	-336 ± 40	[31]
		Corn	64 ± 41	
Chickasha, OK, USA	2012	Switchgrass	-490 ± 59	[33]
		Sorghum	-261 ± 48	
	2013	Switchgrass	-406 ± 24	
		Sorghum	-330 ± 45	
Cadriano, Italy	2014–2016	Switchgrass	-733	[18]
Guelph, ON, Canada	2012	Switchgrass	-380 ± 25	[35]
	2013		-430 ± 30	
Ligonier, PA, USA	2005–2006	Switchgrass	-118	[36]
	2006–2007		-248	
	2007–2008		-189	

Surprisingly, in a study by Zenone et al. [37], the switchgrass field did not exist as a carbon sink but produced CO_2 emissions. However, their measurements were only carried out for 2 years. In contrast, in the 4-year study [18], CO_2 can be fixed each year and NEE stabilized at higher values from the second year, although the cumulative biomass in the first year was relatively low. Zenone et al. [37] and Virgilio et al. [18] conducted studies on a newly established switchgrass field. For mature switchgrass fields, Eichelmann et al. [35] conducted two years of data collection and found that NEE is 106 ± 45 g C m^{-2} in the first year, which was represented as a carbon source, while the NEE in the second year was -59 ± 45 g C m^{-2}, which was manifested as a carbon sink. Previous four-year studies of mature switchgrass fields [36] showed that the first three years of switchgrass forests served as a sink of net CO_2, while the following year became a source of CO_2 emissions. These results suggest that switchgrass may be able to act as a powerful carbon sink in its establishment years, then its benefits will be reduced or even transformed into a carbon source.

4. CH$_4$ Flux as Affected by Switchgrass Cultivation

The soil can be either a sink or a source of CH_4. Some studies [38,39] have shown that forest and grassland soils are the primary consumers of CH_4 thanks to methane oxidation bacteria in soils. However, some agronomic and fertilization measures reduce the function of CH_4 oxidation in soils [40]. This is because these measures can change the N state of the soil, temperature, water content, and other factors [41].

There is less research on CH_4 emissions from switchgrass cultivation but more on the biogas production of switchgrass. However, a recent study [28] shows that under the condition of planting switchgrass, CH_4 consumption per year is 39–47% less than that of unplanted plots. As far as CH_4 is concerned, planting switchgrass is detrimental to CH_4 emission reductions. After all, the GWP of CH_4 in 100 years is about 25 times that of CO_2. The CH_4 flux produced by planting switchgrass should be counted in the carbon budget. In future studies, more measurements of CH_4 data will be needed to pay more attention to CH_4 emissions.

5. N$_2$O Emission from Switchgrass Soil

5.1. N$_2$O Emission of Switchgrass Soil with N Addition

Using winter legumes as nitrogen (N) sources is an N addition measure in agriculture. However, research shows that winter legumes will not increase the yield, cellulose, lignin, and hemicellulose concentration of switchgrass [42]. It may be that legumes are not conducive to use as the main nitrogen source of switchgrass. The primary source of N_2O is the microbial processes of nitrification and denitrification in the soil and it is easy to increase N_2O emissions through N input to the soil. Crutzen et al. [43] argue that N_2O emissions caused by N-fertilizers required for the production of energy plants may offset the effects of energy plants in reducing the greenhouse effect or even exacerbate the greenhouse effect. Qin et al. [44] estimated the potential greenhouse gas emissions of the bioenergy ecosystem using the biogeochemical model AgTEM, a generic agroecosystem model with vegetation specific parameters characterizing specific crop structures and processes [45]. The results show that the N_2O flux of switchgrass and *Miscanthus* in the United States is equivalent to that of corn (Figure 3). According to the crop type and nitrogen application rate, the N_2O flux is about 0.05–0.11 g N m^{-2} per year [44].

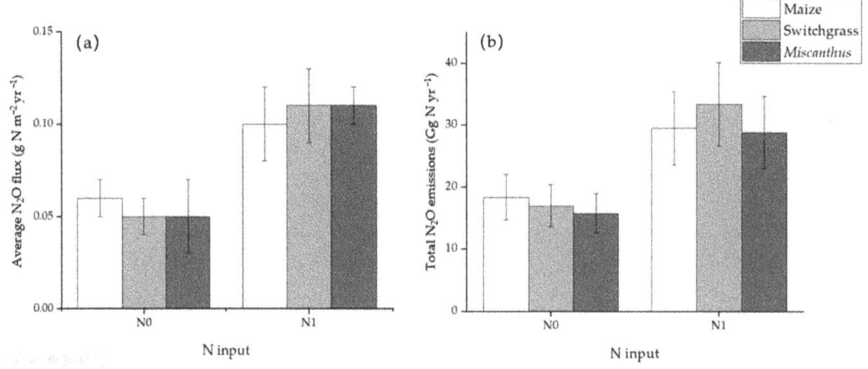

Figure 3. Estimated average N_2O fluxes (**a**) and total N_2O emissions (**b**) at different N input levels (N0: 0 g N ha^{-1} yr^{-1} N1: 67 g N ha^{-1} yr^{-1}) in the conterminous United States. Data were replotted from Qin et al. [44].

However, Wile et al. [46] studied greenhouse gas emissions such as N_2O from N applications to biofuel plants. The results showed that the annual cumulative N_2O emissions of switchgrass cultivation systems were low and would not offset the benefits of using these biofuel feedstocks instead of fossil fuel energy. Wile et al. [46] argue that N fertilizers increase N_2O emissions, but that increases in plant biomass can offset these increases. Similarly, Nikiema et al. [47] found that N fertilizer (0 to 112 kg N ha^{-1}) had no effect on the N_2O emission of switchgrass but increased its yield. This indicated that the N application reduced GHG emissions per unit plant biomass. Schmer et al. [48] determined the greenhouse gas fluxes of switchgrass during the growing season in the Great Plains north of Mantan and found that the application of N fertilizer affected the N_2O flux during the growing season but did not affect the flux of CO_2 and CH_4. However, Ruan et al. [49] demonstrated that applying N fertilizer to mature switchgrass had little effect on yield but increased N_2O emissions (Table 4). McGowan et al. [50] applied different levels of N treatment to switchgrass. The results showed that N fertilizer application higher than switchgrass demand could lead to large N_2O emissions, negatively affecting GHG emissions. Therefore, how to apply N fertilizer reasonably is a crucial problem. It is necessary to ensure that the benefits of N application on climate change mitigation will not be reduced by the N_2O it generates. A meta-analysis from Wullschleger et al. [51] of switchgrass at 39 sites in 19 states of the United States found that the optimal N application amount of switchgrass was about 100 kg N ha^{-1}.

Table 4. Biomass yield of switchgrass and N_2O emissions with different N addition treatments.

Location	Year	N Source	N Treatment (kg N ha^{-1} yr^{-1})	Yield (t ha^{-1} yr^{-1})	N_2O Emissions (g N ha^{-1} yr^{-1})	Citation
Truro, NS, Canada	2009	NH$_4$NO$_3$	0	7.1	463	[46]
			40	6.6	345	
			120	7	933	
Mandan, ND, USA	2010	Urea	0	3.67	58.94	[48]
			67	4.47	184.29	
MI, USA	2009–2011	Urea	0	5.95	374.32	[49]
			28	6.91	512.34	
			56	7.85	698.45	
			84	7.62	964.03	
			112	7.72	1321.62	
			140	8.26	1806.78	
			168	7.82	2486.41	
			196	8.03	2867	

The utilization efficiency of single and mixed cultivation of switchgrass is different. Duran et al. [52] showed that, compared with the mixed planting of switchgrass and local perennial grasses, the single planting of switchgrass increased N_2O emissions and the potential nitrate–nitrogen leaching capacity of fertilized switchgrass plots. This may be because different varieties of herbs have different or partially overlapped demands for N, which provides a reasonable combination for a more effective use of N. Although perennial biofuel plants such as switchgrass will produce N_2O during production, in general, perennial biofuel crops emit less N_2O than annual crops during their establishment. According to Oates et al. [53], perennial systems produce much lower N_2O emissions per unit of ground than annual cropping systems.

5.2. Microbial Mechanism of N_2O Emission from Switchgrass Field

Soil microbial activities related to N_2O emission mainly include nitrification and denitrification [54]. Ammonia oxidizing archaea (AOA) and ammonia oxidizing bacteria (AOB) are closely related to the first step of nitrification. The transformation of dissolved N into gaseous N in denitrification is mainly related to *nirK*, *nirS*, and *nosZ* genes [55]. AOA are more abundant than AOB in agricultural soil [56]. Pannu et al. [57] confirmed that AOB abundance in switchgrass fertilized plots was positively correlated with N_2O emissions. They found that applying N fertilizer increased the quantity and activity of AOB, which would lead to an increase in N_2O emissions from the fertilized plots. Similarly, mycorrhizal fungi, AOA, and AOB increased with N input [58]. The *nirS* and *nosZ* genes are related indicators of denitrification. The expression of *nirS* and *nosZ* genes in the N application area was significantly higher than that in the non-nitrogen application area, indicating that fertilization in these systems may change the denitrification activity and may lead to related nitrogen loss, without yield return [59].

5.3. Environmental Factors Affecting Soil N_2O Emissions with Switchgrass

Switchgrass is a perennial plant that needs to be managed, and switchgrass production can be affected by changes in temperature and precipitation space [60]. Similarly, its production can be affected by temporal changes in climate. Behrman et al. [61] estimated the productivity of current and future switchgrass in the central and eastern United States. They predicted that future climate change would significantly affect the spatial distribution and productivity of switchgrass.

Not only is nitrogen added directly, but many environmental factors will also affect N_2O emissions from switchgrasssuch as temperature and precipitation. For example, under the condition of fertilization, a large amount of precipitation and a high-temperature climate can create a substrate-rich environment with limited oxygen for microorganisms, which is a good promotion of some anaerobic microbial processes such as denitrification.

Duncan et al. [62] used quantile regression to evaluate the correlation of four environmental factors—NH_4^+, NO_3^-, soil temperature, and water-filled pore space (WFPS)—to the upper limit of N_2O emissions from switchgrass soil. The results showed that these four factors were significantly and positively correlated with the upper limit of N_2O flux. However, the regression slope of non-fertilized plots was generally lower than that of fertilized plots. Soil moisture is one of the main factors driving N_2O emissions from soils. A study has shown that N_2O is emitted optimally in the WFPS range of 70–80% [63]. In addition, changes in soil oxygen concentration caused by soil temperature also make soil denitrification extremely sensitive to increasing temperature [64].

Intercropping also affected N_2O emissions from switchgrass. Pannu et al. [57] showed that intercropping alfalfa (70:30, switchgrass: alfalfa) reduced dry matter yield but increased N_2O flux.

6. Application of Switchgrass Cultivation in Degraded Land

Land degradation is an important topic in the 21st century due to its impact on agricultural productivity, the environment, and food security. If the degraded land can be used for biofuel crops, it will benefit agriculture development. Switchgrass is found to have good tolerance to drought and flooding, so it is suitable for marginal land. Slessarev et al. [65] conducted a 10-year (2008–2018) study on degraded land after sandstorms in the United States to assess the impact of switchgrass on the deep organic carbon storage of the three marginal soils. The carbon storage of topsoil (approximately 0–30 cm depth) and subsoil (approximately 30–100 cm depth) in switchgrass areas were significantly higher than those in reference ($p < 0.01$). Moreover, the switchgrass cultivation can increase the operational taxonomic unit (OTU) richness of marginal land [28]. They found that when the marginal land was converted into switchgrass land, the Shannon index increased significantly over time and the community composition changed [28]. This result may indicate that switchgrass has caused the improvement in soil quality.

Long-term soil toxic trace metal pollution will change the soil organic matter and microbial community, thus destroying the ecosystem [66]. Phytoremediation refers to affecting pollutants through plant extraction, which concentrates pollutants (such as toxic trace metals) in the environment into plant tissues [67]. Phytoremediation has been used to repair degraded soil. For example, tomatoes (*Solanum lycopersicum*) are used to repair cadmium (Cd) contaminated soil. Caesar-137 and strontium-90 were removed from power using sunflowers (*Helianthus annuus*) after the Chernobyl accident [68]. Switchgrass is a metal accumulator used in agriculture as a phytoremediation strategy as well. This strategy has the advantage of disposing of contaminated sites without excavation. Switchgrass in situ promotes environmental pollutants' decomposition, fixation, and removal. Switchgrass can accelerate the degradation of atrazine and other herbicides, can absorb toxic trace metals in soil, and has good agronomic characteristics and high biomass [69,70]. A large amount of biomass can be harvested through the annual harvest in several seasons. Therefore, it is feasible to use switchgrass for in situ extraction of toxic trace metals. Finally, the amount of toxic metals will be reduced so that the affected land can restore the natural ecosystem or be used for crops productively. Balsamo et al. [69] studied the enrichment of lead (Pb) by switchgrass and timothy grass (*Pheum pretense*) and found that when the soil Pb concentration was 120 mg kg^{-1}, the Pb content in switchgrass leaves was $0.028 \pm 10\%$ of the dry weight of leaves. In other words, assuming the soil Pb concentration is the same, based on the 7.5 t ha^{-1} harvest yield (Table 3), about 0.02 t lead can be removed by switchgrass every year, which is a considerable number. A study showed that switchgrass has medium tolerance to Cd and that a low concentration of Cd (100–175 μM) promoted the growth of switchgrass [71]. Fertilization can be used to improve the absorption capacity of switchgrass to toxic trace metals. The study showed that plants receiving high nitrogen had significantly the largest leaf dry mass and the highest Pb concentration [72]. Chelating agents can also promote the absorption of Pb by switchgrass. When NTA (nitrogenous acid) and APG (aluminum polyglucoside) were applied together, the Pb concentration in switchgrass leaves was more than doubled [73].

Moreover, the contaminated biomass harvested can be used as raw material for biofuel production. Cellulose, hemicellulose, and pectin can be decomposed into glucose or other sugars by enzymes and then bio-ethanol can be produced by yeast fermentation [74].

7. Conclusions and Future Prospects

Overall, most studies believe that changes in SOC or NEE caused by switchgrass cultivation have a positive effect on climate change. Although short-term research shows that SOC can be significantly increased by perennial biomass production, a long-term measurement is required to assess the dynamics of SOC.

N_2O emissions from switchgrass are lower than from most other perennial grasses and annual crops. For the GHGs (mainly N_2O) directly emitted during switchgrass cultivation, more effective fertilizer utilization strategies must be developed and used. N_2O emissions can be reduced by estimating crop N demand and by improving N-use efficiency through timely fertilization.

Growing energy demands and concerns about climate change drive the use of energy plants, but, even so, the biofuel plant land cannot be developed unbridled. It is not suitable to develop biofuel plants with land that could be planted with large amounts of food, thus posing a danger to food security. Furthermore, more consideration should be given to using some marginal land to develop and grow biofuel plants such as switchgrass. It can be considered to establish switchgrass on some lands with highly toxic trace metals to simultaneously achieve the goal of carbon sequestration and soil restoration.

To develop an optimal cultivation strategy, future studies need to pay more attention to the relationship between fertilization, yield, and C and N loss. It is appropriate to consider the efficient breeding of switchgrass in order to establish switchgrass fields in a shorter time. Meanwhile, CH_4 is a non-negligible carbon debt, which should be taken into account when calculating carbon loss data. Moreover, a comprehensive GHG budget and explicit spatial modeling of soil and plant carbon stocks should be considered to fully assess the impact of the large-scale transformation of these prairie sites.

Author Contributions: Conceptualization, J.B., N.W. and L.Z.; writing—original draft preparation, J.B.; writing—review and editing, L.L., A.L., X.L., X.Z., Y.Y., X.Z., H.W., N.W. and L.Z. All authors have read and agreed to the published version of the manuscript.

Funding: This research received no external funding.

Institutional Review Board Statement: Not applicable.

Informed Consent Statement: Not applicable.

Data Availability Statement: Not applicable.

Conflicts of Interest: The authors declare no conflict of interest.

References

1. Lynas, M.; Houlton, B.Z.; Perry, S. Greater than 99% consensus on human caused climate change in the peer-reviewed scientific literature. *Environ. Res. Lett.* **2021**, *16*, 114005. [CrossRef]
2. Lu, X.; Li, Y.; Wang, H.; Singh, B.P.; Hu, S.; Luo, Y.; Li, J.; Xiao, Y.; Cai, X.; Li, Y. Responses of soil greenhouse gas emissions to different application rates of biochar in a subtropical Chinese chestnut plantation. *Agric. For. Meteorol.* **2019**, *271*, 168–179. [CrossRef]
3. Forster, P.; Storelvmo, T. Chapter 7: The Earths energy budget, climate feedbacks, and climate sensitivity. In *Climate Change 2021: The Physical Science Basis*; Masson-Delmotte, V., Zhai, P., Pirani, A., Connors, S.L., Eds.; Cambridge University Press: Cambridge, Britain, 2021; pp. 923–1025. Available online: https://www.ipcc.ch/report/ar6/wg1/downloads/report/IPCC_AR6_WGI_Chapter07.pdf (accessed on 8 July 2022).
4. Robertson, G.P.; Grace, P.R. Greenhouse gas fluxes in tropical and temperate agriculture: The need for a full-cost accounting of global warming potentials. In *Tropical Agriculture in Transition—Opportunities for Mitigating Greenhouse Gas Emissions?* Springer: Dordrecht, The Netherlands, 2004; pp. 51–63. [CrossRef]

5. McCarthy, J.J.; Canziani, O.F.; Leary, N.A.; Dokken, D.J.; White, K.S. *Climate Change 2001: Impacts, Adaptation, and Vulnerability: Contribution of Working Group II to the Third Assessment Report of the Intergovernmental Panel on Climate Change*; Cambridge University Press: Cambridge, UK, 2001; pp. 75–913.

6. Frank, S.; Havlík, P.; Stehfest, E.; van Meijl, H.; Witzke, P.; Pérez-Domínguez, I.; van Dijk, M.; Doelman, J.C.; Fellmann, T.; Koopman, J.F. Agricultural non-CO_2 emission reduction potential in the context of the 1.5 °C target. *Nat. Clim. Chang.* **2019**, *9*, 66–72. [CrossRef]

7. Kole, C.; Joshi, C.P.; Shonnard, D.R. *Handbook of Bioenergy Crop Plants*; CRC Press: Boca Raton, FL, USA, 2012; pp. 3–119.

8. Eggleston, H.S.; Buendia, L.; Miwa, K.; Ngara, T.; Tanabe, K. 2006 IPCC Guidelines for National Greenhouse Gas Inventories. 2006. Available online: https://www.ipcc-nggip.iges.or.jp/meeting/pdfiles/Washington_Report.pdf (accessed on 1 September 2022).

9. Porter, C.L., Jr. An analysis of variation between upland and lowland switchgrass, *Panicum virgatum* L., in central Oklahoma. *Ecology* **1966**, *47*, 980–992. [CrossRef]

10. Gonulal, E.; Soylu, S.; Sahin, M. Effects of different water stress levels on biomass yield and agronomic traits of switchgrass (*Panicum virgatum* L.) cultivars under arid and semi-arid conditions. *Turkish J. Field Crop.* **2021**, *26*, 25–34. [CrossRef]

11. Bransby, D.I.; McLaughlin, S.B.; Parrish, D.J. A review of carbon and nitrogen balances in switchgrass grown for energy. *Biomass Bioenergy* **1998**, *14*, 379–384. [CrossRef]

12. McLaughlin, S.B.; Kszos, L.A. Development of switchgrass (*Panicum virgatum*) as a bioenergy feedstock in the United States. *Biomass Bioenergy* **2005**, *28*, 515–535. [CrossRef]

13. Lemus, R.; Lal, R. Bioenergy crops and carbon sequestration. *Crit. Rev. Plant Sci.* **2005**, *24*, 1–21. [CrossRef]

14. Kucharik, C.J.; Brye, K.R.; Norman, J.M.; Foley, J.A.; Gower, S.T.; Bundy, L.G. Measurements and modeling of carbon and nitrogen cycling in agroecosystems of southern Wisconsin: Potential for SOC sequestration during the next 50 years. *Ecosystems* **2001**, *4*, 237–258. [CrossRef]

15. Ussiri, D.A.; Lal, R. Long-term tillage effects on soil carbon storage and carbon dioxide emissions in continuous corn cropping system from an alfisol in Ohio. *Soil Tillage Res.* **2009**, *104*, 39–47. [CrossRef]

16. Ma, Z.; Wood, C.W.; Bransby, D.I. Soil management impacts on soil carbon sequestration by switchgrass. *Biomass Bioenergy* **2000**, *18*, 469–477. [CrossRef]

17. Liebig, M.A.; Johnson, H.A.; Hanson, J.D.; Frank, A.B. Soil carbon under switchgrass stands and cultivated cropland. *Biomass Bioenergy* **2005**, *28*, 347–354. [CrossRef]

18. Di Virgilio, N.; Facini, O.; Nocentini, A.; Nardino, M.; Rossi, F.; Monti, A. Four-year measurement of net ecosystem gas exchange of switchgrass in a Mediterranean climate after long-term arable land use. *Glob. Chang. Biol. Bioenergy* **2019**, *11*, 466–482. [CrossRef]

19. Anderson Teixeira, K.J.; Davis, S.C.; Masters, M.D.; Delucia, E.H. Changes in soil organic carbon under biofuel crops. *Glob. Chang. Biol. Bioenergy* **2009**, *1*, 75–96. [CrossRef]

20. Hong, C.O.; Owens, V.N.; Bransby, D.; Farris, R.; Fike, J.; Heaton, E.; Kim, S.; Mayton, H.; Mitchell, R.; Viands, D. Switchgrass response to nitrogen fertilizer across diverse environments in the USA: A regional feedstock partnership report. *Bioenergy Res.* **2014**, *7*, 777–788. [CrossRef]

21. Omonode, R.A.; Vyn, T.J. Vertical distribution of soil organic carbon and nitrogen under warm-season native grasses relative to croplands in west-central Indiana, USA. *Agr. Ecosyst. Environ.* **2006**, *117*, 159–170. [CrossRef]

22. Follett, R.F.; Vogel, K.P.; Varvel, G.E.; Mitchell, R.B.; Kimble, J. Soil carbon sequestration by switchgrass and no-till maize grown for bioenergy. *Bioenergy Res.* **2012**, *5*, 866–875. [CrossRef]

23. Lee, D.K.; Owens, V.N.; Doolittle, J.J. Switchgrass and soil carbon sequestration response to ammonium nitrate, manure, and harvest frequency on conservation reserve program land. *Agron. J.* **2007**, *99*, 462–468. [CrossRef]

24. Mulkey, V.R.; Owens, V.N.; Lee, D.K. Management of switchgrass-dominated conservation reserve program lands for biomass production in South Dakota. *Crop Sci.* **2006**, *46*, 712–720. [CrossRef]

25. Al-Kaisi, M.M.; Grote, J.B. Cropping systems effects on improving soil carbon stocks of exposed subsoil. *Soil Sci. Soc. Am. J.* **2007**, *71*, 1381–1388. [CrossRef]

26. Zan, C.S.; Fyles, J.W.; Girouard, P.; Samson, R.A. Carbon sequestration in perennial bioenergy, annual corn and uncultivated systems in southern Quebec. *Agr. Ecosyst. Environ.* **2001**, *86*, 135–144. [CrossRef]

27. Tulbure, M.G.; Wimberly, M.C.; Boe, A.; Owens, V.N. Climatic and genetic controls of yields of switchgrass, a model bioenergy species. *Agr. Ecosyst. Environ.* **2012**, *146*, 121–129. [CrossRef]

28. Bates, C.T.; Escalas, A.; Kuang, J.; Hale, L.; Wang, Y.; Herman, D.; Nuccio, E.E.; Wan, X.; Bhattacharyya, A.; Fu, Y. Conversion of marginal land into switchgrass conditionally accrues soil carbon but reduces methane consumption. *ISME J.* **2022**, *16*, 10–25. [CrossRef] [PubMed]

29. Zhang, Z.; Zhang, R.; Cescatti, A.; Wohlfahrt, G.; Buchmann, N.; Zhu, J.; Chen, G.; Moyano, F.; Pumpanen, J.; Hirano, T. Effect of climate warming on the annual terrestrial net ecosystem CO_2 exchange globally in the boreal and temperate regions. *Sci. Rep.* **2017**, *7*, 3108. [CrossRef]

30. Zeri, M.; Anderson-Teixeira, K.; Hickman, G.; Masters, M.; DeLucia, E.; Bernacchi, C.J. Carbon exchange by establishing biofuel crops in Central Illinois. *Agr. Ecosyst. Environ.* **2011**, *144*, 319–329. [CrossRef]

31. Eichelmann, E.; Wagner-Riddle, C.; Warland, J.; Deen, B.; Voroney, P. Comparison of carbon budget, evapotranspiration, and albedo effect between the biofuel crops switchgrass and corn. *Agr. Ecosyst. Environ.* **2016**, *231*, 271–282. [CrossRef]

32. Wagle, P.; Kakani, V.G. Seasonal variability in net ecosystem carbon dioxide exchange over a young switchgrass stand. *Glob. Chang. Biol. Bioenergy* **2014**, *6*, 339–350. [CrossRef]

33. Wagle, P.; Kakani, V.G.; Huhnke, R.L. Net ecosystem carbon dioxide exchange of dedicated bioenergy feedstocks: Switchgrass and high biomass sorghum. *Agric. For. Meteorol.* **2015**, *207*, 107–116. [CrossRef]

34. Wagle, P.; Kakani, V.G.; Huhnke, R.L. Evapotranspiration and ecosystem water use efficiency of switchgrass and high biomass sorghum. *Agron. J.* **2016**, *108*, 1007–1019. [CrossRef]

35. Eichelmann, E.; Wagner Riddle, C.; Warland, J.; Deen, B.; Voroney, P. Carbon dioxide exchange dynamics over a mature switchgrass stand. *Glob. Chang. Biol. Bioenergy* **2016**, *8*, 428–442. [CrossRef]

36. Skinner, R.H.; Adler, P.R. Carbon dioxide and water fluxes from switchgrass managed for bioenergy production. *Agr. Ecosyst. Environ.* **2010**, *138*, 257–264. [CrossRef]

37. Zenone, T.; Gelfand, I.; Chen, J.; Hamilton, S.K.; Robertson, G.P. From set-aside grassland to annual and perennial cellulosic biofuel crops: Effects of land use change on carbon balance. *Agric. For. Meteorol.* **2013**, *182*, 1–12. [CrossRef]

38. Tlustos, P.; Willison, T.W.; Baker, J.C.; Murphy, D.V.; Pavlikova, D.; Goulding, K.; Powlson, D.S. Short-term effects of nitrogen on methane oxidation in soils. *Biol. Fertil. Soils* **1998**, *28*, 64–70. [CrossRef]

39. Dutaur, L.; Verchot, L.V. A global inventory of the soil CH_4 sink. *Glob. Biogeochem. Cycles* **2007**, *21*, GB4013. [CrossRef]

40. Chan, A.; Parkin, T.B. Effect of land use on methane flux from soil. *J. Environ. Qual.* **2001**, *30*, 786–797. [CrossRef]

41. Bodelier, P.L.; Laanbroek, H.J. Nitrogen as a regulatory factor of methane oxidation in soils and sediments. *FEMS Microbiol. Ecol.* **2004**, *47*, 265–277. [CrossRef]

42. Sutradhar, A.K.; Miller, E.C.; Arnall, D.B.; Dunn, B.L.; Girma, K.; Raun, W.R. Switchgrass forage yield and biofuel quality with no-tillage interseeded winter legumes in the southern Great Plains. *J. Plant. Nutr.* **2017**, *40*, 2382–2391. [CrossRef]

43. Crutzen, P.J.; Mosier, A.R.; Smith, K.A.; Winiwarter, W. *N_2O Release from Agro-Biofuel Production Negates Global Warming Reduction by Replacing Fossil Fuels*; Springer: Berlin/Heidelberg, Germany, 2016; pp. 227–238. [CrossRef]

44. Qin, Z.; Zhuang, Q.; Zhu, X. Carbon and nitrogen dynamics in bioenergy ecosystems: 2. Potential greenhouse gas emissions and global warming intensity in the conterminous United States. *Glob. Chang. Biol. Bioenergy* **2015**, *7*, 25–39. [CrossRef]

45. Qin, Z.; Zhuang, Q.; Zhu, X. Carbon and nitrogen dynamics in bioenergy ecosystems: 1. Model development, validation and sensitivity analysis. *Glob. Chang. Biol. Bioenergy* **2014**, *6*, 740–755. [CrossRef]

46. Wile, A.; Burton, D.L.; Sharifi, M.; Lynch, D.; Main, M.; Papadopoulos, Y.A. Effect of nitrogen fertilizer application rate on yield, methane and nitrous oxide emissions from switchgrass (*Panicum virgatum* L.) and reed canarygrass (*Phalaris arundinacea* L.). *Can. J. Soil. Sci.* **2014**, *94*, 129–137. [CrossRef]

47. Nikièma, P.; Rothstein, D.E.; Min, D.; Kapp, C.J. Nitrogen fertilization of switchgrass increases biomass yield and improves net greenhouse gas balance in northern Michigan, USA. *Biomass Bioenergy* **2011**, *35*, 4356–4367. [CrossRef]

48. Schmer, M.R.; Liebig, M.A.; Hendrickson, J.R.; Tanaka, D.L.; Phillips, R.L. Growing season greenhouse gas flux from switchgrass in the northern great plains. *Biomass Bioenergy* **2012**, *45*, 315–319. [CrossRef]

49. Ruan, L.; Bhardwaj, A.K.; Hamilton, S.K.; Robertson, G.P. Nitrogen fertilization challenges the climate benefit of cellulosic biofuels. *Environ. Res. Lett.* **2016**, *11*, 64007. [CrossRef]

50. McGowan, A.R.; Min, D.H.; Williams, J.R.; Rice, C.W. Impact of nitrogen application rate on switchgrass yield, production costs, and nitrous oxide emissions. *J. Environ. Qual.* **2018**, *47*, 228–237. [CrossRef] [PubMed]

51. Wullschleger, S.D.; Davis, E.B.; Borsuk, M.E.; Gunderson, C.A.; Lynd, L.R. Biomass production in switchgrass across the United States: Database description and determinants of yield. *Agron. J.* **2010**, *102*, 1158–1168. [CrossRef]

52. Duran, B.E.; Duncan, D.S.; Oates, L.G.; Kucharik, C.J.; Jackson, R.D. Nitrogen fertilization effects on productivity and nitrogen loss in three grass-based perennial bioenergy cropping systems. *PLoS ONE* **2016**, *11*, e151919. [CrossRef]

53. Oates, L.G.; Duncan, D.S.; Gelfand, I.; Millar, N.; Robertson, G.P.; Jackson, R.D. Nitrous oxide emissions during establishment of eight alternative cellulosic bioenergy cropping systems in the North Central United States. *Glob. Chang. Biol. Bioenergy* **2016**, *8*, 539–549. [CrossRef]

54. Hassan, M.U.; Aamer, M.; Mahmood, A.; Awan, M.I.; Barbanti, L.; Seleiman, M.F.; Bakhsh, G.; Alkharabsheh, H.M.; Babur, E.; Shao, J.; et al. Management strategies to mitigate N_2O emissions in agriculture. *Life* **2022**, *12*, 439. [CrossRef]

55. Wei, W.; Isobe, K.; Nishizawa, T.; Zhu, L.; Shiratori, Y.; Ohte, N.; Koba, K.; Otsuka, S.; Senoo, K. Higher diversity and abundance of denitrifying microorganisms in environments than considered previously. *ISME J.* **2015**, *9*, 1954–1965. [CrossRef]

56. Leininger, S.; Urich, T.; Schloter, M.; Schwark, L.; Qi, J.; Nicol, G.W.; Prosser, J.I.; Schuster, S.C.; Schleper, C. Archaea predominate among ammonia-oxidizing prokaryotes in soils. *Nature* **2006**, *442*, 806–809. [CrossRef]

57. Pannu, M.W.; Meinhardt, K.A.; Bertagnolli, A.; Fransen, S.C.; Stahl, D.A.; Strand, S.E. Nitrous oxide emissions associated with ammonia-oxidizing bacteria abundance in fields of switchgrass with and without intercropped alfalfa. *Environ. Microbiol. Rep.* **2019**, *11*, 727–735. [CrossRef] [PubMed]

58. Cha, G.; Meinhardt, K.A.; Orellana, L.H.; Hatt, J.K.; Pannu, M.W.; Stahl, D.A.; Konstantinidis, K.T. The influence of alfalfa-switchgrass intercropping on microbial community structure and function. *Environ. Microbiol.* **2021**, *23*, 6828–6843. [CrossRef] [PubMed]

59. Thompson, K.A.; Deen, B.; Dunfield, K.E. Impacts of surface-applied residues on N-cycling soil microbial communities in miscanthus and switchgrass cropping systems. *Appl. Soil. Ecol.* **2018**, *130*, 79–83. [CrossRef]

60. Casler, M.D.; Boe, A.R. Cultivar× environment interactions in switchgrass. *Crop Sci.* **2003**, *43*, 2226–2233. [CrossRef]

61. Behrman, K.D.; Kiniry, J.R.; Winchell, M.; Juenger, T.E.; Keitt, T.H. Spatial forecasting of switchgrass productivity under current and future climate change scenarios. *Ecol. Appl.* **2013**, *23*, 73–85. [CrossRef] [PubMed]

62. Duncan, D.S.; Oates, L.G.; Gelfand, I.; Millar, N.; Robertson, G.P.; Jackson, R.D. Environmental factors function as constraints on soil nitrous oxide fluxes in bioenergy feedstock cropping systems. *Glob. Chang. Biol. Bioenergy* **2019**, *11*, 416–426. [CrossRef]

63. Davidson, E.A.; Keller, M.; Erickson, H.E.; Verchot, L.V.; Veldkamp, E. Testing a conceptual model of soil emissions of nitrous and nitric oxides: Using two functions based on soil nitrogen availability and soil water content, the hole-in-the-pipe model characterizes a large fraction of the observed variation of nitric oxide and nitrous oxide emissions from soils. *Bioscience* **2000**, *50*, 667–680. [CrossRef]

64. Schaufler, G.; Kitzler, B.; Schindlbacher, A.; Skiba, U.; Sutton, M.A.; Zechmeister Boltenstern, S. Greenhouse gas emissions from European soils under different land use: Effects of soil moisture and temperature. *Eur. J. Soil Sci.* **2010**, *61*, 683–696. [CrossRef]

65. Slessarev, E.W.; Nuccio, E.E.; McFarlane, K.J.; Ramon, C.E.; Saha, M.; Firestone, M.K.; Pett Ridge, J. Quantifying the effects of switchgrass (Panicum virgatum) on deep organic C stocks using natural abundance ^{14}C in three marginal soils. *Glob. Chang. Biol. Bioenergy* **2020**, *12*, 834–847. [CrossRef]

66. Zamulina, I.V.; Gorovtsov, A.V.; Minkina, T.M.; Mandzhieva, S.S.; Burachevskaya, M.V.; Bauer, T.V. Soil organic matter and biological activity under long-term contamination with copper. *Environ. Geochem. Health* **2022**, *44*, 387–398. [CrossRef]

67. Raskin, I.; Salt, D.E.; Smith, R.D. Phytoremediation. *Plant Mol. Biol.* **1998**, *49*, 643–668. [CrossRef]

68. Cooney, C.M. News: Sunflowers remove radionuclides from water in ongoing phytoremediation field tests. *Environ. Sci. Technol.* **1996**, *30*, 194A. [CrossRef]

69. Balsamo, R.A.; Kelly, W.J.; Satrio, J.A.; Ruiz-Felix, M.N.; Fetterman, M.; Wynn, R.; Hagel, K. Utilization of grasses for potential biofuel production and phytoremediation of heavy metal contaminated soils. *Int. J. Phytoremediation* **2015**, *17*, 448–455. [CrossRef] [PubMed]

70. Murphy, I.J.; Coats, J.R. The capacity of switchgrass (*Panicum virgatum*) to degrade atrazine in a phytoremediation setting. *Environ. Toxicol. Chem.* **2011**, *30*, 715–722. [CrossRef]

71. Wang, Q.; Gu, M.; Ma, X.; Zhang, H.; Wang, Y.; Cui, J.; Gao, W.; Gui, J. Model optimization of cadmium and accumulation in switchgrass (*Panicum virgatum* L.): Potential use for ecological phytoremediation in Cd-contaminated soils. *Environ. Sci. Pollut. Res.* **2015**, *22*, 16758–16771. [CrossRef]

72. Greipsson, S.; McElroy, T.; Koether, M. Effects of supplementary nutrients (soil-nitrogen or foliar-iron) on switchgrass (*Panicum virgatum* L.) grown in Pb-contaminated soil. *J. Plant. Nutr.* **2022**, *45*, 2919–2930. [CrossRef]

73. Hart, G.; Koether, M.; McElroy, T.; Greipsson, S. Evaluation of chelating agents used in phytoextraction by switchgrass of lead contaminated soil. *Plants* **2022**, *11*, 1012. [CrossRef]

74. David, K.; Ragauskas, A.J. Switchgrass as an energy crop for biofuel production: A review of its ligno-cellulosic chemical properties. *Energ. Environ. Sci.* **2010**, *3*, 1182–1190. [CrossRef]

Review

Management Strategies to Mitigate N$_2$O Emissions in Agriculture

Muhammad Umair Hassan [1], Muhammad Aamer [1], Athar Mahmood [2], Masood Iqbal Awan [3], Lorenzo Barbanti [4], Mahmoud F. Seleiman [5,6], Ghous Bakhsh [7], Hiba M. Alkharabsheh [8], Emre Babur [9], Jinhua Shao [1], Adnan Rasheed [10] and Guoqin Huang [1,*]

[1]   Research Center on Ecological Sciences, Jiangxi Agricultural University, Nanchang 330045, China; muhassanuaf@gmail.com (M.U.H.); muhammadaamer@jxau.edu.cn (M.A.); jinhuashao1@gmail.com (J.S.)

[2]   Department of Agronomy, University of Agriculture, Faisalabad 38040, Pakistan; athar.mahmood@uaf.edu.pk

[3]   Department of Agronomy, Sub-Campus Depalpur, Okara, University of Agriculture, Faisalabad 38040, Pakistan; masood.awan@uaf.edu.pk

[4]   Department of Agriculture, Food Sciences University of Bologna, 40127 Bologna, Italy; lorenzo.barbanti@unibo.it

[5]   Plant Production Department, College of Food and Agriculture Sciences, King Saud University, Riyadh 11451, Saudi Arabia; mseleiman@ksu.edu.sa

[6]   Department of Crop Sciences, Faculty of Agriculture, Menoufia University, Shibin El-Kom 32514, Egypt

[7]   Training and Publicity, Agriculture Extension, Dera Allah Yar 79000, Pakistan; ghouskhosa2353@gmail.com

[8]   Department of Water Resources and Environmental Management, Faculty of Agricultural Technology, Al Balqa Applied University, Salt 19117, Jordan; drhibakh@bau.edu.jo

[9]   Department of Forest Engineering, Faculty of Forestry, Kahramanmaras Sutcu Imam University, Kahramanmaras 46050, Turkey; emrebabur@ksu.edu.tr

[10] Key Laboratory of Crops Physiology, Ecology and Genetic Breeding, Ministry of Education/College of Agronomy, Jiangxi Agricultural University, Nanchang 330045, China; adnanbreeder@yahoo.com

*   Correspondence: hgqjxauhgq@jxau.edu.cn or hgqjxes@sina.com

Citation: Hassan, M.U.; Aamer, M.; Mahmood, A.; Awan, M.I.; Barbanti, L.; Seleiman, M.F.; Bakhsh, G.; Alkharabsheh, H.M.; Babur, E.; Shao, J.; et al. Management Strategies to Mitigate N$_2$O Emissions in Agriculture. *Life* **2022**, *12*, 439. https://doi.org/10.3390/life12030439

Academic Editor: Dmitry L. Musolin

Received: 24 January 2022
Accepted: 7 March 2022
Published: 17 March 2022

Publisher's Note: MDPI stays neutral with regard to jurisdictional claims in published maps and institutional affiliations.

Abstract: The concentration of greenhouse gases (GHGs) in the atmosphere has been increasing since the beginning of the industrial revolution. Nitrous oxide (N$_2$O) is one of the mightiest GHGs, and agriculture is one of the main sources of N$_2$O emissions. In this paper, we reviewed the mechanisms triggering N$_2$O emissions and the role of agricultural practices in their mitigation. The amount of N$_2$O produced from the soil through the combined processes of nitrification and denitrification is profoundly influenced by temperature, moisture, carbon, nitrogen and oxygen contents. These factors can be manipulated to a significant extent through field management practices, influencing N$_2$O emission. The relationships between N$_2$O occurrence and factors regulating it are an important premise for devising mitigation strategies. Here, we evaluated various options in the literature and found that N$_2$O emissions can be effectively reduced by intervening on time and through the method of N supply (30–40%, with peaks up to 80%), tillage and irrigation practices (both in non-univocal way), use of amendments, such as biochar and lime (up to 80%), use of slow-release fertilizers and/or nitrification inhibitors (up to 50%), plant treatment with arbuscular mycorrhizal fungi (up to 75%), appropriate crop rotations and schemes (up to 50%), and integrated nutrient management (in a non-univocal way). In conclusion, acting on N supply (fertilizer type, dose, time, method, etc.) is the most straightforward way to achieve significant N$_2$O reductions without compromising crop yields. However, tuning the rest of crop management (tillage, irrigation, rotation, etc.) to principles of good agricultural practices is also advisable, as it can fetch significant N$_2$O abatement vs. the risk of unexpected rise, which can be incurred by unwary management.

Keywords: N$_2$O emissions; denitrification; nitrification; C:N ratio; integrated nutrient management

1. Introduction

The sustainability of agricultural activities involves supporting crop yields under adverse natural conditions [1–9]. Many countries across the globe have adopted intensive agricultural practices to assure food security under the rapid increase in world population [10,11]. However, scaling up the level of crop intensiveness has devastating impacts on the environment [12]. Agriculture is a major contributor to greenhouse gases (GHGs) (namely, CO_2, N_2O and CH_4) released into the atmosphere and accounts for 10–12% of the total GHGs produced globally by anthropogenic activities [13,14]. These GHGs are a major source of global warming and climate change across the globe and pose a serious threat to global food security [15,16].

N_2O is a powerful and long-lasting GHG, has a global warming potential (GWP) 298 times as high as that of CO_2 and can contribute to the depletion of the stratospheric ozone layer [17]. Moreover, it is a very reactive gas, which catalyzes the production of the tropospheric ozone, exerting adverse impacts on humans and crop production [18,19]. Agriculture is responsible for about 60% of the global N_2O production, owing to the heavy usage of mineral N and the sustained use of legumes as cover and main crops releasing N at the end of their life cycle [20–22]. For example, from 1990 to 2005, agricultural emissions have increased by 14%, with an average increase of 49 Mt CO_2 per year [23]. Based on another source, during the last decade, approximately 80% of the world's total N_2O emissions were related to agricultural activities, with the concentration in atmosphere increasing from 270 ppb to 319 ppb [24]. Moreover, N_2O emissions are expected to increase by 35–60% in the near future, largely due to poor manure management and increased application of chemical fertilizers [24]. Additionally, excessive use and inappropriate timing of N application can lead to N leaching that affects water quality [25], resulting in increased N_2O emission from the landscape-draining waterways [26].

In soils, N_2O is mainly produced by transformation of reactive N through the microbes [25–29]. When N enters the soil, either from organic or mineral fertilizers in the form of NH_4^+ and NO_3^-, there are different processes that can result in N_2O formation. However, their relative prominence is still not well understood [30,31]. Three main processes, namely nitrification, denitrification and dissimilatory nitrate reductions, are considered the main contributors to N_2O emissions [27]. The contribution of each process to N_2O emission depends upon soil texture, organic C, soil pH, microbial activities and environmental conditions, including precipitation and temperature [28]. The quality and intricacy of N_2O production pathways, and their spatial as well as temporal variability, make the reduction in N_2O from soils quite challenging to interpret [32]. Crop management practices, including tillage and irrigation, N fertilizers, biochar, lime, nitrification inhibitors, slow-releasing fertilizers, arbuscular mycorrhizal fungi (AMF), suitable cultivars, appropriate crop rotations and integrated nutrient management (INM) can significantly influence soil properties, which in turn affect N_2O emissions [33–39]. Therefore, it is generally sensed that emissions can be mitigated by the suitable management of tillage and irrigation practices, reducing the overall N application and using biochar, lime, organic amendments, manures, nitrification inhibitors, fermented fertilizers, AMF, suitable crop rotations and INM (Figure 1).

To better appreciate the extent of these effects, organize in a comprehensive way the multiple contributions on this topic and discuss the variable results obtained in the quest to curb N_2O emission, we set out to review the potential of different management options to reduce N_2O emission on the basis of the available data. It is generally acknowledged that the adoption of suitable practices can play a significant role in restraining N_2O emission, but the extent to which the atmospheric equilibrium and agricultural production will benefit from these efforts is still questioned.

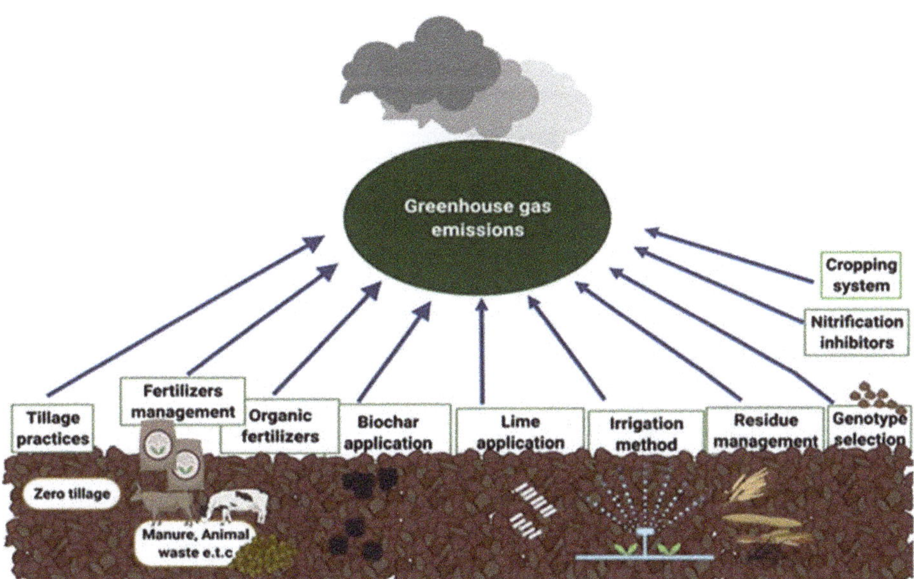

Figure 1. Management practices influencing N_2O emissions to the atmosphere. The adoption of several measures in each specific management sector can contribute to mitigate N_2O emission from agricultural soils.

2. N_2O Production and Emission

Nitrous oxide is produced in the process of nitrification, consisting of the microbial conversion of ammonia (NH_3) to nitrate (NO_3^-). Nitrification (NF) is considered the main process involved in the global N cycle. Most of the transformation of N during nitrification is mediated by autotrophic micro-organisms. The first step in nitrification is NH_3 oxidation to the hydroxylamine (NH_2OH). Both ammonia-oxidizing archaea (AOA) and ammonia-oxidizing bacteria (AOB) mediate this process.

In various soils, the quantity of AOA is higher than AOB, which supports the hypothesis that the abundance of AOA can better control nitrification rates, in turn leading to lower N_2O emission compared to soils with higher AOB [40,41]. This is especially true in the acidic soils, where AOA prevail as a result of their unique adaptation [42]. Nonetheless, the degree to which AOA vs. AOB can affect N_2O emission is still uncertain [43] and might depend on the NH_2OH fate. The metabolic and enzymatic pathways lead to decomposition of NH_2OH into NO_2^- and nitrogen oxide (NO) [44]. NO_2^- is further volatilized into HONO, but NO_2^- may be converted into NO, N_2O and N_2 via nitrifier denitrification [45,46].

In contrast to nitrification, denitrification (DNF) is a reduction process involved in the conversion of NO_3^- to N_2, mediated by facultative anaerobic bacteria [47]. This process can be completed up to N_2 production, but if it remains incomplete, it results in N release in the form of NO and N_2O [48].

The microbial processes of NF and DNF are responsible for 70% of global N_2O emission [49,50]. However, the above description of the two processes as sources of N_2O is a simplification, owing to the fact that the main process pathway can provide a wealth of collateral processes that either form or use N_2O. Moreover, other metabolic processes can contribute to N_2O production in soils:

- The decomposition of hydroxylamine during the process of autotrophic as well as heterotrophic nitrification;
- The chemical DNF of soil NO_2^- and abiotic decomposition of ammonium nitrate in the presence of light, humidity and reacting surfaces;

- The production of N_2O by nitrifier denitrification within the same nitrifying micro-organisms;
- The coupled nitrification–denitrification by different micro-organisms (the nitrite oxidizers produce nitrate, which is denitrified by denitrifiers in situ);
- The DNF conducted by microbes capable of using nitrogen oxides as alternative electron acceptors under O_2 limited conditions;
- The co-denitrification of organic N compounds with NO and nitrate ammonification or dissimilatory nitrate reduction to ammonium [51].

3. Environmental and Anthropic Factors Affecting N_2O Emission from Agricultural Soils

3.1. Soil pH

Soil pH is one of the main factors that can affect N_2O emission (Figure 2). The increase in soil pH can reduce the emission of N_2O [52,53], although some other source reports increased N_2O emission at increasing pH [54], which is consistent with denitrifying bacteria thriving on relatively high pH for their activities. Alkaline pH is considered responsible for enhancing the rates of both NF and DNF processes [55,56]. In general, soil pH influences the microbial population and activity, which directly impact N_2O emission [57].

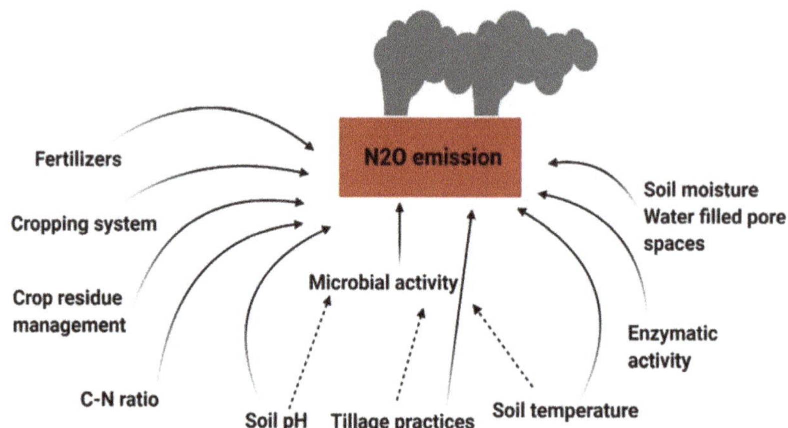

Figure 2. Factors and management practices responsible for N_2O emission from agricultural soils.

3.2. Soil Moisture and Temperature

Large quantities of N_2O are produced under high water-filled pore space (WFPS), owing to the fact that soil moisture controls N_2O emission through organic matter (OM) decomposition. Soil moisture can enhance organic C mineralization, which can control microbial metabolism and activities [58,59]. Thus, higher C stimulates the activities of micro-organisms by increasing substrate availability, which in turn increases N_2O emission. Moist soils enhance N_2O emission over long periods, owing to increased availability of C substrate for microbial activities. Moreover, no tillage (NT) can increase the WFPS compared to conventional tillage (CT), which can be a reason for increased N_2O emission under NT conditions. Soil temperature interacts with moisture in regulating N_2O production. Bacterial populations increase with increasing temperature up to a certain range (25–35 °C) [60,61], and the activities of both nitrifying and denitrifying bacteria are equally enhanced at higher soil temperatures [62].

3.3. Application of Crop Residues

The addition of crop residues and straw provides a source of easily available C and N, henceforth, a potential source of N_2O emission [63]. Nitrogen mineralized from crop residues is quite easily dispersed in the form of N_2O [64]. The release of N and C from mineralization of crop residues largely depends on the C:N ratio of the specific residues [65]. The rate of DNF depends on the amount of C that is made easily available to the pool of denitrifying bacteria [66]. High N_2O emission from loamy soil was observed following the incorporation of straw with low C:N ratio [65], while low N_2O emission from sandy soil was noticed with the addition of cereal straw with higher C:N compared to vegetable residues with lower C:N [67]. Therefore, the characteristics of crop residues incorporated into the soil can be a significant factor in N_2O emissions [68].

3.4. Nitrogen Application

Before 1950, less than 50% of N_2O emission was caused by N fertilizers in the agricultural sector. Nonetheless, most of the N_2O emissions were linked to animal rearing and related activities [69]. However, with the increase in human population and food demand, increased application of N fertilizers was also needed. Agriculture is responsible for more than 60% of N_2O emission [21,22]. Nitrogen fertilizers have high mobility in soil solution: after application, they enter the soil, undergoing diverse reactions resulting in N leaching, immobilization, volatilization and DNF [70]. Therefore, N fertilizers have significant impact on N_2O emission, leading to differentiated emissions according to fertilizer type [71]. The method and timing of N application also have substantial impact on N_2O emission [72]. Among the application methods, the N applied as side banding significantly reduced N_2O emission compared to broadcasting [73]. Similarly, the time of N application is very crucial, and the selection of suitable timing can contribute to N loss reduction. The available ammonium (NH_4^+) and nitrate (NO_3^-) are major sources of N_2O emission from soils [74], and N fertilizers, which more or less directly supply the two N forms, are largely implied in N_2O production and emission [75–77]. The deep placement of fertilizers has been seen to substantially improve crop growth compared to shallow and surface placement [78]. Plant roots tend to proliferate around the fertilizer area; therefore, deep placement considerably increased root density, N and water uptake from deeper layers in various cereals [78,79]. Moreover, in deep placement, a thicker layer must be crossed by diffusing N_2O, which prolongs the residence time and favors the ultimate reduction of N_2O to N_2 in the upper topsoil where no fertilizer N was placed [78], resulting in significant reduction in N_2O emission [79].

3.5. Soil Micro-Organisms

An increase in soil depth considerably decreases microbial biomass and activity. Microbial occurrence is imperative for NO_3^- and NO_2^- reduction to NO, N_2O or N_2; this reaction is coupled with electron transport in the DNF process [77]. Denitrifying bacteria have the ability to reduce NO_3^-, NO_2^- and NO under soil anaerobic conditions. They catch the energy from sunlight and organic or inorganic substrates, and are consequently known as phototroph, organotroph or lithotroph. Moreover, some enzymes, including ammonia monooxygenase, hydroxylamine oxidoreductase and nitrite oxidoreductase, are involved in the NF, and these enzymes either increase or decrease N_2O emission by affecting the rate of NF [80]. In a similar way, other enzymes, including nitrate reductase, nitrite reductase, nitric oxide reductase and nitrous oxide reductase, are involved in the DNF process. The occurrence and amount of these enzymes remarkably influence DNF rate and, consequently, N_2O soil emission [80]. The amount of soil organic carbon positively influences N_2O production and emission [81], also in association with soil moisture [82]. In fact, soil organic C provides the substrate for microbial growth that is needed for both NF and DNF processes [83].

3.6. Soil Characteristics

Fine textured soils emit more N_2O [84], owing to the fact that they have more capillary pores within soil aggregates compared to sandy soils [85]. The pores present in fine soils hold more water, leading to anaerobic conditions, which are maintained for a longer time, resulting in significant increase in N_2O emission compared to sandy soils [86]. The DNF process is also considerably increased, as soil texture becomes finer and WFPS increase [85]. When WFPS decrease, the DNF process is slowed. In fact, it was reported that in clayey soils, N_2O emission was considerably increased with increasing WFPS, up to 40%, and reached its maximum extent at WFPS higher than 70% [85]. Generally, soil texture affects N_2O emission by determining how likely it is for anaerobic vs. aerobic soil conditions to prevail [87,88]. Moreover, soil texture also affects N_2O emission owing to differences in soil N availability, the amount of organic carbon and microbial population [89]. Site exposure influences soil temperature and moisture, in turn affecting N_2O emission, as does field surface morphology; N_2O emission was recorded maximum in depressions vs. ridges and sloped lands, owing to higher moisture content present in depressed areas [90,91]. Lastly, lower air pressure at high altitudes also favors higher N_2O emissions due to a reduction in the counter pressure exerted on the soil [90,91].

4. Management Options to Mitigate N_2O Emission

4.1. Modification of Irrigation Pattern

Irrigation is an important factor in N_2O emission [92]. The amount of water supplied and the method of distribution affect soil moisture spatially and temporally [93], and significantly impact on the N cycle. This includes the processes of NF and DNF on which N_2O production depends [94,95].

Flood irrigation (FI) is the most common irrigation method in developing countries, such as India, Pakistan, Bangladesh and large parts of Africa. In FI, high volumes of water are applied to crops, resulting in fertilizers being strongly diluted and easily absorbed [94]. However, large irrigation volumes determine the anaerobic conditions conducive to N_2O production and nitrate leaching [96]. To prevent this, a precise water application technique, such as alternate wetting and drying (AWD), could be useful to save water while concurrently reducing GHG emissions. However, contrasting results are reported about the effect of AWD on N_2O emission and grain yields even in paddy rice, one of the crops most suited for AWD. On the one hand, Lahue et al. [97] found that AWD vs. FI curbed CH_4 emission by 80% in a clay-loamy soil, while significantly increasing the final yield; on the other hand, Lagomarsino et al. [98] reported that AWD saved water by 70% and decreased CH_4 emission by 97%, but it increased N_2O emission by five times in a clayey soil.

Generally, AWD inhibits CH_4 emission [77]; however, soil moisture during AWD cycles remains high, which can create anaerobic conditions [92] and favor N_2O emission. Soils produce large quantities of N_2O when WFPS fluctuates around 45–90% [99].

Under aerobic soil conditions, NF becomes the dominant N_2O production pathway when WFPS increases up to 60–70% [100]. Conversely, DNF becomes a dominant pathway for N_2O production when WFPS exceeds 60–70% [100]. However, the production of N_2O may still be limited with WFPS around 50–60% as a result of dissimilatory nitrate reduction to ammonia [101]. For instance, continuous flooding in rice releases less N_2O to the atmosphere [102,103], owing to water saturated conditions favoring ultimate NO_3^- reduction to N_2 by denitrifiers [51]. Conversely, AWD may be responsible for increased N_2O emission when it determines soil cracks; stronger aeration at deeper layers increases NF and provides substrate for N_2O emission [104,105].

Similarly, modifications in the irrigation method can play a crucial role in the amount of water used and N_2O emission. Different patterns of water infiltration and redistribution result in variable time trends of soil water content and water infiltration depths; all this has a great impact on soil N_2O emission and its spatial and temporal occurrence [106]. The surface layer in a field irrigated by sprinkler irrigation (SI) is relatively loose compared to FI. Therefore, in such soils, the NO_3-N and NH_4-N ions are less leached and remain more concentrated in the root zone [106,107], which makes them more easily absorbed by plant roots and, therefore, less prone to be turned into N_2O [107,108]. SI is a water-saving approach, and soil conditions during SI, as well as drip irrigation (DI), favor NF in both cases. Enhanced NF provides the substrate for N_2O emission [104–106]; however, SI is associated with modest WFPS, resulting more likely in reduced N_2O emissions [109–113].

It is therefore evinced that more advanced irrigation methods, such as SI and DI, lead to a contained risk of N_2O emission with respect to FI. A controversial role is played by AWD, which is proposed as an advanced version of FI: despite undeniable benefits in terms of water saving and crop performance, the unstable moisture conditions associated with AWD may be conducive to stronger N_2O production. In general, it is sensed that the irrigation practice should be directed toward higher water use efficiency, either by replacing less efficient methods or better tuning the existing ones, as a premise for more balanced moisture conducive to less N_2O emission.

4.2. Tillage Practices

Tillage practices influence crop productivity [114] as well as GHG emission, as they substantially affect soil properties [115]. Tillage disturbs the soil and increases CO_2 emission by aerating the soil and breaking soil aggregates, which release the organic carbon that favors microbial activities responsible for GHG emission [116].

It is not easy to univocally identify which tillage practices could reduce GHG emission [117], as contrasting results have been reported in the literature. In rice fields, Xiao et al. [118] and Liang et al. [119] noticed a substantial reduction in N_2O under no tillage (NT) compared to conventional tillage (CT). Conversely, a meta-analysis conducted by Mei et al. [120], including rice and other arable crops (wheat, maize, others), showed that conservation tillage increases N_2O emission by an average 17.8% compared to CT. Lastly, another meta-analysis conducted by Feng et al. [121] pointed out the advantage for NT in terms of N_2O and CH_4 reduction (-6.6% compared to CT). In this last source [121], special emphasis is given to the interactions of tillage with other crop management practices and land use patterns in triggering/mitigating GHG emission from agricultural soils. Despite the uncertainties in N_2O effects, NT practices appreciably offset GHG emissions owing to C sequestration [122] and reduction in CH_4 emissions. This results in the global warming potential (GWP) of NT being remarkably lower than that of CT [123,124]. In turn, this suggests that NT is beneficial for GHG emission and C-smart agriculture, and must be generally promoted in cropping systems. No tillage reduces the losses of OM and significantly increases soil bulk density (BD) [125]. The long-term use of NT can improve soil structure and reduce soil temperature, owing to the residues present on soil surface reflecting the incoming radiation and acting as a barrier between soil surface and atmospheric air [126]. This, in turn, may lead to reduced N_2O emission compared to CT [127,128]. In another case, CT increased water-holding capacity, WFPS and the availability of substrate for microbial activities, potentially leading to increased N_2O emission [129]. However, the effect of tillage practices can vary according to climate type. For instance, Van Kessel et al. [130] conducted a meta-analysis and found that dry warm climate significantly increases N_2O emissions. Rainfall and temperature are considered key factors affecting N_2O emissions. Higher rainfall increases soil moisture contents, which reduce the soil oxygen availability, which ultimately increases the NF and DNF and results in significant increase in N_2O emissions. Therefore, in warm dry regions, NT can be an important practice to reduce N_2O emission as compared to conventional tillage practices, which can increase N_2O emission due to decomposition of organic matter and increase in microbial activities [131]. Recently, Shakoot et al. [132] (2022) also found that NT reduces N_2O emissions in irrigated areas, whereas it increases N_2O emission in rain-fed areas.

The contrast among studies for N_2O emissions could be ascribed to different soil characteristics, ambient conditions and time at which tillage practices are carried out in a specific soil. However, despite the non-univocal effects on N_2O production, reduced and no tillage are associated with a beneficial effect, in general, in GHG mitigation. Therefore, as in the case of irrigation practices, it appears that more advanced tillage practices provide a more favorable background for the containment of GHG emission.

4.3. Crop Residue Management

Crop residues (CR) return to the soils is widely popular, owing to its benefits in increasing agricultural production and soil fertility [133,134]. Moreover, CR return also influences N_2O emissions by regulating the microbial activities, and C and N availability [135,136]. At a global level, it is estimated that CR return produces 0.4 million metric tons of N_2O-N/year [137]. Nonetheless, contrasting results have been shown in the literature concerning the effects of CR return on N_2O emission from agricultural soils, depending on several CR and soil characteristics.

Various authors noted that returning CR can increase N_2O emissions by increasing C and N availability for microbial activities and modifying soil aeration by improving soil aggregation and microbial demand, which is considered a major factor mediating soil NF and DNF for N_2O production [126,135,138,139]. Conversely, other authors reported that the addition of CR has an inhibitory effect on N_2O emission, depending on soil properties and C/N ratio of crop residues [140,141]. Additional soil characteristics influencing CR effects on N_2O emission are soil pH, texture, water content and residue C and N input to soil [142–144]. Soil pH affects CR decomposition, and C and N availability for NF, as well as DNF [145]. Similarly, soil texture affects soil permeability and water conditions and, therefore, CR decomposition and N transformation processes [146]. Thus, it is important to consider the above-discussed soil and CR properties when estimating N_2O emission from CR.

The return of CR can serve as a source of carbon for microbial growth, stimulating the N assimilation by micro-organisms. This action can prompt a strong competition for NH_4^+ between heterotrophic micro-organisms and autotrophic nitrifiers [147], resulting in N_2O production. Additionally, CRs serve as source of energy for denitrifiers, enhancing DNF and, resultantly, N_2O emission under aerobic conditions. In those agricultural systems where CRs are soil incorporated, they provide N and C for NF. For instance, coarse textured soils have low DNF owing to the limited availability of organic carbon [148,149], and the addition of CRs can result in increased N_2O emission. Moreover, in fine textured soils, CR addition improves soil properties and increases substrate availability and microbial activities; therefore, the addition of CRs with low C:N ratio increases N_2O emission from these soils [139]. Lastly, CRs from mature crops have higher C:N ratio and tend to immobilize N and reduce NO_3^- availability, thus limiting N_2O emission from agricultural soils [149]. The decomposition of CR interacts with soil water content in determining the O_2 status in organic hotspots. For instance, CR significantly increased the N_2O emission at 30 and 60% WFPS; however, after heavy rainfall and increase in WPFS at 90%, N_2O emission was reduced by CR owing to a shift in $N_2O:N_2$ product ratio of DNF due to more reducing conditions [150]. Lastly, residue incorporation during the spring season following N addition of N fertilizers increases the potential interactions between external N source and decomposition of CR, which can increase the DNF and, subsequently, N_2O emissions [151]. Even in CR management, it is perceived that no univocal behavior can be detected with respect to N_2O emission. Several features, including CR characteristics and ambient conditions, must be considered to enhance smart CR management and its contribution to reduced N_2O emission. The trade-offs for successful management are, nevertheless, undeniable.

4.4. Fertilizer Management

4.4.1. Adjusting Fertilizer Dose and Matching N Supply with Demand

The application of optimum levels of N and P fertilizers ensures higher yield and reduces background GHG emissions. N_2O emissions from soils are influenced by fertilizer type, amount and application time [152]. The containment of N doses at the lowest non-limiting levels decreases the soil N availability and, consequently, the N_2O emission [153].

Many experiments demonstrate a substantial increase in N_2O emission with application of N fertilizers however, N_2O emissions also varied according to source of N application (Table 1). In rice, N_2O emission increased with an increase in N rate [154], which is supported by another experiment where a 33% reduction in the reference N application resulted in −28% N_2O emissions [155]. In another study, it was noted that application of N (200 kg ha^{-1}) reduced the methane emissions by 25–30% from rice crop as compared to application of N (400 kg ha^{-1}) [156]. The N application method can also affect N_2O production. In fact, N placement near the roots increased the nitrogen use efficiency (NUE) and reduced N_2O emissions [157]. Moreover, optimizing N fertilizer use to better match nutrient supply with crop demand significantly reduced the soil amount of residual N, curbing N_2O emission [158]. From a practical viewpoint, split fertilizer applications at different crop stages ensure uninterrupted N availability, which in turn improves NUE and reduces N_2O emission [159].

4.4.2. Time of Fertilizer Application

The time of fertilizer application is in tight connection with the amount of fertilizer application from the perspective of reducing N_2O emission. Fertilizer application weeks after sowing instead of prior to sowing increases the chances that applied N will end up in crop tissues instead of getting lost to atmosphere and ground water. For instance, in maize, the side dressing of N at V-6 stage increased NUE and reduced N losses in the form of N_2O [160,161]. Contrarily to this, the autumn application of fertilizers or manure enhanced nitrate and N_2O losses [162,163].

Table 1. Effect of different sources of N fertilizers on N_2O emissions.

Crop	N Sources	N_2O Emission (kg ha^{-1})	References
Rice	Control (no fertilizers)	0.04	[164]
	AS (100 kg ha^{-1})	0.17	
	Urea (100 kg ha^{-1})	0.15	
Rice	Control (no fertilizers)	0.67	[116]
	NPK (210:105:240 kg ha^{-1})	6.51	
Rice	Control (no fertilizers)	0.64	[165]
	Urea (300 kg ha^{-1})	1.39	
Maize	Control (no fertilizers)	1.53 (kg N Mg^{-1})	[77]
	UAN (150 kg ha^{-1})	1.92 (kg N Mg^{-1})	
	CAN (150 kg ha^{-1})	1.81 (kg N Mg^{-1})	
Maize	Control (no fertilizers)	0.16	[166]
	Urea (145 kg ha^{-1})	0.30	
	AN (145 kg ha^{-1})	0.29	

UAN: Urea-ammonium nitrate, AS: Ammonium sulfate, CAN: calcium ammonium nitrate, AN: Ammonium nitrate, NPK: Nitrogen, phosphorus and potassium fertilizer.

4.4.3. Improving N Fertilizer Placement

The deep placement of N fertilizers compared to conventional application ensures effective nutrient availability at later growth stages [167]. The placement of N closer to the plants considerably decreases N_2O emission, as in the case of urea band application instead of broadcasting. Similarly, the side banding in wheat and canola, rather than the banded mid-row, appreciably reduced N_2O emission [168]. In another study, the deeper placement of N fertilizer in maize resulted in a reduction in N_2O emission compared to the shallow placement [168]. The site-specific N application according to field variability improves NUE by tailoring the applied N to soil spatial variability. In maize, site-specifically applied N reduced the overall N use by 25 kg/ha and resulted in a substantial reduction in N_2O emission [169].

Deep placement of fertilizers is potentially useful to reduce N_2O emissions [170]. In lowland rice [171], the deep placement of N fertilizers determined an 80% lower N_2O emission than the conventional surface spreading. In another rice study [172], deep N placement substantially reduced N_2O emission, owing to the fact that a large portion of N was retained in soil for a longer time. Moreover, Chapuis-Lardy et al. [173] argued that deep placement reduces N_2O emission as a result of microbial consumption of N_2O. Rutkowska et al. [174] also noticed a substantial reduction in soil N_2O emission from sandy soils with deep placement of N fertilizers. Conversely, some other authors noted no significant difference in N_2O emission with deep vs. broadcast application of fertilizers [175], and some others noted that deep placement of N fertilizers led to higher N_2O emission [176]. It is sensed that these variations in N_2O emission with deep placement vs. shallow placement or surface spreading can be attributed to differences in N source, the applied amount and interactions amid the soil and weather conditions [11].

4.4.4. Selection of Suitable Fertilizers

Fertilizer type can influence N_2O emission (Table 1) in association with time and amount of fertilizer application [177]. Fertilizers affect N_2O emission because of different content of NH_4^+, NO_3^- and organic C. Grave et al. [178] studied the impact of various N sources on N_2O emission in a maize–wheat rotation. They noted that urea and slurry application increased N_2O emission by 33% and 46%, respectively, as compared to the control plots. Bordoloi et al. [179] studied the impact of different levels of urea on N_2O emissions in a wheat cropping system and found that N_2O emission increased in parallel with urea increase, up to +174% N_2O emission with 100 kg N ha^{-1} from urea. Moreover, Lebender et al. [180] studied the impact of N source (calcium-ammonium-nitrate (CAN; range 0–400 kg ha^{-1})) on N_2O emission from the wheat crop. They noted that over the years, N_2O emission from 400 kg N ha^{-1} was significantly higher as compared to 200 kg N ha^{-1}.

The experimental results reported in Table 1 clearly show the differences among fertilizer sources for N_2O emission. Large differences can be seen among fertilizer forms [172,181]. Specifically, higher N_2O fluxes and losses occur more quickly from ammonium nitrate compared to urea [182,183]. The application of calcium ammonium nitrate, especially in wet soils with high OM, results in higher N_2O emissions [184]. In another work, Nayak et al. [185] reported that replacing urea with ammonium sulphate increases the N_2O and decreases the CH_4 emissions. However, further differences among N fertilizers for N_2O emission can be due to soil properties, such as texture, BD, pH, organic carbon, N and microbial population [186].

Overall, fertilizer management is the premier domain of intervention to mitigate N_2O emissions, as N fertilizers supply the nutrient that, to a varying degree (1.25% on average, according to the IPCC [153]), fuels N_2O emission from agricultural soils. However, N fertilizers are a powerful tool to boost agricultural productions and are, therefore, indispensable to the present level of world food production. More efficient ways of supplying this nutrient, i.e., determining the right amount, time and place of supply, are the only strategy to pursue the increase in agricultural production necessitated by a growing population, while concurrently restraining N_2O emission. Time and place of N application are the least controversial fields to achieve a significant containment of N_2O emission at no cost to potential yield. The higher level of N application significantly increased N_2O emissions [187,188]. The application of higher levels of N significant increases the DNF, which, resultantly, increases N_2O emissions. Moreover, fertilizers and type of N also influence NF and DNF and, resultantly, N_2O emissions. For instance, the application of anhydrous ammonia significantly increased N_2O emissions [189]. Environmental conditions also significantly affect N_2O emissions. The application of heavy doses of N can increase N_2O emissions in warm temperate regions due to favorable microbial activities [190]. The tropical and sub-tropical zones also favor the microbial NF and DNF, which are linked to CO_2 and N_2O emissions [191] (Xu et al., 2012). Therefore, the application of heavy doses of N must be avoided in these regions. Moreover, Muller et al. (2003) [192] also observed N_2O emission observed between −1 and 10 °C, and maximum N_2O emissions occurred near the 0 °C owing to increasing activity of N_2O reductase.

4.5. Biochar Application

Biochar is a C-rich product resulting from the pyrolysis of various sources of organic matter. Soil incorporation of biochar sequesters C and improves soil properties [41,193,194], involving physical, chemical and biochemical changes (Figure 3), influencing N_2O production [195]. The application of biochar can mitigate GHGs emissions from soils [196]. Because of slow degradation, biochar is considered as the best option for long-term carbon seizure in soils [197]. Biochar produced from plant biomass has a significant quantity of carbon that can be sequestered for up to 2000 years of mean residence time in soil [198]. Biochar application hinders GHGs emissions, therefore reducing global warming [197]. The application of biochar could reduce the emission of N_2O and NH_3 by 16.10% and 89.60%, respectively, as compared to control in rice crop [199].

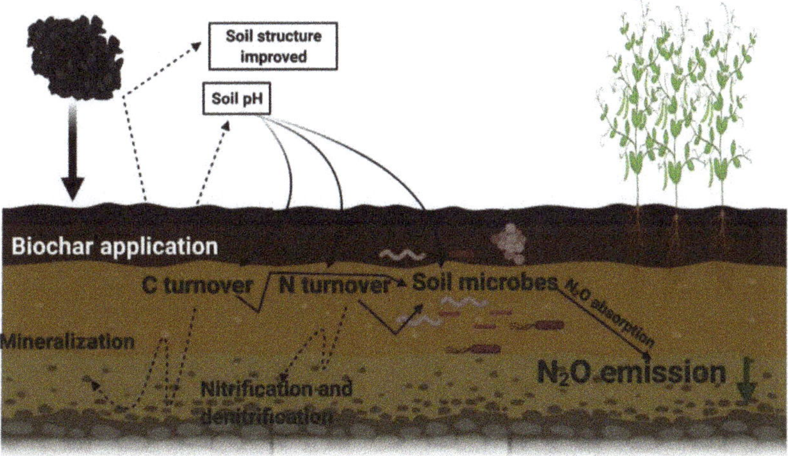

Figure 3. Mechanisms related to the role of biochar in mitigating N_2O emission.

The application of biochar increases soil pH and drives N_2O complete reduction to N_2, thus curbing N_2O emission (Table 2) [200]. However, the impact of biochar on N_2O emission varies according to biochar amount and soil properties, including pH, C:N ratio, organic carbon, water status, microbial and enzymatic activities. The biochar-mediated reduction in N_2O emission is made possible by biotic and abiotic pathways [53]. The main effects of biochar, modification of soil pH, aeration and water-holding capability, are those responsible for reduced N_2O emission [201]. However, biochar also directly absorbs N_2O, which further contributes to reduced emission [202].

An enzyme, N_2OR, catalyzes N_2O transformation into N_2 during the DNF. Under low soil pH, the assembly and functioning of this enzyme are constrained [202]; the application of biochar, by increasing soil pH, restores N_2OR functioning, which explains the relevant reduction in N_2O emission following biochar application [203]. The increase in aeration and O_2 availability resulting from biochar application contributes to further reduction in N_2O emissions by creating adverse conditions for microbial DNF [203].

Table 2. Effect of biochar on N_2O mitigation potential compared to no biochar application.

Biochar Application	N_2O Mitigation Potential (%)	Reference
BBC: 5 tons/ha	38	
BBC: 10 tons/ha	48	[204]
BBC: 15 tons/ha	61	
RCHBC: 50 tons/ha	36	[205]
MSBC: 16.77 tons/ha	10.8	[206]
BBC: 5 tons/ha	24.25	[207]
BBC: 15 tons/ha	30.7	
RSBC: 22.4 tons/ha	72.95	[208]
RSBC: 44.8 tons/ha	235.1	
RSBC: 36 tons/ha	50	[209]
RSBC: 72 tons/ha	83	
WSBC: 10 tons/ha	101.68	[210]
CSBC: 9 tons/ha	46.3	[211]
CSBC:13 tons/ha	33.3	
RSBC: 1% (w/w)	82.28	[212]
RSBC: 5% (w/w)	185.21	

GHBC: Grain husk biochar, BBC: Bamboo biochar, RCHBC: Rice and cotton husk biochar, MSBC: Maize stalk biochar, RSBC: Rice straw biochar, WSBC: Wood shaving biochar, CSBC: Cotton stalk biochar.

Additionally, biochar has a good adsorption potential, resulting in a considerable adsorption on its surface of NH_4^+ and NO_3^- [213], which reduces the N availability for N_2O production [214]. Biochar application also influences soil gene abundance, including nirK and nosZ [215]. These genes are highly sensitive to acidic pH, and they are involved in the process of DNF. The nosZ gene is linked to N_2O reductase, which catalyzes the reduction of N_2O to N_2 [216]. This is a further reason for biochar application resulting in substantial reduction in N_2O emission [217,218]. The application of biochar not only increases the SOC, crop yield and soil fertility, but also influences N_2O emissions. Many authors noted that biochar application reduced N_2O emission from agricultural soils [199]. However, environmental and soil conditions are significant factors that affect N_2O emissions. A meta-analysis conducted by Shakoor et al. [219] showed that application of biochar to fine textured soils significantly increased N_2O and CO_2 emissions. However, biochar application to coarse textured soils had no impact or reduced N_2O [219]. Under all circumstances, these effects can be best predicted by soil moisture and environmental conditions. Therefore, biochar application as a long-term approach to reduce N_2O emission appears quite promising, owing to the fact that the literature does not report any controversy in biochar's final effects. However, detailed mechanisms need to be further elucidated in order to assure higher reliability and, therefore, profitability of this practice.

4.6. Lime Application

Lime application modifies soil pH, which regulates different soil processes, including OM mineralization, NF and DNF, which in turn affect soil N_2O production [57,220]. However, contradictory reports have been issued regarding the impact of lime on N_2O emission, as the increased C and N mineralization, the latter resulting in higher NH_4^+ and NO_3^- contents, are the premise for enhanced NF and DNF, potentially leading to N_2O emission [52,200]. Conversely, other studies pointed out a significant reduction in N_2O with lime application, thanks to the increased N_2O reductase activity, resulting in more N_2 in exchange for less N_2O as the ultimate reduction product [221–223].

Soil N_2O emission is regulated by pH; N_2O emission decreases linearly with increased pH in a pH range of 4–7, irrespective of soil type [224]. The liming material also has great impact on the mineral N content. The addition of lime reduces NH_4^+ and speeds up the NF process, increasing NO_3^- content. The higher NO_3^- content at high pH stimulates micro-organisms to consume N_2O as electron acceptor in lieu of NO_3^- [225]. Thus, lime potentially ensures the complete DNF and promotes N_2O conversion to N_2. The increase in dissolved organic carbon associated with liming serves as a readily available C source for microbial growth, further contributing to N_2O abatement [226].

It is, therefore, evinced that liming acidic soils has an intrinsically favorable role in containing N_2O emissions; yet, the increase in readily available N forms, namely nitrates, is a potential source of N_2O, which deserves to be directed toward plant nutrition in the first instance or needs to be ultimately denitrified to N_2 in the second instance. In other words, the undeniable benefits of liming need to be carefully exploited in order to limit NO_3^- residual amounts which, under unfavorable conditions, fuel N_2O emission.

4.7. Use of Nitrification Inhibitors or Slow-Release Fertilizers

Nitrification inhibitors (NI) or slow-release N fertilizers can reduce both N_2O and CH_4 emissions [227]. The NI reduces N_2O emission directly, by inhibiting NF, as well as indirectly, by reducing NO_3^- availability for DNF [228], without compromising yield [229,230]. The chemical compounds present in the NI deactivate the enzymes responsible for the first step of NF (ammonia mono-oxygenase; AMO), maintaining NH_4^+ for longer periods in soils [231,232]. As a result, the NI decreases the rates of NF and the availability of substrates for denitrifiers, in turn reducing N_2O emission from fertilizers [233]. Various authors noticed a significant reduction in N_2O emission with application of different NI, including dicyandiamide, hydroquinol, nitropyrimidine and benzoic acid [234,235]. Lastly, plant-derived products, such as neem oil, neem cakes and karanja seed extract, can be used to inhibit NF; however, the exact mechanisms behind NF reduction induced by these products are still unclear.

The quest for NUE improvement is oriented toward the utilization of slow-release fertilizers, in order to reduce N_2O emission and the effects of global warming [94]. Slow-release fertilizers are mainly represented by controlled-release fertilizers (CRF) [236]. The CRF are granule-coated fertilizers, which slowly release the nutrients in order to improve nutrient uptake efficiency [237], reducing N losses by delaying the initial N supply and gradually providing the nutrient to the plants [238]. The application of CRF is recommended for those areas where the vulnerability to N losses is very high [239]. In paddy rice, the application of CRF significantly reduced N_2O losses and N application rate by 26–50%, without compromising yield [240]. The application of CRF can be seen as an effective approach to mitigate the N losses in combination [241] or as an alternative to urea [242].

It may be concluded that NI and CRF application is a promising approach to curb N_2O emission and other pathways of N loss, while concurrently improving crop production and NUE. The gradual release of nitrogen determined by both types of products ensures no peak of N supply responsible for increased N_2O emission. The main constraint in the use of NI and CRF is represented by their cost, which needs to be carefully evaluated in view of the expected return.

4.8. Use of Organic Amendments

Organic amendments (OA), including CR and animal wastes (i.e., manures and slurries), have been widely used to reduce N fertilizer application, improve soil fertility and alleviate environmental deterioration [3,14,243,244]. The effects of OA on N_2O emission have been documented in both lab and field studies. Some researchers demonstrated that OA enhance N_2O emission through DNF by serving as energy source for denitrifiers, favoring the formation of anaerobic micro-sites within soil aggregates [245,246]. Conversely, other researchers showed that OA reduce N_2O emission by increasing N microbial assimilation, thus limiting the availability of N substrates for the production of N_2O through NF and DNF [247,248]. The difference between these two contrasting behaviors could be due to differences in OA application, soil and climatic conditions, and fertilization history in the respective studies [249,250]. A long-term study showed that the amount of OA is critical for the accumulation of organic carbon and subsequent impact on N_2O

emission [251]. Moreover, it is assumed that the substitution ratio of synthetic fertilizers by OA is an important feature regulating N_2O emissions [251].

Therefore, OA are a viable alternative to mineral N fertilizers, in whose respect they do not provide clear advantages, as OA denote potential benefits as well as drawbacks in terms of N_2O emissions, depending on specific cases. Based on this, it is not easy to trace a consistent behavior for N_2O abatement through OA; it may only be concluded that a sensitive use of OA can contribute to an alleviation of the N_2O problem, whereas an unconsidered use of OA may result in aggravating the N_2O problem. Generally, NF is considered to be a major source of N_2O emission under limited moisture conditions; however, optimum moisture conditions in irrigated soils can induce anaerobic conditions, which promote the DNF [132]. Manure application ensures quick availability of C-substrates that promote the activity of DNF bacteria and increase the development of micro-sites due to higher moisture contents, which promote N_2O production and emissions [252,253]. Therefore, the application of organic manures in areas with higher rainfall and the application of heavy irrigation could increase N_2O emissions as compared to dry areas.

4.9. Fermented Organic Manures

The incorporation of fermented manures to soil can reduce GHG emission owing to rapid depletion of the pools of OM during fermentation [254]. The application of fermented CR significantly reduced CH_4 emission by 52% compared to application of fresh residues in a lab experiment [255]. A huge difference has been documented among GHG emissions triggered by fresh and pre-fermented materials [256]. For instance, the application of fermented biogas residues increased the CH_4 emission by 42%, while the unfermented material increased the CH_4 emission by more than 110% [234]. In another investigation, Nayak et al. [185] found that composted manure application significantly decreased N_2O and increased C sequestration and CH_4 emission. In rice, Zhang et al. [76] reported that compost application reduced N_2O emission by more than 50% compared to urea. The application of organic material produced as a result of aerobic composting of rice straw considerably reduced GHG emissions (CH_4 and N_2O) compared to fresh straw [255], suggesting that this approach is environmentally friendly.

It appears, therefore, that OA obtained from organic matter fermentation do not show harmful effects in the literature, possibly in association with more controlled doses with respect to OA originating from animal slurries and manures. Higher N_2O emissions in manure-amended and irrigated soils are a major concern in the climate-resilient agroecosystems [132] (Shakoor et al., 2022). Generally, the application of manures to irrigated lands increases N_2O emissions due to substrate availability and increasing micro-sites and microbial activities [252,253]. Higher rainfalls can also induce a significant increase in N_2O emissions following the application of fermented manures. Therefore, it could be suggested that manure application in irrigated soils and areas facing higher rainfall be dealt with cautiously to ensure better production and lower N_2O emissions.

4.10. Composting

Fermentation refers to a breakdown of organic substances into energy and by-products under anaerobic conditions, whereas composting involves the degradation of organic materials into value-added products under aerobic conditions. The application of composted materials has been widely practiced in crop production [6,256–258]. The dissolved organic carbon (DOC) released from composted animal manures can be a source of available C for microbial use in DNF, and the cumulative N_2O emission is directly related to the concentration of DOC in soil [72]. Vermi-composting is a promising approach that involves the conversion of organic materials into compost in the presence of earthworms [208,259]. The material produced as result of their activity has good structure and microbial activity associated with the abundance of liable resources. In a study on rice, the application of vermi-compost decreased the transfer of NH_4^+ and NO_3^- to water [260].

However, extensive use of vermi-compost might increase N_2O gaseous losses, owing to higher N availability, stimulating microbial activity. In fact, the combined use of vermi-compost and inorganic fertilizers increased N_2O emission by increasing the NO_3^- concentration with respect to unamended soil [261]. Conversely, the combined application of biochar and vermi-compost influenced soil properties through the C:N ratio and by increasing the abundance of *nosZ* genes; all of this led to reduced N_2O emission [262]. Therefore, the combined application of biochar and vermi-compost may be a promising approach to reduce N_2O emission. However, more studies are needed on a large scale to determine the influence and interaction of biochar and vermi-compost on N_2O emission and the mechanisms lying behind the reduction in N_2O emission due to these products.

Therefore, as in the case of fermented manures, the application of composted materials appears to be a promising strategy to improve soil properties and the general fertility. This, in turn, will likely result in restrained N_2O emission.

4.11. Role of Arbuscular Mycorrhizal Fungi

The understanding of the N_2O production pathway has been significantly improved recently by the development of isotopic methods for tracing the sources of N_2O [263,264]. N_2O production rate from soils is controlled by the available N, soil pH, OC, N, microbial activity and oxygen availability [26,265]. Arbuscular mycorrhizal fungi (AMF) are a key group of micro-organisms that form symbiotic relationships with most plants [38,39]. It is generally acknowledged that AMF play a role in the N cycle, as they can acquire this nutrient for host plants and have N requirements for themselves [39,40,266]. It has also been documented that AMF reduce NO_3^- leaching [267,268]. In general, these fungi reduce the availability of N sources in NF and DNF for the production of N_2O. AMF are able to acquire both NH_4^+ and NO_3^-; nonetheless, they prefer the more energetically attractive NH_4^+ [38,39,269]. The competition of these fungi with other micro-organisms for inorganic N reduces the N availability for N_2O producers and the consequential N_2O emission [270]. Another study highlighted a significant reduction in N_2O emission from soils affected by AMF-colonized roots compared to soils influenced only by root activity [271]. Similarly, another research outlined a reduction in N_2O fluxes in the rice crop by means of AMF [272]. The above-mentioned studies suggest that AMF alter N_2O emission; however, it has not been determined whether AMF induce N_2O reduction by physiological changes in the AMF-colonized roots or as direct result of the AMF themselves. Recently, it has been noticed that AMF directly reduce N_2O emission [249]. Additionally, AMF also affect the N cycling by capturing the nutrient and transferring some portions to host plants [273]. The availability of N and C are the factors that control NF and DNF [274]. Thus, it is not possible to separate the AMF and root fluxes of N_2O in the mycorrhizosphere without first separating the AMF hyphae from plant roots. Additionally, there is a positive association between the presence of AMF and reduced NF [38,39]. Likewise, the presence of AMF reduces the abundance of *nirk* genes, which are considered responsible for N_2O production [274] Thus, a decrease in N_2O in the presence of AMF can be due to lower NF rates [274]. Additionally, AMF reduce NH_4^+ in the hyphosphere, resulting in a reduction in ammonia-oxidizing bacteria (AOB) population. Since AOB are considered the main producers of N_2O, this may explain the reduction in N_2O emissions owing to AMF activity [274].

It is definitely evinced that AMF, by interacting with the host plant and the soil environment, can play a relevant role in restraining N_2O emission. Specifically, AMF activity buffers the content of available N forms in soil profile, which in turn results in lower amounts of NO_3^- prone to DNF. All the consulted sources are consistent with a potentially beneficial role exerted by AMF in restraining N_2O emission.

4.12. Selection of Plant Genotypes

The selection of suitable cultivars is a prerequisite to obtain the desirable crop production [275–280], while concurrently playing a role in GHG reduction. The variations amid the rice cultivars for CH_4 emission can be related to differences in CH_4 production, oxidation and transport [281]. The mechanisms explaining the differences among plant species for N_2O emission are often unclear [282]; however, numerous prospects can be envisioned. In the case of the rice plant, active pathways exist for N_2O transport through aerenchyma cells to soil submerged with water [283], and during daytime, N_2O is transported from roots to shoots via the transpiration stream and is subsequently lost through stomata [284]. In *Brachiaria humidicola*, a tropical grass, there are cultivars able to produce the chemicals that directly inhibit NF [285], substantially reducing N_2O emission [286]. In another study, it was noticed that the lowest N_2O emission was linked with a plant strategy characterized by higher N uptake [287]. In fact, plant cultivars with higher N uptake were shown able to reduce the N pool, especially NO_3^-, resulting in lower availability of substrate for denitrifiers and subsequently lower N_2O emission. The variation amid cultivars for N_2O emission had also been reported in the intercropping of cereals and legumes [288]. In another study, researchers noticed a significant contribution of plants to N_2O emission and suggested that in the soil-crop system, N_2O emission is markedly influenced by plant characteristics [289].

Therefore, it appears that the breeding of crop plants could be directed, among other things, to the release of cultivars, enabling N_2O containment. All plant strategies conducive to earlier and stronger N uptake deplete soil reserves and leave less NO_3^- exposed to the risk of N_2O production. In this respect, a relevant goal from a productive viewpoint can be associated with breeding with an equally relevant goal from an environmental viewpoint.

4.13. Modifying Cropping Schemes and Crop Rotations

In rice, switching from conventional puddled transplanted rice (TPR) system to directly seeded rice (DSR) may contribute to reducing GHG emissions. In fact, it was noticed that DSR increased N_2O emission when the redox potential (RP) crossed 250 mV [290]. It was concluded that water should be applied in such a way that RP be kept at a range of 100–200 mV to reduce both N_2O and CH_4 emissions. Since DSR system offsets N_2O emission, it is an encouraging production system, thanks to the lower GWP [230]. The DSR has 53% less GWP in terms of N_2O, CH_4 and CO_2 components as compared to traditional TSP [291]. Further, Ahmad et al. [112] stated that GWP of DSR can be further decreased by shifting toward no tillage (NT). The lower GWP and higher production of DSR suggest that DSR would decrease both CH_4 and N_2O emissions. Nonetheless, more detailed studies involving the measurements of GHGs under the concurring effects of factors including water, tillage, nutrients and biochar, are direly needed to support DSR as a suitable system that also reduces the environmental burden.

Few studies investigated the impact of crop rotation diversity on GHG emissions from diverse plant species within the rotation. The GHG fluxes were investigated under a maize–soybean rotation for three years, and it was noticed that maize and soybean emitted a similar amount of CH_4 [292]. In another study, authors reported non-significant differences in N_2O emission from different species, including cowpea, wheat and soybean, in a four-year rotation [293]. In some other works, the authors compared N_2O emissions from crops sown in rotation and mono-cropping; corn sown in rotation decreased the N_2O and CO_2 emissions compared with continuous corn [294,295], owing to the application of large amount of N fertilizers in mono-cropping. However, some authors noticed that wheat grown in rotation and in mono-cropping emitted the same amount of N_2O [296]. In another study, maize staged the same N_2O emissions when grown as continuous crop and in maize–soybean and soybean–wheat–maize rotations [297]. Crops entering a cropping system must be chosen properly because they significantly affect N_2O emissions [219]. For instance, grasslands significantly increased N_2O emissions, whereas maize crop showed a negative impact on N_2O emissions [219]. Intensive grasslands can increase global N_2O emissions owing to the application of manures and animal excreta deposited on the surface of grasslands [298].

Such differences suggest that the effect of crop rotation diversity on GHG emission can vary owing to soil and climate conditions, and crop diversity. Since there is no univocal effect exerted by cropping schemes and rotations, the amount of N_2O emissions and their potential abatement appear to be linked to specific issues in crop management, such as the planting system or N fertilization, whose effects have already been surveyed in the specific sub-sections.

4.14. Integrated Nutrient Management

Integrated nutrient management (INM) involves the combined use of OA and inorganic fertilizers to increase NUE and reduce N losses by synchronizing crop demand with soil nutrient availability [35,299]. A few reports are available about the effects of INM on GHG emission. Some authors compared the effects of NPK fertilizer, compost and their combination on N_2O emission [299,300]. They noted that a combined application of NPK and compost reduced N_2O emission compared to the sole use of compost or NPK. Additionally, they suggested that the application of composted material with C:N ratio lower than 20 significantly reduced N_2O emission, owing to the release of a lower amount of N during decomposition in soil [299,300]. Moreover, one research work measured the impact of INM (cattle manure and AN) on N_2O emission during one growing season for maize and wheat. These authors noted that INM increased N_2O emission compared to cattle manure, whereas it decreased the emission compared to AN. This reduction in N_2O emission with the application of OA was due to slower decomposition of C and N, and slower release of mineralized N [301]. Huang et al. [59] noticed the reduction in N_2O emission with plant amendments at increasing C:N ratio and found that this relation becomes stronger with the addition of inorganic N. Nonetheless, in this study, the treatment featuring highest N_2O emission was associated with the greatest N supply, indicating that the N dose effect remains of paramount importance. In accordance with the previous results, another study suggested that a reduction in N_2O emission occurs when OA with lower C:N ratio are applied alone or when OA with higher C:N ratio are applied together with inorganic fertilizers [302].

Nonetheless, rare field studies are available about the effect of the C:N ratio of OA on N_2O emission. As the micro-organisms involved in the NF and NDF processes depend on C supply, the application of OA with a C:N > 20 tends, under no synthetic N supply, to result in nutrient microbial immobilization, in turn reducing the available N for DNF [303]. Conversely, OA with lower C:N ratio are more quickly mineralized by microbial activity and result in the release of C and N, which increases the microbial activities and, resultantly, N_2O emission [303]. Nonetheless, the microbially induced N_2O emission from INM not only depends on C:N ratio but also on the amount of synthetic N fertilizers added to soil. The application of N fertilizers with OA with a large quantity of labile C further increases DNF, leading to higher N_2O emission [304]. A summary of studies indicates that the INM leads to a reduction in N_2O emission (Table 3).

Table 3. Effect of organic/inorganic nutrients and integrated nutrient management (INM) on N_2O emission.

Crop Rotation	Total Rate of N (kg ha^{-1})	N_2O Emission Trend	References
Maize–wheat	Organic: 150 composted manure (CM), Inorganic: 150 urea, INM: 75 CM + 75 Urea	No significant difference was recorded	[300]
Maize–wheat	Organic: 150 CM, Inorganic: urea, INM: 75 CM + 75 urea	No significant difference was recorded	[301]
Maize–wheat	Organic: 150 CM, Inorganic: urea, INM: 75 CM + 75 urea	INM, Organic < Inorganic	[302]
Rapeseed	Organic: 97.5 cattle manure, Inorganic: ammonium nitrate (AN) 120, INM: 65 cattle manure + 60 AN	INM < Organic, Inorganic	[303]

Table 3. *Cont.*

Crop Rotation	Total Rate of N (kg ha^{-1})	N$_2$O Emission Trend	References
Maize–wheat	Organic: 120 cattle manure, Inorganic: AN 120, INM: 60 CM + 60 AN	Organic < INM < Inorganic	[304]
Maize–wheat	Inorganic: 100% NPK, INM: 100% NPK + FYM	Inorganic < INM	[305]
Rice	Inorganic: 120 kg urea, INM: Compost (30 kg/ha + urea 90 kg/ha	INM < Inorganic	[306]

The total rate of N applied from OA and inorganic fertilizers also explains the amount of N$_2$O emission [300,307]. It is not surprising that the INM approach of combining organic and synthetic N sources at higher N rates results in higher N$_2$O emission compared to their alternative use at lower N rates [308]. Conversely, when half of the suitable N rate was applied from organic and half from inorganic sources, this resulted in reduction in N$_2$O emission compared to the sole application of organic or inorganic N sources at the same N rate [308–310]. It is evinced, therefore, that combining OA with inorganic fertilizers does not assure reduction in N$_2$O emission. However, a meta-analysis conducted by Graham et al. [295] suggests that the application of amendments with very low C:N (<8) ratio in a substitutive strategy of N application (proportional reduction in N rate from each N source) has a good potential to mitigate N$_2$O emission. Therefore, the integrated use of inorganic N with OA at lower C:N ratio helps to avoid two processes, namely rapid mineralization of inorganic N (low C:N ratio) and stimulation of microbial activity through the addition of excessively C-rich substrates (high C:N), which together contribute to N$_2$O emission.

It is perceived, in general, that only a shrewd application of INM can make this approach successful in the quest for mitigating N$_2$O emission. It is equally sensed that none of the crop practices previously surveyed, nor INM alone, can positively contribute to alleviating this problem, unless N$_2$O abatement is considered a major goal in crop production and practices in crop management are directed toward its achievement.

5. Role of Regulatory Authorities in Implementing Environment-Friendly Management Practices to Reduce GHGs Emissions

The intensity of GHGs has substantially increased in recent time, which has in turn increased climate change and global warming [311–316]. Globally, various policies, measures and strategies are being deployed by governments to limit GHGs emissions. Different approaches, including standards, incentives and different permissions, are used to encourage environmentally friendly approaches to restrict GHGs emissions [317,318]. However, these approaches may vary at the national and sub-national levels according to each country. GHGs are major drivers of climate change, and diverse international negotiations have taken place in the last two decades to curb GHG emissions and counter climate change and global warming. Many countries have followed various development cycles since the 1990s to reduce GHGs emission. Initial efforts were made in reducing GHG emissions from developed and industrialized nations, which eventually became the Annex-1 group of the Kyoto Protocol [319]. Similarly, the 27 member states of the European Union (EU-27) and the United Kingdom have signed commitments to become carbon-neutral economies by the end of 2050 [320]. Moreover, the European Commission also proposes to reduce GHG emission by 55% compared to 1991 by the end of year 2030 [321,322]. However, the simultaneous implementation of climate change policies in the EU-27, UK and USA has also put a major focus on heavy industries as the main source of national gross domestic product [323]. By contrast, some medium to large countries have also gone through unprecedented economic growth as a result of industrialization, and they are also experiencing a substantial increase in population growth [324]. The socio-economic and demographic transformations combined with technology are designed to restrict climate change and GHG emissions in a framework of market conditions. An important practice adopted around the globe is the use of renewable energy sources accompanied by the decrease in use of coal and petroleum and the development of efficient energy production and consumption practices [325,326].

During the 1997 UNFCC conference of parties in Kyoto, a protocol was adopted, and it was enforced in 2005. This Kyoto Protocol invented the GHG emission commitments for developed nations for a period of five years (2008–2012). The Kyoto Protocol defined four emission-saving units, including those obtained: (1) by clean development mechanism projects, (2) through joint implementation of projects, (3) through the trading of unused assigned emissions between protocol parties and (4) through reforestation-related projects. Moreover, during the year 2012, an amendment was made to the Kyoto Protocol, and a second commitment period was determined for another seven years (2013–2020) to reduce GHG emission. The proposed amendment targeted a reduction of 18% in GHG emission as compared to 1990 levels [327].

Nowadays, it has been recognized that environmental protection is an essential part of business processes [328]. Environmental protection can yield many benefits, including cost and resource savings, and it can increase satisfaction and loyalty in people [329]. The European Commission developed the European Union (EU) Eco-Management and Audit Scheme (EMAS) for companies and other sectors to adopt the environmentally friendly approaches to restrict environmental footprint [330]. The Environmental Management Systems (EMS), such as ISO (International Organization for Standardization) or EMAS (Eco-Management and Audit Scheme), have been also designed for ensuring higher environmental protection and competitive advantage of organizations resulting from the introduced improvements. Corporate social responsibility (CSR) is another important concept in performing business activities according to which companies still make a profit in strict compliance with the law, and they take into account the impact of their operations on the environment in their business decisions [328]. The application of such approaches improves the quality of life and ensures a sustainable development.

6. Conclusions and Future Prospects

The mushrooming population and rapidly increasing food demand have raised the concern all around the globe of stabilizing the atmospheric greenhouse gases concentration for mitigating the ongoing climate change. Here, we presented comprehensive information about management practices designed to reduce N_2O emission.

The adoption of all the practices reviewed here is expected to mitigate N_2O emission without comprising productivity. The discussion of the literature allowed us to outline the role of management options that can be adopted either alone or in association, in the quest to reduce N_2O emission. Prioritizing the use of fertilizers associated with low N_2O emission, such as ammonium fertilizers, leads to less N_2O compared to nitrate fertilizers. Similarly, the deep placement of N fertilizers should be promoted to reduce N_2O emissions. Plant-breeding activities should be aimed at releasing genotypes with better N uptake, nitrogen fixation and the ability to capitalize those C–N interactions in the rhizosphere, which can be helpful to reduce N_2O emission. Promoting sustainable crop intensification, which can be done by using higher-yielding crop varieties, reducing the use of external inputs, improving nitrogen use efficiency, using biochar and lime to counter acidic soil pH and adopting agroecological practices, can help to mitigate the impact of current management systems on N_2O emissions. The selection of suitable irrigation methods is an important strategy to save water and maintain yields. However, future studies are needed to study irrigation effects on soil hydraulic properties, which affect water distribution and, therefore, N_2O emission. Additionally, these systems are often combined with fertilizer applications, thus future work is required to evaluate the impact of rate, frequency and types of N fertilizer on N_2O emission under sprinkler and drip irrigation systems.

Moreover, to further understand the impact of C:N ratio on N_2O emission, integrated nutrient management studies should be conducted by including a wider range of C:N ratios in organic amendments, along with the application of inorganic fertilizers. In parallel to this, different organic amendments with similar C:N ratio should be applied with constant rates of nitrogen to better appraise the impacts of organic amendment properties beside C:N ratio on N_2O emission. In arbuscular mycorrhizal fungi, future studies should be conducted to explore their interaction with microbial communities, including ammonia-oxidizing archaea and bacteria, nitrifying communities and non-denitrifying N_2O reducers.

A better understanding of successful N_2O mitigation strategies requires studies related to N_2O fluxes in agroecosystems to account for the wide range of biotic and abiotic factors, including ecosystem state factors, such as soil characteristics, climate and topography, which interact with management practices to influence soil N_2O emission. Nonetheless, only few of the above-mentioned studies consider the interactions between eco-system state factors and management practices. Therefore, interdisciplinary and cross scale studies should be run to understand how we can successfully reduce N_2O emission in crop production systems. Finally, at the field level, N_2O measurements and agronomic information can be used to design N_2O mitigation approaches that should reduce the carbon footprint and maximize monetary paybacks of cultivation efforts.

Author Contributions: Conceptualization, M.U.H., M.A., G.H. and L.B.; writing—original draft preparation, M.U.H., L.B. and G.H.; writing—review and editing, M.A., A.M., M.I.A., M.F.S., J.S., G.B., A.R., H.M.A. and E.B. All authors have read and agreed to the published version of the manuscript.

Funding: This work was supported by the National Key R&D Program of China (2016YFD0300208); National Natural Science Foundation of China (41661070); and Key disciplines (construction) of ecology in the 13th Five-Year Plan of Jiangxi Agricultural University.

Institutional Review Board Statement: Not applicable.

Informed Consent Statement: Not applicable.

Data Availability Statement: Not applicable.

Conflicts of Interest: The authors declare no conflict of interest.

References

1. Seleiman, M.F.; Santanen, A.; Stoddard, F.L.; Mäkelä, P. Feedstock quality and growth of bioenergy crops fertilized with sewage sludge. *Chemosphere* **2012**, *89*, 1211–1217. [CrossRef] [PubMed]
2. Seleiman, M.F.; Santanen, A.; Jaakkola, S.; Ekholm, P.; Hartikainen, H.; Stoddard, F.L.; Mäkelä, P.S.A. Biomass yield and quality of bioenergy crops grown with synthetic and organic fertilizers. *Biomass Bioenerg.* **2013**, *59*, 477–485. [CrossRef]
3. Seleiman, M.F.; Alotaibi, M.A.; Alhammad, B.A.; Alharbi, B.M.; Refay, Y.; Badawy, S.A. Effects of ZnO nanoparticles and biochar of rice straw and cow manure on characteristics of contaminated soil and sunflower productivity, oil quality, and heavy metals uptake. *Agronomy* **2020**, *10*, 790. [CrossRef]
4. Seleiman, M.F.; Kheir, A.M.S.; Al-Dhumri, S.; Alghamdi, A.G.; Omar, E.S.H.; Aboelsoud, H.M.; Abdella, K.A.; Abou El Hassan, W.H. Exploring optimal tillage improved soil characteristics and productivity of wheat irrigated with different water qualities. *Agronomy* **2019**, *9*, 233. [CrossRef]
5. Seleiman, M.F.; Santanen, A.; Mäkelä, P.S.A. Recycling sludge on cropland as fertilizer—Advantages and risks. *Resour. Conserv. Recycl.* **2020**, *155*, 104647. [CrossRef]
6. Ding, Z.; Ali, E.F.; Elmahdy, A.M.; Ragab, K.E.; Seleiman, M.F.; Kheir, A.M.S. Modeling the combined impacts of deficit irrigation, rising temperature and compost application on wheat yield and water productivity. *Agric. Water Manag.* **2021**, *244*, 106626. [CrossRef]
7. Taha, R.S.; Seleiman, M.F.; Alotaibi, M.; Alhammad, B.A.; Rady, M.M.; Mahdi, A.H.A. Exogenous potassium treatments elevate salt tolerance and performances of Glycine max L. By boosting antioxidant defense system under actual saline field conditions. *Agronomy* **2020**, *10*, 1741. [CrossRef]
8. Hassan, M.U.; Aamer, M.; Chattha, M.U.; Haiying, T.; Shahzad, B.; Barbanti, L.; Nawaz, M.; Rasheed, A.; Afzal, A.; Liu, Y.; et al. The critical role of zinc in plants facing the drought stress. *Agriculture* **2020**, *10*, 396. [CrossRef]
9. Malyan, S.K.; Bhatia, A.; Tomer, R.; Harit, R.C.; Jain, N.; Bhowmik, A.; Kaushik, R. Mitigation of yield-scaled greenhouse gas emissions from irrigated rice through Azolla, Blue-green algae, and plant growth–promoting bacteria. *Environ. Sci. Poll. Res.* **2021**, *28*, 51425–51439. [CrossRef] [PubMed]
10. Tilman, D.; Cassman, K.G.; Matson, P.A.; Naylor, R.; Polasky, S. Agricultural sustainability and intensive production practices. *Nature* **2002**, *418*, 671–677. [CrossRef] [PubMed]

11. Rasheed, A.; Hassan, M.U.; Aamer, M.; Batool, M.; Fang, S.; WU, Z.; LI, H. A Critical Review on the Improvement of Drought Stress Tolerance in Rice (*Oryza sativa* L.). *Not. Bot. Horti Agrobot. Cluj-Napoca* **2020**, *48*, 1756–1788. [CrossRef]

12. Hassan, M.U.; Chattha, M.U.; Khan, I.; Chattha, M.B.; Barbanti, L.; Aamer, M.; Iqbal, M.M.; Nawaz, M.; Mahmood, A.; Ali, A.; et al. Heat stress in cultivated plants: Nature, impact, mechanisms, and mitigation strategies—A review. *Plant Biosyst.* **2021**, *155*, 211–234. [CrossRef]

13. Tellez-Rio, A.; Vallejo, A.; García-Marco, S.; Martin-Lammerding, D.; Tenorio, J.L.; Rees, R.M.; Guardia, G. Conservation Agriculture practices reduce the global warming potential of rainfed low N input semi-arid agriculture. *Eur. J. Agron.* **2017**, *84*, 95–104. [CrossRef]

14. Malyan, S.K.; Bhatia, A.; Fagodiya, R.K.; Kumar, S.S.; Kumar, A.; Gupta, D.K.; Tomer, R.; Harit, R.C.; Kumar, V.; Jain, N.; et al. Plummeting global warming potential by chemicals interventions in irrigated rice: A lab to field assessment. *Agric. Ecosyst. Environ.* **2021**, *319*, 107545. [CrossRef]

15. Sekoai, T.T.; Yoro, K.O. Biofuel development initiatives in sub-Saharan Africa: Opportunities and challenges. *Climate* **2016**, *4*, 33. [CrossRef]

16. Liu, J.; Xu, H.; Jiang, Y.; Zhang, K.; Hu, Y.; Zeng, Z. Methane emissions and microbial communities as influenced by dual cropping of Azolla along with early rice. *Sci. Rep.* **2017**, *7*, 40635. [CrossRef] [PubMed]

17. Yoro, K.O.; Daramola, M.O. CO_2 emission sources, greenhouse gases, and the global warming effect. In *Advances in Carbon Capture*; Woodhead Publishing: Sawston, UK, 2020; pp. 3–28.

18. Intergovernmental Panel on Climate Change. *Climate Change 2014 Mitigation of Climate Change*; Intergovernmental Panel on Climate Change: Geneva, Switzerland, 2014; pp. 1–164.

19. Anenberg, S.C.; Schwartz, J.; Shindell, D.; Amann, M.; Faluvegi, G.; Klimont, Z.; Janssens-Maenhout, G.; Pozzoli, L.; van Dingenen, R.; Vignati, E.; et al. Global air quality and health co-benefits of mitigating near-term climate change through methane and black carbon emission controls. *Environ. Health Perspect.* **2012**, *120*, 831–839. [CrossRef]

20. Avnery, S.; Mauzerall, D.L.; Liu, J.; Horowitz, L.W. Global crop yield reductions due to surface ozone exposure: 1. Year 2000 crop production losses and economic damage. *Atmos. Environ.* **2011**, *45*, 2284–2296. [CrossRef]

21. Davidson, E.A. The contribution of manure and fertilizer nitrogen to atmospheric nitrous oxide since 1860. *Nat. Geosci.* **2009**, *2*, 659–662. [CrossRef]

22. Stocker, T.F.; Qin, D.; Plattner, G.-K.; Tignor, M.; Allen, S.K.; Boschung, J.; Nauels, A.; Xia, Y.; Bex, V.; Midgley, P.M. *Climate Change 2013: The Physical Science Basis: Working Group I Contribution to the Fifth Assessment Report of the Intergovernmental Panel on Climate Change*; Stocker, T.F., Qin, D., Plattner, G.-K., Tignor, M., Allen, S.K., Boschung, J., Nauels, A., Xia, Y., Bex, V., Midgley, P.M., Eds.; Cambridge University Press: Cambridge, UK; New York, NY, USA, 2014. Available online: https://www.ipcc.ch/site/assets/uploads/2018/02/WG1AR5_Chapter08_FINAL.pdf (accessed on 5 May 2021).

23. Mannina, G.; Capodici, M.; Cosenza, A.; Ditrapani, D.; Vanloosdrecht, M.C.J.J. Nitrous oxide emission in a University of Cape Town membrane bioreactor: The effect of carbon to nitrogen ratio. *J. Clean. Prod.* **2017**, *149*, 180–190. [CrossRef]

24. Haider, A.; Bashir, A.; Husnain, M.I. Impact of agricultural land use and economic growth on nitrous oxide emissions: Evidence from developed and developing countries. *Sci. Total Environ.* **2020**, *741*, 140421. [CrossRef] [PubMed]

25. Kammann, C.; Ippolito, J.; Hagemann, N.; Borchard, N.; Cayuela, M.L.; Estavillo, J.M.; Fuertes-Mendizabal, T.; Jeffery, S.; Kern, J.; Novak, J.; et al. Biochar as a tool to reduce the agricultural greenhouse-gas burden–knowns, unknowns and future research needs. *J. Environ. Eng. Landsc. Manag.* **2017**, *25*, 114–139. [CrossRef]

26. Ding, Y.; Liu, Y.-X.; Wu, W.-X.; Shi, D.-Z.; Yang, M.; Zhong, Z.-K. Evaluation of biochar effects on nitrogen retention and leaching in multi-layered soil columns. *Water Air Soil Pollut.* **2010**, *213*, 47–55. [CrossRef]

27. Tian, L.; Zhu, B.; Akiyama, H. Seasonal variations in indirect N_2O emissions from an agricultural headwater ditch. *Biol. Fertil. Soils* **2017**, *53*, 651–662. [CrossRef]

28. Baggs, E.M. Soil microbial sources of nitrous oxide: Recent advances in knowledge, emerging challenges and future direction. *Curr. Opin. Environ. Sustain.* **2011**, *3*, 321–327. [CrossRef]

29. Thomson, A.J.; Giannopoulos, G.; Pretty, J.; Baggs, E.M.; Richardson, D.J. Biological sources and sinks of nitrous oxide and strategies to mitigate emissions. *Phil. Trans. R. Soc. B* **2012**, *367*, 1157–1168. [CrossRef]

30. Aamer, M.; Hassan, M.U.; Shaaban, M.; Rasul, F.; Haiying, T.; Qiaoying, M.; Batool, M.; Rasheed, A.; Chuan, Z.; Qitao, S.; et al. Rice straw biochar mitigates N_2O emissions under alternate wetting and drying conditions in paddy soil. *J. Saudi Chem. Soc.* **2021**, *25*, 101172. [CrossRef]

31. Fernandes, S.O.; Bonin, P.C.; Michotey, V.D.; Garcia, N.; LokaBharathi, P.A. Nitrogen-limited mangrove ecosystems conserve N through dissimilatory nitrate reduction to ammonium. *Sci. Rep.* **2012**, *2*, 419. [CrossRef]

32. Zhu, X.; Burger, M.; Doane, T.A.; Horwath, W.R. Ammonia oxidation pathways and nitrifier denitrification are significant sources of N_2O and NO under low oxygen availability. *Proc. Natl. Acad. Sci. USA* **2013**, *110*, 6328–6333. [CrossRef]

33. Venterea, R.T.; Halvorson, A.D.; Kitchen, N.; Liebig, M.A.; Cavigelli, M.A.; Del Grosso, S.J.; Motavalli, P.P.; Nelson, K.A.; Spokas, K.A.; Singh, B.P. Challenges and opportunities for mitigating nitrous oxide emissions from fertilized cropping systems. *Front. Ecol. Environ.* **2012**, *10*, 562–570. [CrossRef]

34. Seleiman, M.; Abdel-Aal, M. Response of growth, productivity and quality of some Egyptian wheat cultivars to different irrigation regimes. *Egypt. J. Agron.* **2018**, *40*, 313–330. [CrossRef]

35. Seleiman, M.; Abdel-Aal, M. Effect of organic, inorganic and bio-fertilization on growth, yield and quality traits of some chickpea (*Cicer arietinum* L.) varieties. *Egypt. J. Agron.* **2018**, *40*, 105–117. [CrossRef]

36. Seleiman, M.F.; Kheir, A.M.S. Maize productivity, heavy metals uptake and their availability in contaminated clay and sandy alkaline soils as affected by inorganic and organic amendments. *Chemosphere* **2018**, *204*, 514–522. [CrossRef] [PubMed]

37. Seleiman, M.F.; Hardan, A.N. Importance of mycorrhizae in crop productivity. In *Mitigating Environmental Stresses for Agricultural Sustainability in Egypt*; Awaad, H., Abu-hashim, M., Negm, A., Eds.; Springer Nature: Cham, Switzerland, 2021; pp. 471–484.

38. Seleiman, M.F.; Hafez, E.M. Optimizing inputs management for sustainable agricultural development. In *Mitigating Environmental Stresses for Agricultural Sustainability in Egypt*; Awaad, H., Abu-hashim, M., Negm, A., Eds.; Springer Nature: Cham, Switzerland, 2021; pp. 487–507.

39. Seleiman, M.F.; Refay, Y.; Al-Suhaibani, N.; Al-Ashkar, I.; El-Hendawy, S.; Hafez, E.M. Hafez integrative effects of rice-straw biochar and silicon on oil and seed quality, yield and physiological traits of *Helianthus annuus* L. grown under water deficit stress. *Agronomy* **2019**, *9*, 637. [CrossRef]

40. Seleiman, M.F.; Santanen, A.; Kleemola, J.; Stoddard, F.L.; Mäkelä, P.S.A. Improved sustainability of feedstock production with sludge and interacting mycorrhiza. *Chemosphere* **2013**, *91*, 1236–1242. [CrossRef] [PubMed]

41. Prosser, J.I.; Nicol, G.W. Archaeal and bacterial ammonia-oxidisers in soil: The quest for niche specialisation and differentiation. *Trends Microbiol.* **2012**, *20*, 523–531. [CrossRef]

42. Shen, J.P.; Zhang, L.M.; Di, H.J.; He, J.Z. A review of ammonia-oxidizing bacteria and archaea in Chinese soils. *Front. Microbiol.* **2012**, *3*, 296. [CrossRef]

43. Lehtovirta-Morley, L.E.; Sayavedra-Soto, L.A.; Gallois, N.; Schouten, S.; Stein, L.Y.; Prosser, J.I.; Nicol, G.W. Identifying potential mechanisms enabling acidophily in the ammonia-oxidizing archaeon "*Candidatus* Nitrosotalea devanaterra". *Appl. Environ. Microbiol.* **2016**, *82*, 2608–2619. [CrossRef] [PubMed]

44. Conrad, R. Metabolism of nitric oxide in soil and soil microorganisms and regulation of flux into the atmosphere. In *Microbiology of Atmospheric Trace Gases*; Murrell, J.C., Kelly, D.P., Eds.; Springer: Berlin/Heidelberg, Germany, 1996; pp. 167–203.

45. Mushinski, R.M.; Phillips, R.P.; Payne, Z.C.; Abney, R.B.; Jo, I.; Fei, S.; Pusede, S.E.; White, J.R.; Rusch, D.B.; Raff, J.D. Microbial mechanisms and ecosystem flux estimation for aerobic NO_y emissions from deciduous forest soils. *Proc. Natl. Acad. Sci. USA* **2019**, *116*, 2138–2145. [CrossRef]

46. Abeliovich, A. The Nitrite-oxidizing bacteria introduction. *Prokaryotes* **2006**, *5*, 861–872.

47. Pilegaard, K. Processes regulating nitric oxide emissions from soils. *Philos. Trans. R. Soc. B* **2013**, *368*, 20130126. [CrossRef]

48. Moreira, F.M.S.; Siqueira, J.O. *Microbiology and Soil Bio-Chemistry*, 2nd ed.; Academic Press: Cambridge, MA, USA, 2006.

49. Syakila, A.; Kroeze, C. The global nitrous oxide budget revisited. *Greenh. Gas Meas. Manag.* **2011**, *1*, 17–26. [CrossRef]

50. Braker, G.; Conrad, R. Diversity, structure, and size of N_2O-producing microbial communities in soils-what matters for their functioning? *Adv. Appl. Microbiol.* **2011**, *75*, 33–70.

51. Butterbach-Bahl, K.; Baggs, E.M.; Dannenmann, M.; Kiese, R.; Zechmeister-Boltenstern, S. Nitrous oxide emissions from soils: How well do we understand the processes and their controls? *Philos. Trans. R. Soc. B* **2013**, *368*, 20130122. [CrossRef] [PubMed]

52. Čuhel, J.; Šimek, M.; Laughlin, R.J.; Bru, D.; Chèneby, D.; Watson, C.J.; Philippot, L. Insights into the effect of soil pH on N_2O and N_2 emissions and denitrifier community size and activity. *Appl. Environ. Microbiol.* **2010**, *76*, 1870–1878. [CrossRef]

53. Sun, P.; Zhuge, Y.; Zhang, J.; Cai, Z. Soil pH was the main controlling factor of the denitrification rates and N_2/N_2O emission ratios in forest and grassland soils along the northeast China transect. *Soil Sci. Plant Nutr.* **2012**, *58*, 517–525. [CrossRef]

54. Baggs, E.M.; Smales, C.L.; Bateman, E.J. Changing pH shifts the microbial source as well as the magnitude of N_2O emission from soil. *Biol. Fertil. Soils* **2010**, *46*, 793–805. [CrossRef]

55. Khan, S.; Clough, T.J.; Goh, K.M.; Sherlock, R.R. Influence of soil pH on NO_x and N_2O emissions from bovine urine applied to soil columns. *N. Z. J. Agric. Res.* **2011**, *54*, 285–301. [CrossRef]

56. Groffman, P.M.; Altabet, M.A.; Böhlke, J.K.; Butterbach-Bahl, K.; David, M.B.; Firestone, M.K.; Giblin, A.E.; Kana, T.M.; Nielsen, L.P.; Voytek, M.A. Methods for measuring denitrification: Diverse approaches to a difficult problem. *Ecol. Appl.* **2006**, *16*, 2091–2122. [CrossRef]

57. Tate, K.R.; Ross, D.J.; Saggar, S.; Hedley, C.B.; Dando, J.; Singh, B.K.; Lambie, S.M. Methane uptake in soils from *Pinus radiata* plantations, a reverting shrubland and adjacent pastures: Effects of land-use change, and soil texture, water and mineral nitrogen. *Soil Biol. Biochem.* **2007**, *39*, 1437–1449. [CrossRef]

58. Granli, T.; Bøckman, O.C. Nitrous oxide from agriculture. *Norwegian J. Agric. Sci.* **1994**, *3*, 14–21.

59. Bolan, N.S.; Adriano, D.C.; Kunhikrishnan, A.; James, T.; McDowell, R.; Senesi, N. Dissolved organic matter: Biogeochemistry, dynamics, and environmental significance in soils. *Adv. Agron.* **2011**, *110*, 1–75.

60. Stres, B.; Danevčič, T.; Pal, L.; Fuka, M.M.; Resman, L.; Leskovec, S.; Hacin, J.; Stopar, D.; Mahne, I.; Mandic-Mulec, I. Influence of temperature and soil water content on bacterial, archaeal and denitrifying microbial communities in drained fen grassland soil microcosms. *FEMS Microbiol. Ecol.* **2008**, *66*, 110–122. [CrossRef] [PubMed]

61. Braker, G.; Schwarz, J.; Conrad, R. Influence of temperature on the composition and activity of denitrifying soil communities. *FEMS Microbiol. Ecol.* **2010**, *73*, 134–148. [CrossRef]

62. Szukics, U.; Abell, G.C.J.; Hödl, V.; Mitter, B.; Sessitsch, A.; Hackl, E.; Zechmeister-Boltenstern, S. Nitrifiers and denitrifiers respond rapidly to changed moisture and increasing temperature in a pristine forest soil. *FEMS Microbiol. Ecol.* **2010**, *72*, 395–406. [CrossRef]

63. Lemke, R.L.; Izaurralde, R.C.; Nyborg, M.; Solberg, E.D. Tillage and N source influence soil-emitted nitrous oxide in the Alberta Parkland region. *Can. J. Soil Sci.* **1999**, *79*, 15–24. [CrossRef]

64. Huang, Y.; Zou, J.; Zheng, X.; Wang, Y.; Xu, X. Nitrous oxide emissions as influenced by amendment of plant residues with different C:N ratios. *Soil Biol. Biochem.* **2004**, *36*, 973–981. [CrossRef]

65. Eichner, M.J. Nitrous oxide emissions from fertilized soils: Summary of available data. *J. Environ. Qual.* **1990**, *19*, 272–280. [CrossRef]

66. Patten, D.K.; Bremner, J.M.; Blackmer, A.M. Effects of drying and air-dry storage of soils on their capacity for denitrification of nitrate. *Soil Sci. Soc. Am. J.* **1980**, *44*, 67–70. [CrossRef]

67. Baggs, E.M.; Rees, R.M.; Smith, K.A.; Vinten, A.J.A. Nitrous oxide emission from soils after incorporating crop residues. *Soil Use Manag.* **2000**, *16*, 82–87. [CrossRef]

68. Shelp, M.L.; Beauchamp, E.G.; Thurtell, G.W. Nitrous oxide emissions from soil amended with glucose, alfalfa, or corn residues. *Commun. Soil Sci. Plant Anal.* **2000**, *31*, 877–892. [CrossRef]

69. Rochette, P.; Worth, D.E.; Lemke, R.L.; McConkey, B.G.; Pennock, D.J.; Wagner-Riddle, C.; Desjardins, R.L. Estimation of N_2O emissions from agricultural soils in Canada. I. Development of a country-specific methodology. *Can. J. Soil Sci.* **2008**, *88*, 655–669. [CrossRef]

70. Signor, D.; Cerri, C.E.P. Nitrous oxide emissions in agricultural soils: A review. *Pesqui. Agropecu. Trop.* **2013**, *43*, 322–338. [CrossRef]

71. Stehfest, E.; Bouwman, L. N_2O and NO emission from agricultural fields and soils under natural vegetation: Summarizing available measurement data and modeling of global annual emissions. *Nutr. Cycl. Agroecosyst.* **2006**, *74*, 207–228. [CrossRef]

72. Malhi, S.S.; Lemke, R. Tillage, crop residue and N fertilizer effects on crop yield, nutrient uptake, soil quality and nitrous oxide gas emissions in a second 4-yr rotation cycle. *Soil Tillage Res.* **2007**, *90*, 171–183. [CrossRef]

73. De Boer, W.; Kowalchuk, G.A. Nitrification in acid soils: Micro-organisms and mechanisms. *Soil Biol. Biochem.* **2001**, *33*, 853–866. [CrossRef]

74. Venterea, R.T.; Hyatt, C.R.; Rosen, C.J. Fertilizer management effects on nitrate leaching and indirect nitrous oxide emissions in irrigated potato production. *J. Environ. Qual.* **2011**, *40*, 1103–1112. [CrossRef]

75. Gagnon, B.; Ziadi, N.; Rochette, P.; Chantigny, M.H.; Angers, D.A. Fertilizer source influenced nitrous oxide emissions from a clay soil under corn. *Soil Sci. Soc. Am. J.* **2011**, *75*, 595–604. [CrossRef]

76. Zhang, J.B.; Zhu, T.B.; Cai, Z.C.; Qin, S.W.; Müller, C. Effects of long-term repeated mineral and organic fertilizer applications on soil nitrogen transformations. *Eur. J. Soil Sci.* **2012**, *63*, 75–85. [CrossRef]

77. Bremner, J.M.; Blackmer, A.M. Effects of acetylene and soil water content on emission of nitrous oxide from soils. *Nature* **1979**, *280*, 380–381. [CrossRef]

78. Khalil, M.I.; Baggs, E.M. CH_4 oxidation and N_2O emissions at varied soil water-filled pore spaces and headspace CH_4 concentrations. *Soil Biol. Biochem.* **2005**, *37*, 1785–1794. [CrossRef]

79. Rychel, K.; Meurer, K.H.; Börjesson, G.; Strömgren, M.; Getahun, G.T.; Kirchmann, H.; Kätterer, T. Deep N fertilizer placement mitigated N_2O emissions in a Swedish field trial with cereals. *Nutr. Cycl. Agroecosyst.* **2020**, *118*, 133–148. [CrossRef]

80. Li, L.; Tian, H.; Zhang, M.; Fan, P.; Ashraf, U.; Liu, H.; Chen, X.; Duan, M.; Tang, X.; Wang, Z.; et al. Deep placement of nitrogen fertilizer increases rice yield and nitrogen use efficiency with fewer greenhouse gas emissions in a mechanical direct-seeded cropping system. *Crop J.* **2021**, *9*, 1386–1396. [CrossRef]

81. Wrage, N.; Velthof, G.L.; Van Beusichem, M.L.; Oenema, O. Role of nitrifier denitrification in the production of nitrous oxide. *Soil Biol. Biochem.* **2001**, *33*, 1723–1732. [CrossRef]

82. Brentrup, F.; Kusters, J.; Lammel, J.; Kuhlmann, H. Methods to estimate on-field nitrogen emissions from crop production as an input to LCA studies in the agricultural sector. *Int. J. Life Cycle Assess.* **2000**, *5*, 349–357. [CrossRef]

83. Ciampitti, I.A.; Ciarlo, E.A.; Conti, M.E. Nitrous oxide emissions from soil during soybean [(*Glycine max* (L.) Merrill] crop phenological stages and stubbles decomposition period. *Biol. Fertil. Soils* **2008**, *44*, 581–588. [CrossRef]

84. Bremner, J.M. Sources of nitrous oxide in soils. *Nutr. Cycl. Agroecosyst.* **1997**, *49*, 7–16. [CrossRef]

85. Lesschen, J.P.; Velthof, G.L.; De Vries, W.; Kros, J. Differentiation of nitrous oxide emission factors for agricultural soils. *Environ. Pollut.* **2011**, *159*, 3215–3222. [CrossRef] [PubMed]

86. Parton, W.J.; Mosier, A.R.; Ojima, D.S.; Valentine, D.W.; Schimel, D.S.; Weier, K.; Kulmala, A.E. Generalized model for N_2 and N_2O production from nitrification and denitrification. *Glob. Biogeochem. Cycles* **1996**, *10*, 401–412. [CrossRef]

87. Gaillard, R.; Duval, B.D.; Osterholz, W.R.; Kucharik, C.J. Simulated effects of soil texture on nitrous oxide emission factors from corn and soybean agroecosystems in Wisconsin. *J. Environ. Qual.* **2016**, *45*, 1540–1548. [CrossRef]

88. Charles, A.; Rochette, P.; Whalen, J.K.; Angers, D.A.; Chantigny, M.H.; Bertrand, N. Global nitrous oxide emission factors from agricultural soils after addition of organic amendments: A meta-analysis. *Agric. Ecosyst. Environ.* **2017**, *236*, 88–98. [CrossRef]

89. Meurer, K.H.; Franko, U.; Stange, C.F.; Dalla Rosa, J.; Madari, B.E.; Jungkunst, H.F. Direct nitrous oxide (N_2O) fluxes from soils under different land use in Brazil—A critical review. *Environ. Res. Lett.* **2016**, *11*, 23001. [CrossRef]

90. Xu, Y.; Xu, Z.; Cai, Z.; Reverchon, F. Review of denitrification in tropical and subtropical soils of terrestrial ecosystems. *J. Soils Sediments* **2013**, *13*, 699–710. [CrossRef]

91. Hefting, M.M.; Bobbink, R.; De Caluwe, H. Nitrous oxide emission and denitrification in chronically nitrate-loaded riparian buffer zones. *J. Environ. Qual.* **2003**, *32*, 1194–1203. [CrossRef] [PubMed]

92. Oertel, C.; Matschullat, J.; Zurba, K.; Zimmermann, F.; Erasmi, S. Greenhouse gas emissions from soils—A review. *Geochemistry* **2016**, *76*, 327–352. [CrossRef]

93. Chattha, M.U.; Hassan, M.U.; Khan, I.; Chattha, M.B.; Munir, H.; Nawaz, M.; Mahmood, A.; Usman, M.; Kharal, M. Alternate skip irrigation strategy ensure sustainable sugarcane yield. *J. Anim. Plant Sci.* **2017**, *27*, 1604–1610.

94. Liu, S.; Zhang, Y.; Lin, F.; Zhang, L.; Zou, J. Methane and nitrous oxide emissions from direct-seeded and seedling-transplanted rice paddies in southeast China. *Plant Soil* **2014**, *374*, 285–297. [CrossRef]

95. Naghedifar, S.M.; Ziaei, A.N.; Ansari, H. Simulation of irrigation return flow from a triticale farm under sprinkler and furrow irrigation systems using experimental data: A case study in arid region. *Agric. Water Manag.* **2018**, *210*, 185–197. [CrossRef]

96. Sanz-Cobena, A.; Lassaletta, L.; Aguilera, E.; del Prado, A.; Garnier, J.; Billen, G.; Iglesias, A.; Sánchez, B.; Guardia, G.; Abalos, D.; et al. Strategies for greenhouse gas emissions mitigation in Mediterranean agriculture: A review. *Agric. Ecosyst. Environ.* **2017**, *238*, 5–24. [CrossRef]

97. Qi, L.; Niu, H.D.; Zhou, P.; Jia, R.J.; Gao, M. Effects of biochar on the net greenhouse gas emissions under continuous flooding and water-saving irrigation conditions in paddy soils. *Sustainablity* **2018**, *10*, 1403. [CrossRef]

98. LaHue, G.T.; Chaney, R.L.; Adviento-Borbe, M.A.; Linquist, B.A. Alternate wetting and drying in high yielding direct-seeded rice systems accomplishes multiple environmental and agronomic objectives. *Agric. Ecosyst. Environ.* **2016**, *229*, 30–39. [CrossRef]

99. Lagomarsino, A.; Agnelli, A.E.; Linquist, B.; Adviento-Borbe, M.A.; Agnelli, A.; Gavina, G.; Ravaglia, S.; Ferrara, R.M. Alternate wetting and drying of rice reduced CH_4 emissions but triggered N_2O peaks in a clayey soil of Central Italy. *Pedosphere* **2016**, *26*, 533–548. [CrossRef]

100. Hou, H.; Peng, S.; Xu, J.; Yang, S.; Mao, Z. Seasonal variations of CH_4 and N_2O emissions in response to water management of paddy fields located in Southeast China. *Chemosphere* **2012**, *89*, 884–892. [CrossRef] [PubMed]

101. Ruser, R.; Flessa, H.; Russow, R.; Schmidt, G.; Buegger, F.; Munch, J. Emission of N_2O, N_2 and CO_2 from soil fertilized with nitrate: Effect of compaction, soil moisture and rewetting. *Soil Biol. Biochem.* **2006**, *38*, 263–274. [CrossRef]

102. Friedl, J.; De Rosa, D.; Rowlings, D.W.; Grace, P.R.; Müller, C.; Scheer, C. Dissimilatory nitrate reduction to ammonium (DNRA), not denitrification dominates nitrate reduction in subtropical pasture soils upon rewetting. *Soil Biol. Biochem.* **2018**, *125*, 340–349. [CrossRef]

103. Akiyama, H.; Yan, X.; Yagi, K. Evaluation of effectiveness of enhanced-efficiency fertilizers as mitigation options for N_2O and NO emissions from agricultural soils: Meta-analysis. *Glob. Chang. Biol.* **2010**, *16*, 1837–1846. [CrossRef]

104. Gaihre, Y.K.; Singh, U.; Islam, S.M.M.; Huda, A.; Islam, M.R.; Sanabria, J.; Satter, M.A.; Islam, M.R.; Biswas, J.C.; Jahiruddin, M. Nitrous oxide and nitric oxide emissions and nitrogen use efficiency as affected by nitrogen placement in lowland rice fields. *Nutr. Cycl. Agroecosyst.* **2018**, *110*, 277–291. [CrossRef]

105. Zhou, Z.; Zheng, X.; Xie, B.; Liu, C.; Song, T.; Han, S.; Zhu, J. Nitric oxide emissions from rice-wheat rotation fields in eastern China: Effect of fertilization, soil water content, and crop residue. *Plant Soil* **2010**, *336*, 87–98. [CrossRef]

106. Mitsch, W.J.; Zhang, L.; Anderson, C.J.; Altor, A.E.; Hernández, M.E. Creating riverine wetlands: Ecological succession, nutrient retention, and pulsing effects. *Ecol. Eng.* **2005**, *25*, 510–527. [CrossRef]

107. Zou, J.; Huang, Y.; Zheng, X.; Wang, Y. Quantifying direct N_2O emissions in paddy fields during rice growing season in mainland China: Dependence on water regime. *Atmos. Environ.* **2007**, *41*, 8030–8042. [CrossRef]

108. Yang, W.; Kang, Y.; Feng, Z.; Gu, P.; Wen, H.; Liu, L.; Jia, Y. Sprinkler Irrigation Is Effective in Reducing Nitrous Oxide Emissions from a Potato Field in an Arid Region: A Two-Year Field Experiment. *Atmosphere* **2019**, *10*, 242. [CrossRef]

109. Sun, Z.; Kang, Y.; Jiang, S. Effects of water application intensity, drop size and water application amount on the characteristics of topsoil pores under sprinkler irrigation. *Agric. Water Manag.* **2008**, *95*, 869–876. [CrossRef]

110. Wang, X.J.; Wei, C.Z.; Zhang, J.; Dong, P.; Wang, J.; Zhu, Q.C.; Wang, J.X. Effects of irrigation mode and N application rate on cotton field fertilizer N use efficiency and N losses. *J. Appl. Ecol.* **2012**, *23*, 2751–2758.

111. Lv, G.; Kang, Y.; Li, L.; Wan, S. Effect of irrigation methods on root development and profile soil water uptake in winter wheat. *Irrig. Sci.* **2010**, *28*, 387–398. [CrossRef]

112. Sánchez-Martín, L.; Vallejo, A.; Dick, J.; Skiba, U.M. The influence of soluble carbon and fertilizer nitrogen on nitric oxide and nitrous oxide emissions from two contrasting agricultural soils. *Soil Biol. Biochem.* **2008**, *40*, 142–151. [CrossRef]

113. Sanchez-Martín, L.; Meijide, A.; Garcia-Torres, L.; Vallejo, A. Combination of drip irrigation and organic fertilizer for mitigating emissions of nitrogen oxides in semiarid climate. *Agric. Ecosyst. Environ.* **2010**, *137*, 99–107. [CrossRef]

114. Maqsood, M.; Chattha, M.U.; Chattha, M.B.; Khan, I.; Fayyaz, M.A.; Hassan, M.U.; Zaman, Q.U.; Chattha, M.U. Influence of foliar applied potassium and deficit irrigation under different tillage systems on productivity of hybrid maize. *Pak. J. Agric. Sci.* **2017**, *69*, 317–322.

115. Li, C.; Zhang, Z.; Guo, L.; Cai, M.; Cao, C. Emissions of CH_4 and CO_2 from double rice cropping systems under varying tillage and seeding methods. *Atmos. Environ.* **2013**, *80*, 438–444. [CrossRef]

116. Sainju, U.M. Tillage, cropping sequence, and nitrogen fertilization influence dryland soil nitrogen. *Agron. J.* **2013**, *105*, 1253–1263. [CrossRef]

117. Beare, M.H.; Gregorich, E.G.; St-Georges, P. Compaction effects on CO_2 and N_2O production during drying and rewetting of soil. *Soil Biol. Biochem.* **2009**, *41*, 611–621. [CrossRef]

118. Xiao, X.P.; Wu, F.L.; Huang, F.Q.; Li, Y.; Sun, G.F.; Hu, Q.; He, Y.Y.; Chen, F.; Yang, G.L. Greenhouse air emission under different pattern of rice straw returned to field in double rice area. *Res. Agric. Mod.* **2007**, *28*, 629–632.

119. Liang, W.; Shi, Y.; Zhang, H.; Yue, J.; Huang, G.H. Greenhouse gas emissions from northeast China rice fields in fallow season. *Pedosphere* **2007**, *17*, 630–638. [CrossRef]

120. Mei, K.; Wang, Z.; Huang, H.; Zhang, C.; Shang, X.; Dahlgren, R.A.; Zhang, M.; Xia, F. Stimulation of N_2O emission by conservation tillage management in agricultural lands: A meta-analysis. *Soil Tillage Res.* **2018**, *182*, 86–93. [CrossRef]

121. Feng, J.; Li, F.; Zhou, X.; Xu, C.; Ji, L.; Chen, Z.; Fang, F. Impact of agronomy practices on the effects of reduced tillage systems on CH_4 and N_2O emissions from agricultural fields: A global meta-analysis. *PLoS ONE* **2018**, *13*, e0196703. [CrossRef] [PubMed]

122. Meurer, K.H.; Haddaway, N.R.; Bolinder, M.A.; Kätterer, T. Tillage intensity affects total SOC stocks in boreo-temperate regions only in the topsoil—A systematic review using an ESM approach. *Earth-Sci. Rev.* **2018**, *177*, 613–622. [CrossRef]

123. Nyamadzawo, G.; Wuta, M.; Chirinda, N.; Mujuru, L.; Smith, J.L. Greenhouse gas emissions from intermittently flooded (dambo) rice under different tillage practices in chiota smallholder farming area of Zimbabwe. *Atmos. Clim. Sci.* **2013**, *3*, 13–20. [CrossRef]

124. Ahmad, S.; Li, C.; Dai, G.; Zhan, M.; Wang, J.; Pan, S.; Cao, C. Greenhouse gas emission from direct seeding paddy field under different rice tillage systems in central China. *Soil Tillage Res.* **2009**, *106*, 54–61. [CrossRef]

125. Lampurlanés, J.; Cantero-Martínez, C. Soil bulk density and penetration resistance under different tillage and crop management systems and their relationship with barley root growth. *Agron. J.* **2003**, *95*, 526–536. [CrossRef]

126. Pareja-Sánchez, E.; Plaza-Bonilla, D.; Ramos, M.C.; Lampurlanés, J.; Álvaro-Fuentes, J.; Cantero-Martínez, C. Long-term no-till as a means to maintain soil surface structure in an agroecosystem transformed into irrigation. *Soil Tillage Res.* **2017**, *174*, 221–230. [CrossRef]

127. Grandy, A.S.; Loecke, T.D.; Parr, S.; Robertson, G.P. Long-Term Trends in nitrous oxide emissions, soil nitrogen, and crop yields of till and no-till cropping systems. *J. Environ. Qual.* **2006**, *35*, 1487–1495. [CrossRef]

128. Ball, B.C.; Crichton, I.; Horgan, G.W. Dynamics of upward and downward N_2O and CO_2 fluxes in ploughed or no-tilled soils in relation to water-filled pore space, compaction and crop presence. *Soil Tillage Res.* **2008**, *1*, 20–30. [CrossRef]

129. Elder, J.W.; Lal, R. Tillage effects on gaseous emissions from an intensively farmed organic soil in North Central Ohio. *Soil Tillage Res.* **2008**, *98*, 45–55. [CrossRef]

130. Van Kessel, C.; Venterea, R.; Six, J.; Adviento-Borbe, M.A.; Linquist, B.; Van Groenigen, K.J. Climate, duration, and N placement determine N_2O emissions in reduced tillage systems: A meta-analysis. *Glob. Chang. Biol.* **2013**, *19*, 33–44. [CrossRef]

131. Shakoor, A.; Shahbaz, M.; Farooq, T.H.; Sahar, N.E.; Shahzad, S.M.; Altaf, M.M.; Ashraf, M. A global meta-analysis of greenhouse gases emission and crop yield under no-tillage as compared to conventional tillage. *Sci. Total Environ.* **2021**, *750*, 142299. [CrossRef] [PubMed]

132. Shakoor, A.; Dar, A.A.; Arif, M.S.; Farooq, T.H.; Yasmeen, T.; Shahzad, S.M.; Tufail, M.A.; Ahmed, W.; Albasher, G.; Ashraf, M. Do soil conservation practices exceed their relevance as a countermeasure to greenhouse gases emissions and increase crop productivity in agriculture? *Sci. Total Environ.* **2022**, *805*, 150337. [CrossRef]

133. Memon, M.; Guo, J.; Tagar, A.; Perveen, N.; Ji, C.; Memon, S.; Memon, N. The effects of tillage and straw incorporation on soil organic carbon status, rice crop productivity, and sustainability in the rice-wheat cropping system of eastern China. *Sustainability* **2018**, *10*, 961. [CrossRef]

134. Turmel, M.S.; Speratti, A.; Baudron, F.; Verhulst, N.; Govaerts, B. Crop residue management and soil health: A systems analysis. *Agric. Syst.* **2015**, *134*, 6–16. [CrossRef]

135. Gregorutti, V.C.; Caviglia, O.P. Nitrous oxide emission after the addition of organic residues on soil surface. *Agric. Ecosyst. Environ.* **2017**, *246*, 234–242. [CrossRef]

136. Henderson, S.L.; Dandie, C.E.; Patten, C.L.; Zebarth, B.J.; Burton, D.L.; Trevors, J.T.; Goyer, C. Changes in denitrifier abundance, denitrification gene mRNA levels, nitrous oxide emissions, and denitrification in anoxic soil microcosms amended with glucose and plant residues. *Appl. Environ. Microbiol.* **2010**, *76*, 2155–2164. [CrossRef]

137. Mosier, A.; Kroeze, C.; Nevison, C.; Oenema, O.; Seitzinger, S.; Oswald, C. Closing the global N_2O budget: Nitrous oxide emissions through the agricultural nitrogen cycle. *Nutr. Cycl. Agroecosyst.* **1998**, *52*, 225–248. [CrossRef]

138. Shang, Q.; Yang, X.; Gao, C.; Wu, P.; Liu, J.; Xu, Y.; Shen, Q.; Zou, J.; Guo, S. Net annual global warming potential and greenhouse gas intensity in Chinese double rice-cropping systems: A 3-year field measurement in long-term fertilizer experiments. *Glob. Chang. Biol.* **2011**, *17*, 2196–2210. [CrossRef]

139. Xia, L.; Wang, S.; Yan, X. Effects of long-term straw incorporation on the net global warming potential and the net economic benefit in a rice-wheat cropping system in China. *Agric. Ecosyst. Environ.* **2014**, *197*, 118–127. [CrossRef]

140. Sander, B.O.; Samson, M.; Buresh, R.J. Methane and nitrous oxide emissions from flooded rice fields as affected by water and straw management between rice crops. *Geoderma* **2014**, *235*, 355–362. [CrossRef]

141. Ma, E.; Zhang, G.; Ma, J.; Xu, H.; Cai, Z.; Yagi, K. Effects of rice straw returning methods on N_2O emission during wheat-growing season. *Nutr. Cycl. Agroecosyst.* **2010**, *88*, 463–469. [CrossRef]

142. Chen, H.; Li, X.; Hu, F.; Shi, W. Soil nitrous oxide emissions following crop residue addition: A meta-analysis. *Glob. Chang. Biol.* **2013**, *19*, 2956–2964. [CrossRef]

143. Hu, N.; Chen, Q.; Zhu, L. The responses of soil N_2O emissions to residue returning systems: A meta-analysis. *Sustainability* **2019**, *11*, 748. [CrossRef]

144. Kudo, Y.; Noborio, K.; Shimoozono, N.; Kurihara, R. The effective water management practice for mitigating greenhouse gas emissions and maintaining rice yield in central Japan. *Agric. Ecosyst. Environ.* **2014**, *186*, 77–85. [CrossRef]

145. Shaaban, M.; Wu, Y.; Peng, Q.; Wu, L.; Van Zwieten, L.; Khalid, M.S.; Younas, A.; Lin, S.; Zhao, J.; Bashir, S.; et al. The interactive effects of dolomite application and straw incorporation on soil N_2O emissions. *Eur. J. Soil Sci.* **2018**, *69*, 502–511. [CrossRef]

146. Weitz, A.M.; Linder, E.; Frolking, S.; Crill, P.M.; Keller, M. N_2O emissions from humid tropical agricultural soils: Effects of soil moisture, texture and nitrogen availability. *Soil Biol. Biochem.* **2001**, *33*, 1077–1093. [CrossRef]

147. Burger, M.; Jackson, L.E. Microbial immobilization of ammonium and nitrate in relation to ammonification and nitrification rates in organic and conventional cropping systems. *Soil Biol. Biochem.* **2003**, *35*, 29–36. [CrossRef]

148. Chantigny, M.H.; Rochette, P.; Angers, D.A.; Bittman, S.; Buckley, K.; Massé, D.; Bélanger, G.; Eriksen-Hamel, N.; Gasser, M.-O. Soil nitrous oxide emissions following band-incorporation of fertilizer nitrogen and swine manure. *J. Environ. Qual.* **2010**, *39*, 1545–1553. [CrossRef] [PubMed]

149. Pelster, D.E.; Chantigny, M.H.; Rochette, P.; Angers, D.A.; Rieux, C.; Vanasse, A. Nitrous oxide emissions respond differently to mineral and organic nitrogen sources in contrasting soil types. *J. Environ. Qual.* **2012**, *41*, 427–435. [CrossRef] [PubMed]

150. Li, X.; Hu, F.; Shi, W. Plant material addition affects soil nitrous oxide production differently between aerobic and oxygen-limited conditions. *Appl. Soil Ecol.* **2013**, *64*, 91–98. [CrossRef]

151. Duan, Y.F.; Kong, X.W.; Schramm, A.; Labouriau, R.; Eriksen, J.; Petersen, S.O. Microbial N transformations and N_2O emission after simulated grassland cultivation: Effects of the nitrification inhibitor 3,4-Dimethylpyrazole Phosphate (DMPP). *Appl. Environ. Microbiol.* **2017**, *83*, 0201916. [CrossRef]

152. Pittelkow, C.M.; Adviento-Borbe, M.A.; Hill, J.E.; Six, J.; van Kessel, C.; Linquist, B.A. Yield-scaled global warming potential of annual nitrous oxide and methane emissions from continuously flooded rice in response to nitrogen input. *Agric. Ecosyst. Environ.* **2013**, *177*, 10–20. [CrossRef]

153. IPCC Guidelines for National Greenhouse Gas Inventories. Workbook. In *IPCC Guidelines of National Greenhouse Gas Inventory*; Institute for Global Environmental Strategies: Hayama, Japan, 1997; Volume 2.

154. Huang, S.H.; Jiang, W.W.; Lu, J.; Cao, J.M. Influence of nitrogen and phosphorus fertilizers on N_2O emissions in rice fields. *Zhongguo Huanjing Kexue/China Environ. Sci.* **2005**, *25*, 540–543.

155. Li, C.F.; Kou, Z.K.; Yang, J.H.; Cai, M.L.; Wang, J.P.; Cao, C.G. Soil CO_2 fluxes from direct seeding rice fields under two tillage practices in central China. *Atmos. Environ.* **2010**, *44*, 2696–2704. [CrossRef]

156. Xu, H.; Zhu, B.; Liu, J.; Li, D.; Yang, Y.; Zhang, K.; Jiang, Y.; Hu, Y.; Zeng, Z. Azolla planting reduces methane emission and nitrogen fertilizer application in double rice cropping system in southern China. *Agron. Sustain. Dev.* **2017**, *37*, 29. [CrossRef]

157. Jin, F.; Yang, H.; Zhao, Q.G. Research progress of soil organic carbon reserves and its impacting factors. *Soil* **2000**, *1*, 11–17, (In Chinese, with English abstract).

158. Sharma, L.K.; Sukhwinder, K.B. A review of methods to improve nitrogen use efficiency in agriculture. *Sustainability* **2018**, *10*, 51. [CrossRef]

159. An, H.; Owens, J.; Beres, B.; Li, Y.; Hao, X. Nitrous oxide emissions with enhanced efficiency and conventional urea fertilizers in winter wheat. *Nutr. Cycl. Agroecosyst.* **2021**, *119*, 307–322. [CrossRef]

160. Zebarth, B.J.; Rochette, P.; Burton, D.L.; Price, M. Effect of fertilizer nitrogen management on N_2O emissions in commercial corn fields. *Can. J. Soil Sci.* **2008**, *88*, 189–195. [CrossRef]

161. Sogbedji, J.M.; Es, H.M.; Yang, C.L.; Geohring, L.D.; Magdoff, F.R. Nitrate leaching and nitrogen budget as affected by maize nitrogen rate and soil type. *J. Environ. Qual.* **2000**, *29*, 1813–1820. [CrossRef]

162. Randall, G.W.; Mulla, D.J. Nitrate nitrogen in surface waters as influenced by climatic conditions and agricultural practices. *J. Environ. Qual.* **2001**, *30*, 337–344. [CrossRef] [PubMed]

163. Wagner-Riddle, C.; Thurtell, G.W. Nitrous oxide emissions from agricultural fields during winter and spring thaw as affected by management practices. *Nutr. Cycl. Agroecosyst.* **1998**, *52*, 151–163. [CrossRef]

164. Ghosh, S.; Majumdar, D.; Jain, M.C. Methane and nitrous oxide emissions from an irrigated rice of North India. *Chemosphere* **2003**, *51*, 181–195. [CrossRef]

165. Zhang, A.; Cui, L.; Pan, G.; Li, L.; Hussain, Q.; Zhang, X.; Zheng, J.; Crowley, D. Effect of biochar amendment on yield and methane and nitrous oxide emissions from a rice paddy from Tai Lake plain, China. *Agric. Ecosyst. Environ.* **2010**, *139*, 469–475. [CrossRef]

166. Dell, C.J.; Han, K.; Bryant, R.B.; Schmidt, J.P. Nitrous oxide emissions with enhanced efficiency nitrogen fertilizers in a rainfed system. *Agron. J.* **2014**, *106*, 723–731. [CrossRef]

167. Nkebiwe, M.; Weinmann, M.; Bartal, A.; Müller, T. Fertilizer placement to improve crop nutrient acquisition and yield: A review and meta-analysis. *Field Crop Res.* **2016**, *196*, 389–401. [CrossRef]

168. Breitenbeck, G.A.; Bremner, J.M. Effects of rate and depth of fertilizer application on emission of nitrous oxide from soil fertilized with anhydrous ammonia. *Biol. Fertil. Soils* **1986**, *2*, 201–204. [CrossRef]

169. Sehy, U.; Ruser, R.; Munch, J.C. Nitrous oxide fluxes from maize fields: Relationship to yield, site-specific fertilization, and soil conditions. *Agric. Ecosyst. Environ.* **2003**, *99*, 97–111. [CrossRef]

170. Signor, D.; Cerri, C.E.P.; Conant, R. N_2O emissions due to nitrogen fertilizer applications in two regions of sugarcane cultivation in Brazil. *Environ. Res. Lett.* **2013**, *8*, 015013. [CrossRef]

171. Gaihre, Y.K.; Singh, U.; Huda, A.; Islam, S.M.M.; Islam, M.R.; Biswas, J.C.; DeWald, J. Nitrogen use efficiency, crop productivity and environmental impacts of urea deep placement in lowland rice fields. In Proceedings of the 2016 International Nitrogen Initiative Conference, Melbourne, Australia, 4–8 December 2016; pp. 1–4. Available online: https://www.researchgate.net/profile/Yam-Gaihre/publication/312196435.pdf (accessed on 22 May 2021).

172. Chatterjee, D.; Mohanty, S.; Guru, P.K.; Swain, C.K.; Tripathi, R.; Shahid, M.; Kumar, U.; Kumar, A.; Bhattacharyya, P.; Gautam, P.; et al. Comparative assessment of urea briquette applicators on greenhouse gas emission, N loss and soil enzymatic activities in tropical lowland rice. *Agric. Ecosyst. Environ.* **2018**, *252*, 178–190. [CrossRef]

173. Chapuis-Lardy, L.; Wrage, N.; Metay, A.; Bernoux, M. Soils, a sink for N_2O? A review. *Glob. Chang. Biol.* **2006**, *13*, 1–17. [CrossRef]

174. Rutkowska, B.; Szulc, W.; Szara, E.; Skowro'nska, M.; Jadczyszyn, T. Soil N_2O emissions under conventional and reduced tillage methods and maize cultivation. *Plant Soil Environ.* **2017**, *63*, 242–347.

175. Adviento-Borbe, M.A.A.; Linquist, B. Assessing fertilizer N placement on CH_4 and N_2O emissions in irrigated rice systems. *Geoderma* **2016**, *266*, 40–45. [CrossRef]

176. Linquist, B.; Van Groenigen, K.J.; Adviento-Borbe, M.A.; Pittelkow, C.; Van Kessel, C. An agronomic assessment of greenhouse gas emissions from major cereal crops. *Glob. Chang. Biol.* **2012**, *18*, 194–209. [CrossRef]

177. Schwenke, G.D.; Haigh, B.M. The interaction of seasonal rainfall and nitrogen fertiliser rate on soil N_2O emission, total N loss and crop yield of dryland sorghum and sunflower grown on sub-tropical vertosols. *Soil Res.* **2016**, *54*, 604–618. [CrossRef]

178. Grave, R.A.; Nicoloso, R.D.S.; Cassol, P.C.; da Silva, M.L.B.; Mezzari, M.P.; Aita, C.; Wuaden, C.R. Determining the effects of tillage and nitrogen sources on soil N_2O emission. *Soil Tillage Res.* **2018**, *175*, 1–12. [CrossRef]

179. Bordoloi, N.; Baruah, K.K.; Bhattacharyya, P. Emission estimation of nitrous oxide (N_2O) from a wheat cropping system under varying tillage practices and different levels of nitrogen fertiliser. *Soil Res.* **2016**, *54*, 767–776. [CrossRef]

180. Lebender, U.; Senbayram, M.; Lammel, J.; Kuhlmann, H. Impact of mineral N fertilizer application rates on N_2O emissions from arable soils under winter wheat. *Nutr. Cycl. Agroecosyst.* **2014**, *100*, 111–120. [CrossRef]

181. Bouwman, A.F.; Boumans, L.J.M.; Batjes, N.H. Emissions of N_2O and NO from fertilized fields: Summary of available measurement data. *Glob. Biogeochem. Cycl.* **2002**, *16*, 1058. [CrossRef]

182. Venterea, R.T.; Burger, M.; Spokas, K.A. Nitrogen oxide and methane emissions under varying tillage and fertilizer management. *J. Environ. Qual.* **2005**, *34*, 1467–1477. [CrossRef] [PubMed]

183. Jones, S.K.; Rees, R.M.; Skiba, U.M.; Ball, B.C. Influence of organic and mineral N fertiliser on N_2O fluxes from a temperate grassland. *Agric. Ecosyst. Environ.* **2007**, *121*, 74–83. [CrossRef]

184. Watson, C.; Laughlin, R.; McGeough, K. Modification of nitrogen fertilisers using inhibitors: Opportunities and potentials for improving nitrogen use efficiency. *Proc. Int. Fertil. Soc.* **2009**, *658*, 1–40.

185. Nayak, D.; Saetnan, E.; Cheng, K.; Wang, W.; Koslowski, F.; Cheng, Y.F.; Zhu, W.Y.; Wang, J.K.; Liu, J.X.; Moran, D.; et al. Management opportunities to mitigate greenhouse gas emissions from Chinese agriculture. *Agric. Ecosyst. Environ.* **2015**, *209*, 108–124. [CrossRef]

186. Kelliher, F.M.; Cox, N.; Van Der Weerden, T.J.; De Klein, C.A.M.; Luo, J.; Cameron, K.C.; Di, H.J.; Giltrap, D.; Rys, G. Statistical analysis of nitrous oxide emission factors from pastoral agriculture field trials conducted in New Zealand. *Environ. Pollut.* **2014**, *186*, 63–66. [CrossRef]

187. Zhao, X.; Liu, S.L.; Pu, C.; Zhang, X.Q.; Xue, J.F.; Zhang, R.; Wang, Y.Q.; Lal, R.; Zhang, H.L.; Chen, F. Methane and nitrous oxide emissions under no-till farming in China: A meta-analysis. *Glob. Chang. Biol.* **2016**, *22*, 1372–1384. [CrossRef]

188. Shakoor, A.; Xu, Y.; Wang, Q.; Chen, N.; He, F.; Zuo, H.; Yin, H.; Yan, X.; Ma, Y.; Yang, S. Effects of fertilizer application schemes and soil environmental factors on nitrous oxide emission fluxes in a rice-wheat cropping system, east China. *PLoS ONE* **2018**, *13*, e0202016. [CrossRef]

189. Shakoor, A.; Shakoor, S.; Rehman, A.; Ashraf, F.; Abdullah, M.; Shahzad, S.M.; Farooq, T.H.; Ashraf, M.; Manzoor, M.A.; Altaf, M.M.; et al. Effect of animal manure, crop type, climate zone, and soil attributes on greenhouse gas emissions from agricultural soils-a global meta-analysis. *J. Clean. Prod.* **2021**, *278*, 124019. [CrossRef]

190. Parn, J.; Verhoeven, J.T.A.; Butterbach-Bahl, K.; Dise, N.B.; Ullah, S.; Aasa, A.; Egorov, S.; Espenberg, M.; Jearveoja, J.; Jauhiainen, J.; et al. Nitrogen-rich organic soils under warm well-drained conditions are global nitrous oxide emission hotspots. *Nat. Commun.* **2018**, *9*, 1135. [CrossRef] [PubMed]

191. Xu, R.; Prentice, I.C.; Spahni, R.; Niu, H.S. Modelling terrestrial nitrous oxide emissions and implications for climate feedback. *New Phytol.* **2012**, *196*, 472–488.

192. Müller, C.; Kammann, C.; Ottow, J.C.G.; Jeager, H.J. Nitrous oxide emission from frozen grassland soil and during thawing periods. *J. Plant Nutr. Soil Sci.* **2003**, *166*, 46–53. [CrossRef]

193. Kunhikrishnan, A.; Thangarajan, R.; Bolan, N.S.; Xu, Y.; Mandal, S.; Gleeson, D.B.; Seshadri, B.; Zaman, M.; Barton, L.; Tang, C.; et al. Functional relationships of soil acidification, liming, and greenhouse gas flux. *Adv. Agron.* **2016**, *139*, 1–71.

194. Wu, F.; Jia, Z.; Wang, S.; Chang, S.X.; Startsev, A. Contrasting effects of wheat straw and its biochar on greenhouse gas emissions and enzyme activities in a Chernozemic soil. *Biol. Fertil. Soils* **2013**, *49*, 555–565. [CrossRef]

195. Case, S.D.C.; McNamara, N.P.; Reay, D.S.; Whitaker, J. The effect of biochar addition on N_2O and CO_2 emissions from a sandy loam soil—The role of soil aeration. *Soil Biol. Biochem.* **2012**, *51*, 125–134. [CrossRef]

196. Gupta, D.K.; Gupta, C.K.; Dubey, R.; Fagodiya, R.K.; Sharma, G. Role of biochar in carbon sequestration and greenhouse gas mitigation. In *Biochar Applications in Agriculture and Environment Management*; Springer: Cham, Switzerland, 2020; pp. 141–165.

197. Malyan, S.K.; Kumar, S.S.; Fagodiya, R.K.; Ghosh, P.; Kumar, A.; Singh, R.; Singh, L. Biochar for environmental sustainability in the energy-water-agroecosystem nexus. *Renew. Sustain. Energy Rev.* **2021**, *149*, 111379. [CrossRef]

198. Kuzyakov, Y.; Subbotina, I.; Chen, H.; Bogomolova, I.; Xu, X. Black carbon decomposition and incorporation into soil microbial biomass estimated by 14C labeling. *Soil Biol. Biochem.* **2009**, *41*, 210–219. [CrossRef]

199. He, L.; Xu, Y.; Li, J.; Zhang, Y.; Liu, Y.; Lyu, H.; Wang, Y.; Tang, X.; Wang, S.; Zhao, X.; et al. Biochar mitigated more N-related global warming potential in rice season than that in wheat season: An investigation from ten-year biochar-amended rice-wheat cropping system of China. *Sci. Total Environ.* **2022**, *821*, 153344. [CrossRef]

200. Cornelissen, G.; Rutherford, D.W.; Arp, H.P.H.; Dörsch, P.; Kelly, C.N.; Rostad, C.E. Sorption of pure N_2O to biochars and other organic and inorganic materials under anhydrous conditions. *Environ. Sci. Technol.* **2013**, *47*, 7704–7712. [CrossRef]

201. Bakken, L.R.; Bergaust, L.; Liu, B.; Frostegård, Å. Regulation of denitrification at the cellular level: A clue to the understanding of N_2O emissions from soils. *Philos. Trans. R. Soc. B* **2012**, *367*, 1226–1234. [CrossRef] [PubMed]

202. Obia, A.; Cornelissen, G.; Mulder, J.; Dörsch, P. Effect of soil pH increase by biochar on NO, N_2O and N_2 production during denitrification in acid soils. *PLoS ONE* **2015**, *10*, e0138781. [CrossRef] [PubMed]

203. Van Zwieten, L.; Kimber, S.; Morris, S.; Downie, A.; Berger, E.; Rust, J.; Scheer, C. Influence of biochars on flux of N_2O and CO_2 from Ferrosol. *Aust. J. Soil Res.* **2010**, *48*, 555–568. [CrossRef]

204. Oo, A.Z.; Sudo, S.; Akiyama, H.; Win, K.T.; Shibata, A.; Yamamoto, A.; Sano, T.; Hirono, Y. Effect of dolomite and biochar addition on N_2O and CO_2 emissions from acidic tea field soil. *PLoS ONE* **2018**, *13*, e0192235. [CrossRef]

205. Tan, G.; Wang, H.; Xu, N.; Liu, H.; Zhai, L. Biochar amendment with fertilizers increases peanut N uptake, alleviates soil N_2O emissions without affecting NH_3 volatilization in field experiments. *Environ. Sci. Pollut. Res.* **2018**, *25*, 8817–8826. [CrossRef]

206. Cheng, Q.; Cheng, H.; Wu, Z.; Pu, X.; Lu, L.; Wang, J.; Zhao, J.; Zheng, A. Biochar amendment and *Calamagrostis angustifolia* planting affect sources and production pathways of N_2O in agricultural ditch systems. *Environ. Sci. Process. Impacts* **2019**, *21*, 727–737. [CrossRef]

207. Song, Y.; Li, Y.; Cai, Y.; Fu, S.; Luo, Y.; Wang, H.; Liang, C.; Lin, Z.; Hu, S.; Li, Y.; et al. Biochar decreases soil N_2O emissions in Moso bamboo plantations through decreasing labile N concentrations, N-cycling enzyme activities and nitrification/denitrification rates. *Geoderma* **2019**, *348*, 135–145. [CrossRef]

208. Aamer, M.; Shaaban, M.; Hassan, M.U.; Guoqin, H.; Ying, L.; Hai Ying, T.; Rasul, F.; Qiaoying, M.; Zhuanling, L.; Rasheed, A.; et al. Biochar mitigates the N_2O emissions from acidic soil by increasing the nosZ and nirK gene abundance and soil pH. *J. Environ. Manag.* **2020**, *255*, 109891. [CrossRef]

209. Aamer, M.; Shaaban, M.; Hassan, M.U.; Ying, L.; Haiying, T.; Qiaoying, M.; Munir, H.; Rasheed, A.; Xinmei, L.; Ping, L. N_2O Emissions mitigation in acidic soil following biochar application under different moisture regimes. *J. Soil Sci. Plant Nutr.* **2020**, *20*, 2454–2464. [CrossRef]

210. Barracosa, P.; Cardoso, I.; Marques, F.; Pinto, A.; Oliveira, J.; Trindade, H.; Rodrigues, P.; Pereira, J.L.S. Effect of biochar on emission of greenhouse gases and productivity of cardoon crop (*Cynara cardunculus* L.). *J. Soil Sci. Plant Nutr.* **2020**, *20*, 1524–1531. [CrossRef]

211. Liu, H.; Li, H.; Zhang, A.; Rahaman, M.A.; Yang, Z. Inhibited effect of biochar application on N_2O emissions is amount and time-dependent by regulating denitrification in a wheat-maize rotation system in North China. *Sci. Total Environ.* **2020**, *721*, 137636. [CrossRef] [PubMed]

212. Yang, Z.; Yu, Y.; Hu, R.; Xu, X.; Xian, J.; Yang, Y.; Liu, L.; Cheng, Z. Effect of rice straw and swine manure biochar on N_2O emission from paddy soil. *Sci. Rep.* **2020**, *10*, 10843. [CrossRef] [PubMed]

213. Pereira, J.L.S.; Carranca, C.; Coutinho, J.; Trindade, H. The effect of soil type on gaseous emissions from flooded rice fields in Portugal. *J. Soil Sci. Plant Nutr.* **2020**, *20*, 1732–1740. [CrossRef]

214. Clough, T.; Condron, L.; Kammann, C.; Müller, C. A review of biochar and soil nitrogen dynamics. *Agronomy* **2013**, *3*, 275–293. [CrossRef]

215. Levy-Booth, D.J.; Prescott, C.E.; Grayston, S.J. Microbial functional genes involved in nitrogen fixation, nitrification and denitrification in forest ecosystems. *Soil Biol. Biochem.* **2014**, *75*, 11–25. [CrossRef]

216. Song, Y.; Zhang, X.; Ma, B.; Chang, S.X.; Gong, J. Biochar addition affected the dynamics of ammonia oxidizers and nitrification in microcosms of a coastal alkaline soil. *Biol. Fertil. Soils* **2013**, *50*, 321–332. [CrossRef]

217. Shaaban, M.; Peng, Q.; Hu, R.; Lin, S.; Wu, Y.; Ullah, B.; Zhao, J.; Liu, S.; Li, Y. Dissolved organic carbon and nitrogen mineralization strongly affect CO_2 emissions following lime application to acidic soil. *J. Chem. Soc. Pak.* **2015**, *36*, 875–879.

218. Feng, K.; Yan, F.; Hütsch, B.W.; Schubert, S. Nitrous oxide emission as affected by liming an acidic mineral soil used for arable agriculture. *Nutr. Cycl. Agroecosyst.* **2003**, *13*, 289–298. [CrossRef]

219. Shakoor, A.; Arif, M.S.; Shahzad, S.M.; Farooq, T.H.; Ashraf, F.; Altaf, M.M.; Ahmed, W.; Tufail, M.A.; Ashraf, M. Does biochar accelerate the mitigation of greenhouse gaseous emissions from agricultural soil?—A global meta-analysis. *Environ. Res.* **2021**, *202*, 111789. [CrossRef]

220. Yamulki, S.; Harrison, R.M.; Goulding, K.W.T.; Webster, C.P. N_2O, NO and NO_2 fluxes from a grassland: Effect of soil pH. *Soil Biol. Biochem.* **1997**, *29*, 1199–1208. [CrossRef]

221. Qu, Z.; Wang, J.; Almøy, T.; Bakken, L.R. Excessive use of nitrogen in Chinese agriculture results in high $N_2O/(N_2O+N_2)$ product ratio of denitrification, primarily due to acidification of the soils. *Glob. Chang. Biol.* **2014**, *20*, 1685–1698. [CrossRef] [PubMed]

222. Shaaban, M.; Peng, Q.; Lin, S.; Wu, Y.; Zhao, J.; Hu, R. Nitrous Oxide emission from two acidic soils as affected by dolomite application. *Soil Res.* **2014**, *52*, 841–848. [CrossRef]

223. Senbayram, M.; Budai, A.; Bol, R.; Chadwick, D.; Marton, L.; Gündogan, R.; Wu, D. Soil NO_3^- level and O_2 availability are key factors in controlling N_2O reduction to N_2 following long-term liming of an acidic sandy soil. *Soil Biol. Biochem.* **2019**, *132*, 165–173. [CrossRef]

224. Shaaban, M.; Wu, Y.; Wu, L.; Hu, R.; Younas, A.; Nunez-Delgado, A.; Xu, P.; Sun, Z.; Lin, S.; Xu, X.; et al. The effects of pH change through liming on soil N_2O emissions. *Processes* **2020**, *8*, 702. [CrossRef]

225. Hussain, S.; Peng, S.; Fahad, S.; Khaliq, A.; Huang, J.; Cui, K.; Nie, L. Rice management interventions to mitigate greenhouse gas emissions: A review. *Environ. Sci. Pollut. Res.* **2015**, *22*, 3342–3360. [CrossRef] [PubMed]

226. Lan, T.; Zhang, H.; Han, Y.; Deng, O.; Tang, X.; Luo, L.; Zeng, J.; Chen, G.; Wang, C.; Gao, X. Regulating CH_4, N_2O, and NO emissions from an alkaline paddy field under rice–wheat rotation with controlled release N fertilizer. *Environ. Sci. Pollut. Res.* **2021**, *28*, 18246–18259. [CrossRef]

227. Huérfano, X.; Fuertes-Mendizábal, T.; Fernández-Diez, K.; Estavillo, J.M.; González-Murua, C.; Menéndez, S. The new nitrification inhibitor 3,4-dimethylpyrazole succinic (DMPSA) as an alternative to DMPP for reducing N_2O emissions from wheat crops under humid Mediterranean conditions. *Eur. J. Agron.* **2016**, *80*, 78–87. [CrossRef]

228. Cayuela, M.L.; Aguilera, E.; Sanz-Cobena, A.; Adams, D.C.; Abalos, D.; Barton, L.; Ryals, R.; Silver, W.L.; Alfaro, M.A.; Pappa, V.A. Direct nitrous oxide emissions in Mediterranean climate cropping systems: Emission factors based on a meta-analysis of available measurement data. *Agric. Ecosyst. Environ.* **2017**, *238*, 25–35. [CrossRef]

229. Ruser, R.; Schulz, R. The effect of nitrification inhibitors on the nitrous oxide (N_2O) release from agricultural soils—A review. *J. Plant Nutr. Soil Sci.* **2015**, *178*, 171–188. [CrossRef]

230. Gilsanz, C.; Báez, D.; Misselbrook, T.H.; Dhanoa, M.S.; Cárdenas, L.M. Development of emission factors and efficiency of two nitrification inhibitors, DCD and DMPP. *Agric. Ecosyst. Environ.* **2016**, *216*, 1–8. [CrossRef]

231. Tenuta, M.; Beauchamp, E.G. Nitrous oxide production from granular nitrogen fertilizers applied to a silt loam soil. *Can. J. Soil Sci.* **2003**, *83*, 521–532. [CrossRef]

232. Forster, P.; Ramaswamy, V.; Artaxo, P.; Berntsen, T.; Betts, R.; Fahey, D.W.; Haywood, J.; Lean, J.; Lowe, D.C.; Myhre, J.; et al. *Climate Change 2007: Changes in Atmospheric Constituents and in Radiative Forcing: The Physical Science Basis. Contribution of Working Group I to the Fourth Assessment Report of the Intergovernmental Panel on Climate Change*; Solomon, S., Qin, D., Manning, M., Chen, Z., Marquis, M., Averyt, K.B., Tignor, M., Miller, H.L., Eds.; Cambridge University Press: Cambridge, UK; New York, NY, USA, 2007; Available online: https://www.ipcc.ch/site/assets/uploads/2018/02/ar4-wg1-chapter2-1.pdf (accessed on 11 May 2021).

233. Zebarth, B.J.; Snowdon, E.; Burton, D.L.; Goyer, C.; Dowbenko, R. Controlled release fertilizer product effects on potato crop response and nitrous oxide emissions under rain-fed production on a medium-textured soil. *Can. J. Soil Sci.* **2012**, *92*, 759–769. [CrossRef]

234. Akiyama, H.; Yan, X.; Yagi, K. Estimations of emission factors for fertilizer-induced direct N_2O emissions from agricultural soils in Japan: Summary of available data. *Soil Sci. Plant Nutr.* **2006**, *52*, 774–787. [CrossRef]

235. Zekri, M.; Koo, R.C.J. Evaluation of controlled-release fertilizers for young citrus trees. *J. Am. Soc. Hortic. Sci.* **1991**, *116*, 987–990. [CrossRef]

236. Ji, Y.; Liu, G.; Ma, J.; Xu, H.; Yagi, K. Effect of controlled-release fertilizer on nitrous oxide emission from a winter wheat field. *Nutr. Cycl. Agroecosyst.* **2012**, *94*, 111–122. [CrossRef]

237. Jarosiewicz, A.; Tomaszewska, M. Controlled-release NPK fertilizer encapsulated by polymeric membranes. *J. Agric. Food Chem.* **2003**, *51*, 413–417. [CrossRef] [PubMed]

238. Ji, Y.; Liu, G.; Ma, J.; Zhang, G.; Xu, H.; Yagi, K. Effect of controlled-release fertilizer on mitigation of N_2O emission from paddy field in South China: A multi-year field observation. *Plant Soil* **2013**, *371*, 473–486. [CrossRef]

239. Nawaz, M.; Chattha, M.U.; Chattha, M.B.; Ahmad, R.; Munir, H.; Usman, M.; Hassan, M.U.; Khan, S.; Kharal, M. Assessment of compost as nutrient supplement for spring planted sugarcane (*Saccharum officinarum* L.). *J. Anim. Plant Sci.* **2017**, *27*, 283–293.

240. Nawaz, M.; Khan, S.; Ali, H.; Ijaz, M.; Chattha, M.U.; Hassan, M.U.; Irshad, S.; Hussain, S.; Khan, S. Assessment of Environment-friendly Usage of Spent Wash and its Nutritional Potential for Sugarcane Production. *Commun. Soil Sci. Plant Anal.* **2019**, *50*, 1239–1249. [CrossRef]

241. Chen, H.; Zhao, Y.; Feng, H.; Liu, J.; Si, B.; Zhang, A.; Chen, J.; Cheng, G.; Sun, B.; Pi, X.; et al. Effects of straw and plastic film mulching on greenhouse gas emissions in Loess Plateau, China: A field study of 2 consecutive wheat-maize rotation cycles. *Sci. Total Environ.* **2017**, *579*, 814–824. [CrossRef] [PubMed]

242. Cui, P.; Fan, F.; Yin, C.; Song, A.; Huang, P.; Tang, Y.; Zhu, P.; Peng, C.; Li, T.; Wakelin, S.A.; et al. Long-term organic and inorganic fertilization alters temperature sensitivity of potential N_2O emissions and associated microbes. *Soil Biol. Biochem.* **2016**, *93*, 131–141. [CrossRef]

243. Wei, D.; Xu-Ri; Wang, Y.; Wang, Y.; Liu, Y.; Yao, T. Responses of CO_2, CH_4 and N_2O fluxes to livestock exclosure in an alpine steppe on the Tibetan Plateau, China. *Plant Soil* **2012**, *359*, 45–55. [CrossRef]

244. Frimpong, K.A.; Baggs, E.M. Do combined applications of crop residues and inorganic fertilizer lower emission of N_2O from soil? *Soil Use Manag.* **2010**, *26*, 412–424. [CrossRef]

245. Van Groenigen, J.W.; Huygens, D.; Boeckx, P.; Kuyper, T.W.; Lubbers, I.M.; Rütting, T.; Groffman, P.M. The soil n cycle: New insights and key challenges. *Soil* **2015**, *1*, 235–256. [CrossRef]

246. Jain, N.; Arora, P.; Tomer, R.; Mishra, S.V.; Bhatia, A.; Pathak, H.; Chakraborty, D.; Kumar, V.; Dubey, D.S.; Harit, R.C.; et al. Greenhouse gases emission from soils under major crops in Northwest India. *Sci. Total Environ.* **2016**, *542*, 551–561. [CrossRef]

247. Gu, J.; Yuan, M.; Liu, J.; Hao, Y.; Zhou, Y.; Qu, D.; Yang, X. Trade-off between soil organic carbon sequestration and nitrous oxide emissions from winter wheat-summer maize rotations: Implications of a 25-year fertilization experiment in Northwestern China. *Sci. Total Environ.* **2017**, *595*, 371–379. [CrossRef]

248. Wassmann, R.; Lantin, R.S.; Neue, H.U.; Buendia, L.V.; Corton, T.M.; Lu, Y. Characterization of methane emissions from rice fields in Asia. III. Mitigation options and future research needs. *Nutr. Cycl. Agroecosyst.* **2000**, *58*, 23–36. [CrossRef]

249. Corton, T.M.; Bajita, J.B.; Grospe, F.S.; Pamplona, R.R.; Asis, C.A.; Wassmann, R.; Lantin, R.S.; Buendia, L.V. Methane emission from irrigated and intensively managed rice fields in Central Luzon (Philippines). *Nutr. Cycl. Agroecosyst.* **2000**, *58*, 37–53. [CrossRef]

250. Wassmann, R.; Buendia, L.V.; Lantin, R.S.; Bueno, C.S.; Lubigan, L.A.; Umali, A.; Nocon, N.N.; Javellana, A.M.; Neue, H.U. Mechanisms of crop management impact on methane emissions from rice fields in Los Baños, Philippines. *Nutr. Cycl. Agroecosyst.* **2000**, *58*, 107–119. [CrossRef]

251. Chattha, M.U.; Hassan, M.U.; Barbanti, L.; Chattha, M.B.; Khan, I.; Usman, M.; Ali, A.; Nawaz, M. Composted sugarcane by-product press mud cake supports wheat growth and improves soil properties. *Int. J. Plant Prod.* **2019**, *13*, 241–249. [CrossRef]

252. Grigatti, M.; Barbanti, L.; Hassan, M.U.; Ciavatta, C. Fertilizing potential and CO_2 emissions following the utilization of fresh and composted food-waste anaerobic digestates. *Sci. Total Environ.* **2020**, *698*, 134198. [CrossRef] [PubMed]

253. Ding, Z.; Kheir, A.M.S.; Ali, O.A.M.; Hafez, E.M.; ElShamey, E.A.; Zhou, Z.; Wang, B.; Lin, X.; Ge, Y.; Fahmy, A.E. A vermicompost and deep tillage system to improve saline-sodic soil quality and wheat productivity. *J. Environ. Manag.* **2021**, *277*, 111388. [CrossRef] [PubMed]

254. Doan, T.T.; Bouvier, C.; Bettarel, Y.; Bouvier, T.; Henry-des-Tureaux, T.; Janeau, J.L.; Lamballe, P.; Van Nguyen, B.; Jouquet, P. Influence of buffalo manure, compost, vermicompost and biochar amendments on bacterial and viral communities in soil and adjacent aquatic systems. *Appl. Soil Ecol.* **2014**, *73*, 78–86. [CrossRef]

255. Doan, T.T.; Henry-Des-Tureaux, T.; Rumpel, C.; Janeau, J.L.; Jouquet, P. Impact of compost, vermicompost and biochar on soil fertility, maize yield and soil erosion in Northern Vietnam: A three year mesocosm experiment. *Sci. Total Environ.* **2015**, *514*, 147–154. [CrossRef]

256. Rodriguez, V.; de los Angeles Valdez-Perez, M.; Luna-Guido, M.; Ceballos-Ramirez, J.M.; Franco-Hernández, O.; van Cleemput, O.; Marsch, R.; Thalasso, F.; Dendooven, L. Emission of nitrous oxide and carbon dioxide and dynamics of mineral N in wastewater sludge, vermicompost or inorganic fertilizer amended soil at different water contents: A laboratory study. *Appl. Soil Ecol.* **2011**, *49*, 263–267. [CrossRef]

257. Cayuela, M.L.; van Zwieten, L.; Singh, B.P.; Jeffery, S.; Roig, A.; Sánchez-Monedero, M.A. Biochar's role in mitigating soil nitrous oxide emissions: A review and meta-analysis. *Agric. Ecosyst. Environ.* **2014**, *191*, 5–16. [CrossRef]

258. Mandal, S.; Thangarajan, R.; Bolan, N.S.; Sarkar, B.; Khan, N.; Ok, Y.S.; Naidu, R. Biochar-induced concomitant decrease in ammonia volatilization and increase in nitrogen use efficiency by wheat. *Chemosphere* **2016**, *142*, 120–127. [CrossRef]

259. Barthod, J.; Rumpel, C.; Dignac, M.F. Composting with additives to improve organic amendments. A review. *Agron. Sustain. Dev.* **2018**, *38*, 17. [CrossRef]

260. Nigussie, A.; Kuyper, T.W.; Bruun, S.; De-Neergaard, A. Vermicomposting as a technology for reducing nitrogen losses and greenhouse gas emissions from small-scale composting. *J. Clean. Prod.* **2016**, *139*, 429–439. [CrossRef]

261. Di, W.U.; Yanfang, F.; Lihong, X.; Manqiang, L.I.U.; Bei, Y.; Feng, H.U.; Linzhang, Y. Biochar combined with vermicompost increases crop production while reducing ammonia and nitrous oxide emissions from a paddy soil. *Pedosphere* **2019**, *29*, 82–94.

262. Kool, D.M.; Van Groenigen, J.W.; Wrage, N. Source determination of nitrous oxide based on nitrogen and oxygen isotope tracing, dealing with oxygen exchange. *Methods Enzymol.* **2011**, *496*, 139–160. [PubMed]

263. Ostrom, N.E.; Ostrom, P.H. The isotopomers of nitrous oxide, analytical considerations and application to resolution of microbial production pathways. In *Handbook of Environmental Isotope Geochemistry*; Baskaran, M., Ed.; Springer: Berlin/Heidelberg, Germany, 2011; pp. 453–476.

264. Hino, T.; Matsumoto, Y.; Nagano, S.; Sugimoto, H.; Fukumori, Y.; Murata, T.; Iwata, S.; Shiro, Y. Structural basis of biological N_2O generation by bacterial nitric oxide reductase. *Science* **2010**, *330*, 1666–1670. [CrossRef] [PubMed]

265. Herman, D.J.; Firestone, M.K.; Nuccio, E.; Hodge, A. Interactions between an arbuscular mycorrhizal fungus and a soil microbial community mediating litter decomposition. *FEMS Microbiol. Ecol.* **2012**, *80*, 236–247. [CrossRef] [PubMed]

266. Cavagnaro, T.R.; Bender, S.F.; Asghari, H.R.; van der Heijden, M.G.A. The role of arbuscular mycorrhizas in reducing soil nutrient loss. *Trends Plant Sci.* **2015**, *20*, 283–290. [CrossRef] [PubMed]

267. Köhl, L.; van der Heijden, M.G.A. Arbuscular mycorrhizal fungal species differ in their effect on nutrient leaching. *Soil Biol. Biochem.* **2016**, *94*, 191–199. [CrossRef]

268. Hodge, A.; Storer, K. Arbuscular mycorrhiza and nitrogen: Implications for individual plants through to ecosystems. *Plant Soil* **2014**, *386*, 1–19. [CrossRef]

269. Storer, K.; Coggan, A.; Ineson, P.; Hodge, A. Arbuscular mycorrhizal fungi reduce nitrous oxide emissions from N_2O hotspots. *New Phytol.* **2018**, *220*, 1285–1295. [CrossRef]

270. Bender, S.F.; Plantenga, F.; Neftel, A.; Jocher, M.; Oberholzer, H.R.; Köhl, L.; Giles, M.; Daniell, T.J.; Van Der Heijden, M.G.A. Symbiotic relationships between soil fungi and plants reduce N_2O emissions from soil. *ISME J.* **2014**, *8*, 1336–1345. [CrossRef] [PubMed]

271. Zhang, X.; Wang, L.; Ma, F.; Shan, D. Effects of arbuscular mycorrhizal fungi on N_2O emissions from rice paddies. *Water Air Soil Pollut.* **2015**, *226*, 222. [CrossRef]

272. Thomas, B.W.; Hao, X.; Larney, F.J.; Goyer, C.; Chantigny, M.H.; Charles, A. Nonlegume cover crops can increase non-growing season nitrous oxide emissions. *Soil Sci. Soc. Am. J.* **2017**, *81*, 189–199. [CrossRef]

273. Dungan, R.S.; Leytem, A.B.; Tarkalson, D.D. Greenhouse gas emissions from an irrigated cropping rotation with dairy manure utilization in a semiarid climate. *Agron. J.* **2021**, *113*, 1222–1237. [CrossRef]

274. Thirkell, T.J.; Cameron, D.D.; Hodge, A. Resolving the 'nitrogen paradox' of arbuscular mycorrhizas: Fertilization with organic matter brings considerable benefits for plant nutrition and growth. *Plant. Cell Environ.* **2016**, *39*, 1683–1690. [CrossRef] [PubMed]

275. Hassan, M.U.; Chattha, M.U.; Mahmood, A.; Sahi, S.T. Performance of sorghum cultivars for biomass quality and biomethane yield grown in semi-arid area of Pakistan. *Environ. Sci. Pollut. Res.* **2018**, *25*, 12800–12807. [CrossRef] [PubMed]

276. Hassan, M.U.; Chattha, M.U.; Barbanti, L.; Chattha, M.B.; Mahmood, A.; Khan, I.; Nawaz, M. Combined cultivar and harvest time to enhance biomass and methane yield in sorghum under warm dry conditions in Pakistan. *Ind. Crops Prod.* **2019**, *132*, 84–91. [CrossRef]

277. Hassan, M.U.; Chattha, M.U.; Chattha, M.B.; Mahmood, A.; Sahi, S.T. Chemical composition and methane yield of sorghum as influenced by planting methods and cultivars. *J. Anim. Plant Sci.* **2019**, *29*, 251–259.

278. Hassan, M.U.; Chattha, M.U.; Barbanti, L.; Mahmood, A.; Chattha, M.B.; Khan, I.; Mirza, S.; Aziz, S.A.; Nawaz, M.; Aamer, M. Cultivar and seeding time role in sorghum to optimize biomass and methane yield under warm dry climate. *Ind. Crops Prod.* **2020**, *145*, 111983. [CrossRef]

279. Lou, Y.; Inubushi, K.; Mizuno, T.; Hasegawa, T.; Lin, Y.; Sakai, H.; Cheng, W.; Kobayashi, K. CH_4 emission with differences in atmospheric CO_2 enrichment and rice cultivars in a Japanese paddy soil. *Glob. Chang. Biol.* **2008**, *14*, 2678–2687. [CrossRef]

280. Abalos, D.; van Groenigen, J.W.; De Deyn, G.B. What plant functional traits can reduce nitrous oxide emissions from intensively managed grasslands? *Glob. Chang. Biol.* **2018**, *24*, 248–258. [CrossRef]

281. Xu, H.; Zhang, X.; Han, S.; Wang, Y.; Chen, G. N_2O emissions by trees under natural conditions. *Environ. Sci.* **2001**, *22*, 7–11.

282. Ferch, N.J.; Römheld, V. Release of water-dissolved nitrous oxide by plants: Does the transpiration water flow contribute to the emission of dissolved N_2O by sunflower? In Proceedings of the 14th International Plant Nutrition Colloqium, Beijing, China, 14–17 September 2001; pp. 228–229.

283. Gopalakrishnan, S.; Subbarao, G.V.; Nakahara, K.; Yoshihashi, T.; Ito, O.; Maeda, I.; Ono, H.; Yoshida, M. Nitrification inhibitors from the root tissues of *Brachiaria humidicola*, a tropical grass. *J. Agric. Food Chem.* **2007**, *55*, 1385–1388. [CrossRef]

284. Byrnes, R.C.; Nùñez, J.; Arenas, L.; Rao, I.; Trujillo, C.; Alvarez, C.; Arango, J.; Rasche, F.; Chirinda, N. Biological nitrification inhibition by *Brachiaria* grasses mitigates soil nitrous oxide emissions from bovine urine patches. *Soil Biol. Biochem.* **2017**, *107*, 156–163. [CrossRef]

285. Pappa, V.A.; Rees, R.M.; Walker, R.L.; Baddeley, J.A.; Watson, C.A. Nitrous oxide emissions and nitrate leaching in an arable rotation resulting from the presence of an intercrop. *Agric. Ecosyst. Environ.* **2011**, *141*, 153–161. [CrossRef]

286. Zou, J.; Huang, Y.; Sun, W.; Zheng, X.; Wang, Y. Contribution of plants to N_2O emissions in soil-winter wheat ecosystem: Pot and field experiments. *Plant Soil* **2005**, *269*, 205–211. [CrossRef]

287. Hou, A.X.; Chen, G.X.; Wang, Z.P.; Van Cleemput, O.; Patrick, W.H. Methane and nitrous oxide emissions from a rice field in relation to soil redox and microbiological processes. *Soil Sci. Soc. Am. J.* **2000**, *64*, 2180–2186. [CrossRef]

288. Pathak, H.; Chakrabarti, B.; Bhatia, A.; Jain, N.; Aggarwal, P.K. Potential and cost of low carbon technologies in rice and wheat systems: A case study for the Indo-Gangetic Plains. In *Low Carbon Technologies for Agriculture: A Study on Rice and Wheat Systems in the Indo-Gangetic Plains*; Pathak, H., Aggarwal, P.K., Eds.; Indian Agricultural Research Institute: New Delhi, India, 2012; pp. 12–40.

289. Wegner, B.R.; Chalise, K.S.; Singh, S.; Lai, L.; Abagandura, G.O.; Kumar, S.; Osborne, S.L.; Lehman, R.M.; Jagadamma, S. Response of soil surface greenhouse gas fluxes to crop residue removal and cover crops under a corn–soybean rotation. *J. Environ. Qual.* **2018**, *47*, 1146–1154. [CrossRef]

290. Lehman, R.M.; Osborne, S.L.; Duke, S.E. Diversified no-till crop rotation reduces nitrous oxide emissions, increases soybean yields, and promotes soil carbon accrual. *Soil Sci. Soc. Am. J.* **2017**, *81*, 76–83. [CrossRef]

291. Behnke, G.D.; Zuber, S.M.; Pittelkow, C.M.; Nafziger, E.D.; Villamil, M.B. Long-term crop rotation and tillage effects on soil greenhouse gas emissions and crop production in Illinois, USA. *Agric. Ecosyst. Environ.* **2018**, *261*, 62–70. [CrossRef]

292. Omonode, R.A.; Smith, D.R.; Gál, A.; Vyn, T.J. Soil nitrous oxide emissions in corn following three decades of tillage and rotation treatments. *Soil Sci. Soc. Am. J.* **2011**, *75*, 152–163. [CrossRef]

293. Barton, L.; Murphy, D.V.; Butterbach-Bahl, K. Influence of crop rotation and liming on greenhouse gas emissions from a semi-arid soil. *Agric. Ecosyst. Environ.* **2013**, *167*, 23–32. [CrossRef]

294. Drury, C.F.; Yang, X.M.; Reynolds, W.D.; McLaughlin, N.B. Nitrous oxide and carbon dioxide emissions from monoculture and rotational cropping of corn, soybean and winter wheat. *Can. J. Soil Sci.* **2008**, *88*, 163–174. [CrossRef]

295. Graham, R.F.; Wortman, S.E.; Pittelkow, C.M. Comparison of organic and integrated nutrient management strategies for reducing soil N_2O emissions. *Sustainability* **2017**, *9*, 510. [CrossRef]

296. Ding, W.; Luo, J.; Li, J.; Yu, H.; Fan, J.; Liu, D. Effect of long-term compost and inorganic fertilizer application on background N_2O and fertilizer-induced N_2O emissions from an intensively cultivated soil. *Sci. Total Environ.* **2013**, *465*, 115–124. [CrossRef] [PubMed]

297. Cai, Y.; Ding, W.; Luo, J. Nitrous oxide emissions from Chinese maize-wheat rotation systems: A 3-year field measurement. *Atmos. Environ.* **2013**, *65*, 112–122. [CrossRef]

298. He, W.; Dutta, B.; Grant, B.B.; Chantigny, M.H.; Hunt, D.; Bittman, S.; Tenuta, M.; Worth, D.; VanderZaag, A.; Desjardins, R.L.; et al. Assessing the effects of manure application rate and timing on nitrous oxide emissions from managed grasslands under contrasting climate in Canada. *Sci. Total Environ.* **2020**, *716*, 135374. [CrossRef]

299. Yao, Z.; Zheng, X.; Wang, R.; Liu, C.; Lin, S.; Butterbach-Bahl, K. Benefits of integrated nutrient management on N_2O and NO mitigations in water-saving ground cover rice production systems. *Sci. Total Environ.* **2019**, *646*, 1155–1163.

300. Dittert, K.; Lampe, C.; Gasche, R.; Butterbach-Bahl, K.; Wachendorf, M.; Papen, H.; Sattelmacher, B.; Taube, F. Short-term effects of single or combined application of mineral N fertilizer and cattle slurry on the fluxes of radiatively active trace gases from grassland soil. *Soil Biol. Biochem.* **2005**, *37*, 1665–1674. [CrossRef]

301. Meng, L.; Ding, W.; Cai, Z. Long-term application of organic manure and nitrogen fertilizer on N_2O emissions, soil quality and crop production in a sandy loam soil. *Soil Biol. Biochem.* **2005**, *37*, 2037–2045. [CrossRef]

302. Nyamadzawo, G.; Wuta, M.; Nyamangara, J.; Smith, J.L.; Rees, R.M. Nitrous oxide and methane emissions from cultivated seasonal wetland (dambo) soils with inorganic, organic and integrated nutrient management. *Nutr. Cycl. Agroecosyst.* **2014**, *100*, 161–175. [CrossRef]

303. Nyamadzawo, G.; Shi, Y.; Chirinda, N.; Olesen, J.E.; Mapanda, F.; Wuta, M.; Wu, W.; Meng, F.; Oelofse, M.; De Neergaard, A.; et al. Combining organic and inorganic nitrogen fertilisation reduces N_2O emissions from cereal crops: A comparative analysis of China and Zimbabwe. *Mitig. Adap. Strateg. Glob. Chang.* **2017**, *22*, 233–245.

304. Dhadli, H.S.; Brar, B.S.; Black, T.A. N_2O emissions in a long-term soil fertility experiment under maize–wheat cropping system in Northern India. *Geoderma Reg.* **2016**, *7*, 102–109. [CrossRef]

305. Das, S.; Adhya, T.K. Effect of combine application of organic manure and inorganic fertilizer on methane and nitrous oxide emissions from a tropical flooded soil planted to rice. *Geoderma* **2014**, *213*, 185–192. [CrossRef]

306. Hodge, A.; Robinson, D.; Fitter, A. Are microorganisms more effective than plants at competing for nitrogen? *Trends Plant Sci.* **2000**, *5*, 304–308. [CrossRef]

307. Senbayram, M.; Chen, R.; Budai, A.; Bakken, L.; Dittert, K. N_2O emission and the $N_2O/(N_2O+N_2)$ product ratio of denitrification as controlled by available carbon substrates and nitrate concentrations. *Agric. Ecosyst. Environ.* **2012**, *147*, 4–12. [CrossRef]

308. Shcherbak, I.; Millar, N.; Robertson, G.P. Global metaanalysis of the nonlinear response of soil nitrous oxide (N_2O) emissions to fertilizer nitrogen. *Proc. Natl. Acad. Sci. USA* **2014**, *111*, 9199–9204. [CrossRef] [PubMed]

309. McSwiney, C.P.; Robertson, G.P. Nonlinear response of N_2O flux to incremental fertilizer addition in a continuous maize (*Zea mays* L.) cropping system. *Glob. Chang. Biol.* **2005**, *11*, 1712–1719. [CrossRef]

310. Sarkodie-Addo, J.; Lee, H.C.; Baggs, E.M. Nitrous oxide emissions after application of inorganic fertilizer and incorporation of green manure residues. *Soil Use Manag.* **2006**, *19*, 331–339. [CrossRef]

311. Xu, X.; Yuan, X.; Zhang, Q.; Wei, Q.; Liu, X.; Deng, W.; Wang, J.; Yang, W.; Deng, B.; Zhang, L. Biochar derived from spent mushroom substrate reduced N_2O emissions with lower water content but increased CH_4 emissions under flooded condition from fertilized soils in *Camellia oleifera* plantations. *Chemosphere* **2022**, *287*, 132110. [CrossRef]

312. Deng, B.; Yuan, X.; Siemann, E.; Wang, S.; Fang, H.; Wang, B.; Gao, Y.; Shad, N.; Liu, X.; Zhang, W.; et al. Feedstock particle size and pyrolysis temperature regulate effects of biochar on soil nitrous oxide and carbon dioxide emissions. *Waste Manag.* **2021**, *120*, 33–40. [CrossRef]

313. Xu, X.; He, C.; Yuan, X.; Zhang, Q.; Wang, S.; Wang, B.; Guo, X.; Zhang, L. Rice straw biochar mitigated more N_2O emissions from fertilized paddy soil with higher water content than that derived from ex situ biowaste. *Environ. Poll.* **2020**, *263*, 114477. [CrossRef] [PubMed]

314. Deng, B.; Shi, Y.; Zhang, L.; Fang, H.; Gao, Y.; Luo, L.; Feng, W.; Hu, X.; Wan, S.; Huang, W.; et al. Effects of spent mushroom substrate-derived biochar on soil CO2 and N_2O emissions depend on pyrolysis temperature. *Chemosphere* **2020**, *246*, 125608. [CrossRef] [PubMed]

315. Deng, B.; Fang, H.; Jiang, N.; Feng, W.; Luo, L.; Wang, J.; Wang, H.; Hu, D.; Guo, X.; Zhang, L. Biochar is comparable to dicyandiamide in the mitigation of nitrous oxide emissions from *Camellia oleifera* Abel. fields. *Forests* **2019**, *10*, 1076. [CrossRef]

316. Deng, B.L.; Wang, S.L.; Xu, X.T.; Wang, H.; Hu, D.N.; Guo, X.M.; Shi, Q.H.; Siemann, E.; Zhang, L. Effects of biochar and dicyandiamide combination on nitrous oxide emissions from *Camellia oleifera* field soil. *Environ. Sci. Poll. Res.* **2019**, *26*, 4070–4077. [CrossRef] [PubMed]

317. Hahn, R.W. *A Primer on Environmental Policy Design*; Routledge: London, UK, 2001; pp. 1–35.

318. Sterner, T. *Policy Instruments for Environmental and Natural Resource Management*; Resources for the Future Press: Washington, DC, USA, 2003.

319. UNFCCC. *The Kyoto Protocol to the United Nations Framework Convention on Climate Change*; done at COP3 on 11 December 1997, at Kyoto; UNFCCC: Kyoto, Japan, 1997.

320. Oreggioni, G.D.; Ferraio, F.M.; Crippa, M.; Muntean, M.; Schaaf, E.; Guizzardi, D.; Solazzo, E.; Duerr, M.; Perry, M.; Vignati, E. Climate change in a changing world: Socio-economic and technological transitions, regulatory frameworks and trends on global greenhouse gas emissions from EDGAR v. 5.0. *Glob. Environ. Chang.* **2021**, *70*, 102350. [CrossRef]

321. EC (European Commission). Communication from the Commission to the European Parliament, the European Council, the Council, the Council, the European Economic and Social Committee and the Committee of the Regions: Stepping up Europe's 2030 Climate Ambition. Investing in a Climate-Neutral Future for the Benefit of Our People. 2020. Available online: https://ec.europa.eu/clima/policies/eu-climate-action/2030_ctp_en (accessed on 3 March 2022).

322. House of Commons. The Climate Change Act 2008 (2050 Target Amendment) Order 2019. 2019. Available online: https://www.legislation.gov.uk/ukdsi/2019/9780111187654/pdfs/ukdsi_9780111187654_en.pdf (accessed on 3 March 2022).

323. WB. Data of Industry, Value Added (% of GDP). 2020. Available online: https://data.worldbank.org/indicator/nv.ind.totl.zs (accessed on 3 March 2022).

324. UNDP. *Population Statistics, World Population Prospects (WPP), the 2019 Revision Report United Nations*; Department of Economic and Social Affairs, Population Division: New York, NY, USA, 2019.

325. BP. British Petroleum Statistics of World Energy. 2019. Available online: https://www.bp.com/en/global/corporate/energy-economics/statistical-review-of-world-energy.html (accessed on 3 March 2022).

326. Lapillonne, B.; Sudries, L. Energy Efficiency Trends in the EU: Have We Got off Track? Presentation Given at the Odyssee-Mure Webinar Series on Energy Efficiency Organised by Leonardo ENERGY June 25th 2020. 2020. Available online: https://www.odyssee-mure.eu/events/webinar/energy-efficiency-trends-webinar-june-2020.pdf (accessed on 3 March 2022).

327. UNFCCC. Doha Amendment to the Kyoto Protocol. 2012. Available online: https://treaties.un.org/doc/Publication/CN/2012/CN.718.2012-Eng.pdf (accessed on 24 January 2020).

328. Zelazna, A.; Bojar, M.; Bojar, E. Corporate Social Responsibility towards the Environment in Lublin Region, Poland: A comparative study of 2009 and 2019. *Sustainability* **2020**, *12*, 4463. [CrossRef]

329. Tran, K.T.; Nguyen, P.V. Corporate Social Responsibility: Findings from the Vietnamese Paint Industry. *Sustainability* **2020**, *12*, 1044. [CrossRef]

330. Novelli, V.; Geatti, P.; Bianco, F.; Ceccon, L.; Del Frate, S.; Badin, P. The EMAS registration of the livenza furniture district in the province of Pordenone (Italy). *Sustainability* **2020**, *12*, 898. [CrossRef]

Article

Landscape Analysis of Runoff and Sedimentation Based on Land Use/Cover Change in Two Typical Watersheds on the Loess Plateau, China

Xiaojun Liu [1] and Yi Zhang [2,3,*]

[1] School of Agriculture, Ningxia University, Yinchuan 750021, China
[2] Breeding Base for State Key Laboratory of Land Degradation and Ecological Restoration in Northwest China, Ningxia University, Yinchuan 750021, China
[3] Key Laboratory for Restoration and Reconstruction of Degraded Ecosystem in Northwest China of Ministry of Education, Ningxia University, Yinchuan 750021, China
* Correspondence: 1180411016@stu.xaut.edu.cn; Tel./Fax: +86-0951-2077800

Abstract: Understanding sedimentation and runoff variations caused by land use change have emerged as important research areas, due to the ecological functions of landscape patterns. The aims of this study were to determine the relationship between landscape metrics (LMs), runoff, and sedimentation and explore the crucial LMs in the watersheds on the Loess Plateau. From 1985 to 2010, grassland was the dominant landscape in the Tuweihe (TU) and Gushanchuan (GU) watersheds. Unused land and cropland experienced the greatest transformations. The landscape in the study area tended to become regular, connected, and aggregated, represented by increasing of the Shannon's diversity index and the largest patch index, and decreasing landscape division over time. The landscape stability of the TU watershed was higher than that of the GU watershed. Annual runoff and sedimentation gradually decreased and a significant relationship was found between water and soil loss. Due to larger cropland area and lower landscape stability in the GU watershed, the sedimentation of the two watersheds were similar, even though the runoff in the TU watershed was greater. There were stronger effects of LMs on runoff than that on sedimentation yield. The Shannon's evenness and the patch cohesion index was identified as the key factors of influencing water and soil loss, which had the greatest effects on runoff and sedimentation. Results indicated that regional water and soil loss is sensitive to landscape regulation, which could provide a scientific understanding for the prevention and treatment of soil erosion at landscape level.

Keywords: land use/cover change; landscape; runoff; sedimentation; Loess Plateau

Citation: Liu, X.; Zhang, Y. Landscape Analysis of Runoff and Sedimentation Based on Land Use/Cover Change in Two Typical Watersheds on the Loess Plateau, China. *Life* **2022**, *12*, 1688. https://doi.org/10.3390/life12111688

Academic Editor: Dmitry L. Musolin

Received: 30 August 2022
Accepted: 14 October 2022
Published: 24 October 2022

Publisher's Note: MDPI stays neutral with regard to jurisdictional claims in published maps and institutional affiliations.

1. Introduction

Water and soil loss is a serious problem across the globe and can influence both the biological and physical properties of soil, particularly those related to infiltration rates, nutrient storage, overland flow velocity, and overall soil productivity [1–3]. Environmental services and ecological equilibrium are threatened when soil loss is greater than soil production [4]. About 75 billion tons of soil is eroded every year from terrestrial ecosystems across the world [5], and approximately half of the land is affected by water and soil loss [6]. Water and soil loss is, therefore, an important research area.

Previous studies have shown that vegetation can control soil erosion and help retain runoff. Many studies [7–10] have shown that a high vegetation cover can control water erosion. When rainwater falls on soil, the canopy, roots, and litter components of the vegetation can retain water, weaken the impact of splash erosion, and slow down runoff velocity. These processes of runoff and sediment production are affected by the soil structure, land use type, and vegetation growth patterns [11]. The types and changes of vegetation are the critical factors affecting water and soil loss [12]. A decrease in vegetation cover may result in a growth in erosion problems [13]. For example, deforestation and land reclamation on slopes can accelerate runoff and sedimentation [14]. There is a lot of evidence linking forest clearance and continual cultivation resulting in serious soil erosion [15] because cultivation can change soil properties, such as soil aggregates, permeability, nutrient content, etc., which increases the likelihood of soil erosion [16]. The composition and types of land cover are closely related to runoff process characteristics and sediment yield [17,18]. Excessive land development may weaken the protective action of vegetation on water and soil retention, and encourage runoff and soil erosion. Examples of irrational land uses are planting olive orchards in the Alqueva reservoir region [19], leaving land unused, the inappropriate planting of vineyards [20], and land abandonment. Studies of land use/cover change could help us understand the characteristics of runoff and sedimentation variations, improve eco-environmental stability, and promote the sustainable utilization of water and soil resources.

Water and soil loss is strongly related with land use in landscapes [21]. The relationship between runoff, sedimentation, and vegetation have received attention in recent years [22–24]. The spatial configuration and composition of plant communities has become a vital and widely applied factor in studies of the geomorphological processes related to erosion [25]. Patch level landscape analyses have indicated that forests, shrubland, and grassland patches lead to better soil properties and have consequently reduced runoff and sediment yield [26,27]. Changes in landscape pattern could have a large impact on erosion [28]. The current landscape distributions or variations can be characterized by landscape metrics (LMs), which were classified into three levels that are patch, class, and landscape level. Natural conditions and human disturbances can be remarkably reflected by landscape, including configuration, composition, and topography [29]. Assessing water and soil loss via key environmental parameters and quantifying the respective influence of LMs can facilitate the development of water and soil quality management strategies [30]. For example, Silva [31] found that LMs are sensitive to changes in the soil surface when erosion occurs. In the upper Du River watershed, LMs were found to account for almost 65% and 74% of the variation in soil erosion and sedimentation yield, respectively. In a previous study, four main contributing LMs were highlighted that were closely related to the variations in the erosion modulus [32]. Shi et al. [33] identified several LMs that were the main indices that influenced watershed soil erosion and sediment yield using partial least-squares regression. A recent study identified the largest patch index of farmland and the landscape index of forest as the key LMs for preferred landscape planning to protect the water quality [34]. Therefore, LMs can be used for both geomorphic evaluations and quantifications of water and soil when they are subject to runoff and sedimentation inputs [4]. Although quantitative research has analyzed the impact of LMs on soil and water loss, it is still not clear whether the impact is more significant for LMs on soil loss or water loss. In addition, the reason that caused the influence differences of the LMs on water and soil loss between different regions is subject of debate.

Severe soil erosion and water loss in the Loess Plateau in China has attracted widespread attention, since it restrained local socio-economic development and seriously threatened environmental security [35]. It is particularly challenging to establish the relationship between LMs, and runoff and sedimentation on the Loess Plateau. The semi-arid landscapes of the Loess Plateau are water-limited due to the high evaporation and relatively low rainfall. Therefore, this area is particularly sensitive to a deterioration in environmental quality. Investigating the quantitative relationships between LMs and water and soil loss is crucial if soil erosion is to be prevented in these seasonally affected environments [36]. This is of particular importance when attempting to predict runoff and sedimentation.

2. Materials and Methods

2.1. Research Methodology

By field investigation, data collection, and processing, this paper firstly explored the land use and LM changes of time series in two typical comparative watersheds; secondly, a relation between runoff and sedimentation and LMs at the landscape level was derived by combining ecologically significant LMs; and, finally, the dominant LMs that influence water and soil loss were verified, and the difference exhibited from the two watersheds was discussed.

2.2. Study Area

The Tuweihe and Gushanchuan rivers are tributaries of the Yellow River and are located on the right bank of the middle stream. The two watersheds are located between 109°26′–110°05′ E and 38°18′–39°26′ N, and have areas of 4503.40 and 1263.11 km^2, respectively (Figure 1). Their elevations range from 743 to 1517 m, with a high terrain in the northwest and a low topography in the southeast. They are affected by the northern temperate continental monsoon climate, and the two watersheds are arid and semi-arid regions, with annual mean temperatures of 8.5 and 7.3 °C, respectively. Their annual rainfall amounts are 417.4 and 430 mm, respectively. Concentrated rainfall occurs in the summer, and high evaporation and high intensity storms are the main reasons for the runoff and sedimentation losses in the watersheds. Quaternary loss is widespread in the hilly and gully regions where there is serious wind–water erosion. Two deep fully developed gullies have been cut and their erosion moduli are 2244 and 3299 t km^{-2} a^{-1}, respectively [37,38].

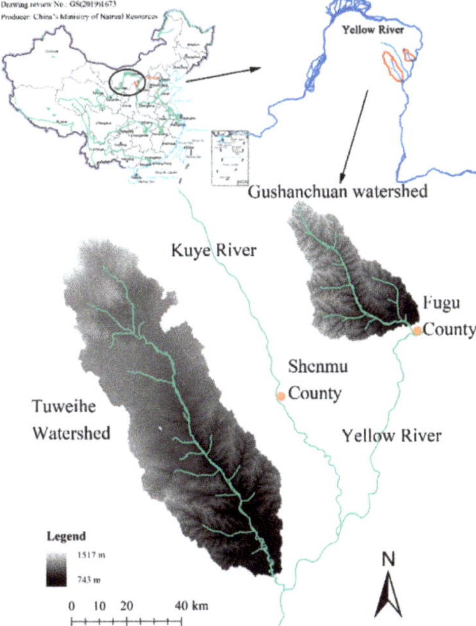

Figure 1. The location of Tuweihe and Gushanchuan watershed, China.

2.3. Research Methods

2.3.1. Data Sources

A digital elevation model (DEM) dataset was provided by the Geospatial Data Cloud, the Computer Network Information Center, Chinese Academy of Sciences (http://www.gscloud.cn, accessed on 28 February 2020). It was processed by the Advanced Spaceborne Thermal Emission and Reflection Radiometer Global Digital Elevation Model (ASTER GDEM) Version 1 and the spatial resolution was 30 m when the Universal Transverse Mercator (UTM, 49N) projection was used. The DEM data were subjected to a mosaic and clipping process, which allowed the study area to be generated. Then, the data were subjected to a depression detention, which meant that the flow generation and drainage network extraction values could be derived and the control watersheds could be created. The hydrological sites at the outlets provided the annual runoff and sedimentation data for 1985–2010.

The land use dataset was provided by the Cold and Arid Regions Science Data Center at Lanzhou, China (http://westdc.westgis.ac.cn, accessed on 8 May 2020). It was funded by major grants from the Chinese Academy of Sciences under the 'National Resources and Environment Survey and Dynamic Monitoring Using Remote Sensing' program (96-B02-01). Researchers experienced in interpreting the spectra, texture, and tone of such images created visual interpretations, which were based on Landsat Multispectral Scanner (MSS), Thematic Mapper (TM), and Enhanced Thematic Mapper (ETM) information. Their results were evaluated by field studies and the precision was as high as 95%.

Due to the large study area, the land use type subcategories were merged into the main categories, which provided a scientific rationality and flexibility for landscape change analysis. Six landscape types were established in the geographic information system (GIS) database: cropland, forest land, grassland, water area, urban and rural land, and unused land. The spatial analyst module in the ArcGIS system and conversion tools were used to transform the vector data for land use into raster data for the following analysis.

2.3.2. Research Methods

Describing the characteristics and variations of the landscape using LMs and identifying relationships between landscape patterns and processes are the most common quantitative methods applied in landscape ecology research [39,40]. The Fragstats 3.3 landscape analysis software was used to determine the relevant LMs according to the Fragstats 3.3 operation manual. The LMs were then used to study the pattern properties of the watershed. Fragstats 3.3 can calculate more than 50 LMs. These metrics were divided into three levels representing three different scales. (1) Patch level: this reflected the structural characteristics of single patches in the landscape and provided the computational basis for the other levels. (2) Class level: this reflected the structural characteristics of multiple patches in the landscape. (3) Landscape level: this reflected all the structural characteristics of the landscape. This study used the landscape-level metrics, number of patches, patch density, the largest patch index, the landscape shape index, the perimeter area fractal dimension, the contagion index, the patch cohesion index, the landscape division index, Shannon's diversity index, and Shannon's evenness index. These indexes were used because they reflect area, density, proximity, diversity, and divergence [30]. The computing method and ecological significance of each metric are listed in Table 1.

Table 1. Description and ecological significance of landscape metrics in this study.

Landscape Index	Formula	Physical Significance	Ecological Significance
Number of patches	N	N: numbers of the patches; Unit: a	the whole numbers of the patches in the landscape
Patch density	$\frac{1}{A}\sum_{j=1}^{M} N_i$	M: types of landscape in the study area; A: total area of the landscape; N: ditto; Unit: a/km^2	degree of fragmentation and spatial heterogeneity; reflecting the human disturbance to a certain extent
Largest patch index	$\frac{Max(a_1,a_2,\ldots a_n)}{A}(100)$	a_i: area of the "i" patch; A: ditto; Unit:%	help to confirm the dominant type; the variation could change the intensity and frequency of the disturbance that reflect the direction and strength of human activities
Landscape shape	$\frac{0.25E}{\sqrt{A}}$	E: perimeter of patches; A: ditto	the bigger the value, the more complicated the patches
Perimeter area fractal dimension	$\dfrac{\left\{n_i\sum_{i=1}^{m}\sum_{j=1}^{n}\left(\ln P_{ij}\times\ln A_{ij}\right)-\left(\sum_{i=1}^{m}\sum_{j=1}^{n}\ln P_{ij}\right)\left(\sum_{i=1}^{m}\sum_{j=1}^{n}\ln A_{ij}\right)\right\}^{2}}{\left(n_i\sum_{i=1}^{m}\sum_{j=1}^{n}\ln P^2_{ij}\right)-\left(\sum_{i=1}^{m}\sum_{j=1}^{n}\ln P_{ij}\right)}$	P_i: proportion of i type in the whole landscape; g_{ik}: number of patches between i and k type; m: ditto	the bigger the value, the greater the fragmentation of patches

Table 1. *Cont.*

Landscape Index	Formula	Physical Significance	Ecological Significance
Contagion	$\left[1 + \dfrac{\sum_{i=1}^{m}\sum_{j=1}^{n}(P_i)\frac{g_{ik}}{\sum_{k=1}^{m}g_{ik}}\times\ln(p_i)\frac{g_{ik}}{\sum_{k=1}^{m}g_{ik}}}{2\ln(m)}\right]\times 100$	M: ditto; Pij: probability of the random selected two adjacent grid belonging to i and j type; Unit: %	the degree of agglomeration and extending tendency of different patch type
Patch cohesion	$\left(1 - \dfrac{\sum_{j=1}^{m}P_{ij}}{\sum_{j=1}^{m}P_{ij}\sqrt{a_{ij}}}\right)\left(1 - \dfrac{1}{\sqrt{A}}\right)^{-1}g(100)$	P_{ij}: perimeter of patch ij; a_{ij}: ditto; A: ditto; Unit: %	spatial connection between a type of patch with the adjacent patches
Landscape division	$\left[1 - \sum_{j=1}^{n}\left(\dfrac{a_{ij}}{A}\right)\right]$	a_{ij}: area of patch j with landscape i; A: ditto	fragmentation of the landscape
Shannon's diversity	$-\sum_{i=1}^{m}[P_i\ln(P_i)]$	P_i: proportion of i type in the whole landscape; i: numbers of patch	complicity and heterogeneity of the landscape, emphasizing the contribution of rare patch to the information
Shannon's evenness	$\dfrac{-\sum_{i=1}^{m}(P_i\ln P_i)}{\ln m}$	ditto	reflecting one or several dominant landscapes

The significance and correlation analyses were undertaken by one-way ANOVA and multiple linear regression using IBM SPSS Statistics Version 2.0 software.

3. Results

3.1. Land Use Changes in the Watersheds

Table 2 describes the change characteristics of the six land use types over the 25-year period. The area of the Tuweihe River watershed is 4503.40 km^2 and grassland represented the greatest proportion of the land cover (between 38.13 and 53.49% over the 25-year period), followed by unused land (between 23.08 and 37.90% over the 25-year period). An analysis of the land-use transfer matrix showed that the unused land variance was largest over the study period at 33.82%. Between 1985 and 2010, 35.12% of unused land was turned into grassland, with 67.28% of the conversion occurring between 1985 and 1996 (Figure 2). Furthermore, 92.93 km^2 of cropland was returned to forest and grassland, with the largest changes occurring between 2000 and 2010 (76.52 km^2). The proportion of land converted from forest to other land uses was lowest at 5.46%.

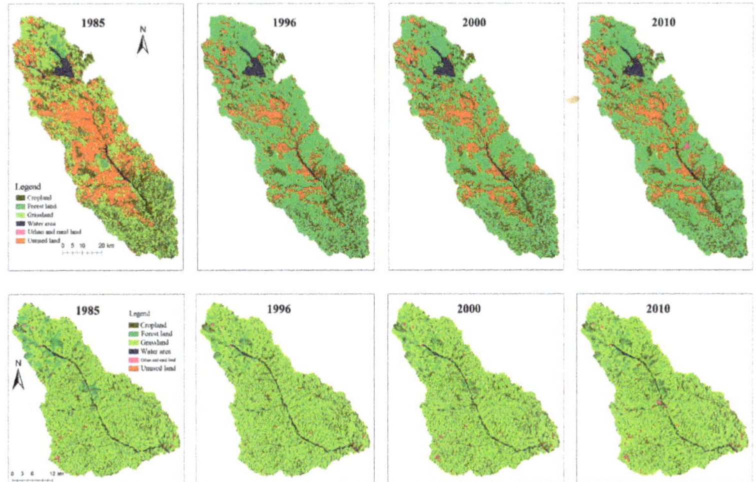

Figure 2. Land use distribution and variations with time in Tu (**up**) and Gu (**down**) watershed.

Table 2. The change variations of land use with time in the study area (km^2).

Land Use	Tuweihe Watershed				Gushanchuan Watershed			
	1985	1996	2000	2010	1985	1996	2000	2010
Cropland	1129.26	1134.52	1116.35	1086.42	410.49	405.94	409.21	383.59
Forest land	203.77	201.87	204.74	212.33	60.47	48.41	64.45	72.91
Grassland	1681.02	2251.95	2124.38	2175.39	772.61	790.87	770.47	785.15
Water area	106.10	105.44	104.98	102.97	12.37	12.82	12.10	12.22
Urban and Rural land	8.70	8.62	9.03	18.65	6.12	4.51	6.32	8.53
Unused land	1374.55	801.00	943.91	909.61	1.05	0.56	0.56	0.55

The GU watershed has an area of 1263.11 km^2. The largest proportion of the wetland was grassland (between 61.00 and 62.61% over the 25-year period), followed by cropland (between 30.37 and 32.50% over the 25-year period). From 1985 to 2010, the cropland and unused land areas gradually decreased, and forest land, grassland, and urban and rural land areas increased. The water area was stable but, in the TU watershed, the water area slowly decreased. The cropland area changed the most (52.85 km^2). Between 2000 and 2010, 49.76 km^2 of cropland was converted into forest and grassland, while unused land had the highest transfer ratio (51.95%) of all the landscape types.

3.2. Land Metrics and Landscape Stability (LS)

Table 3 lists the LMs for the TU and GU watersheds at four time periods. As time progressed, the TU watershed's number of patches, contagion, and patch cohesion values gradually decreased, whereas the largest patch index, the landscape shape index, perimeter area fractal dimension, landscape division, and Shannon's diversity values tended to increase. The patch density values remained almost the same over the 25 years. The LMs in the GU watershed changed over the 25-year period but there was no obvious pattern to the variance. The patch density, the largest patch index, perimeter area fractal dimension, contagion, and patch cohesion were all lower in the TU watershed than in the GU watershed.

Table 3. Annual variations in landscape indices (units: see Table 1).

Watershed	Time	Number of Patches	Patch Density	Largest Patch	Landscape Shape	Perimeter Area Fractal Dimension	Contagion	Patch Cohesion	Division	Shannon's Diversity	Shannon's Eveness
Tuweihe	1985	1393	0.31	20.30	36.48	1.60	36.47	97.79	0.91	1.32	0.734
	1996	1332	0.30	41.04	35.32	1.58	39.82	98.72	0.80	1.24	0.690
	2000	1343	0.30	37.39	36.16	1.58	38.46	98.60	0.83	1.27	0.706
	2010	1340	0.30	34.03	35.94	1.57	38.36	98.44	0.86	1.27	0.707
Gushanchuan	1985	938	0.74	61.00	37.14	1.68	53.34	99.18	0.62	0.89	0.495
	1996	909	0.72	62.50	36.34	1.69	55.22	99.21	0.61	0.85	0.476
	2000	959	0.76	60.81	37.27	1.69	53.07	99.17	0.63	0.89	0.498
	2010	928	0.74	61.79	35.73	1.68	52.59	99.14	0.62	0.91	0.506

The cropland, forest, and WAR land use types had the highest LS values. The grassland had the lowest average LS value, particularly between 2000 and 2010. During this period, the character stability (CS) and density stability (DS) of urban and rural land in the TU watershed was 0.409 and 0.881, respectively, followed by a DS of 0.591 for unused land between 1985 and 1996. In the GU watershed, the lowest LS value occurred for forest land, followed by unused land and urban and rural land, which had the lowest values between 1985 and 1996. In general, the LS values for the TU watershed were higher than those for the GU watershed. The LS values for grassland and unused land increased over time, whereas the cropland and urban and rural land declined.

3.3. Variation in and the Relationship between Annual Runoff and Sedimentation

The Mann–Kendall trend results showed that runoff and sedimentation tended to decrease over time ($p < 0.01$). The peaks of annual runoff and sedimentation were correlated with each other (Figure 3). Up to 2010, the runoff rate in the TU and GU watersheds was 52.52% and 80.95% of the annual runoff in 1956, respectively. The annual reduction in runoff volume was 3.75 and 1.35 million m^3, respectively. Sedimentation in 2010 decreased by 97.26% and 99.77%, respectively, relative to the value in 1985, and the average sedimentation reduction per year was 0.21 and 0.46 million tons in the TU and GU watersheds, respectively. The runoff in the TU watershed, which has a larger area than the GU watershed, was higher than in the GU watershed, but annual sedimentation was much the same (1.40 million tons). The rank-sum test showed that there was a break point for annual runoff and the sedimentation process in the two watersheds. In the TU watershed, the break points for runoff and sedimentation occurred in 1981 and 2001, respectively, while it was 1999 for both runoff and sedimentation in the GU watershed.

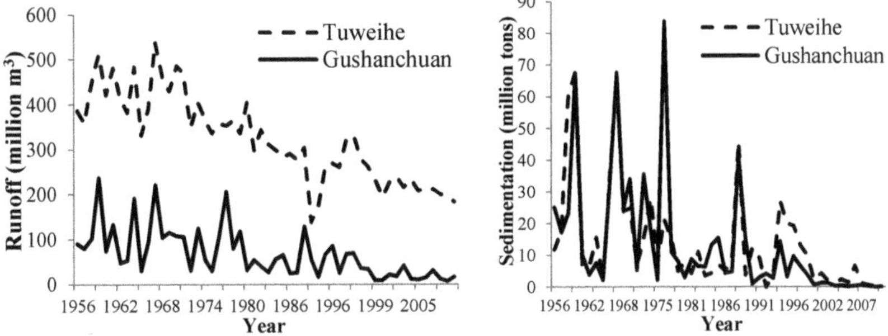

Figure 3. Interannual variation in annual runoff and sedimentation from 1956 to 2010.

The Pearson's correlation analysis revealed that there was a strong positive relationship between runoff and sedimentation in the two watersheds ($p < 0.01$, Figure 4). The coefficient of determination for the TU watershed, which has a higher landscape diversity, was 0.48, whereas the coefficient of determination for the GU watershed was 0.85. The sediment-carrying capacity of the runoff (i.e., the slope of the regression line) in the GU watershed was greater than in the TU watershed carrying capacity. This showed that the sedimentation yields of the GU watershed were similar to those of the TU watershed, even though there was substantially less runoff (19.27% of that in the TU watershed).

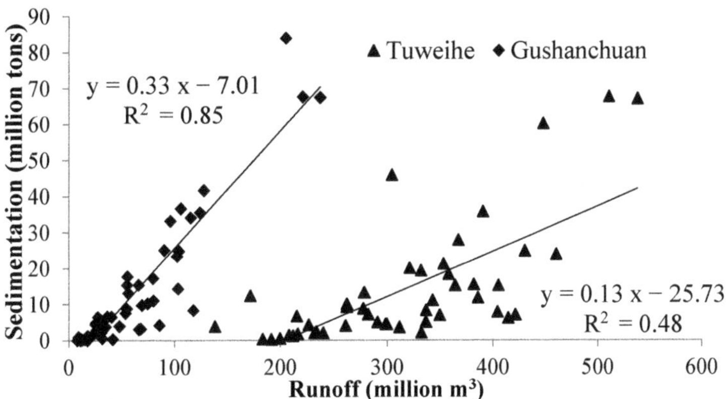

Figure 4. Linear relationship between annual runoff and sedimentation.

3.4. Response Relationships between Runoff, Sedimentation, and LMs

A Pearson's analysis was conducted to determine the effects of landscape on runoff and sedimentation (Table 4). The results showed that there were significant correlations between the factors. More LMs were significantly ($p < 0.05$) or highly significantly ($p < 0.01$) correlated with annual runoff. When patch density, contagion, and patch cohesion rose, the annual runoff declined. In contrast, the LMs related to landscape diversity, such as Shannon's diversity and Shannon's evenness, were positively associated with annual runoff and sedimentation ($p < 0.01$). These relationships implied that the increase in patch density and area led to a decrease in runoff. Furthermore, contagion and patch cohesion had direct impacts ($p < 0.05$) on erosion (coefficients of determination of 0.773 and 0.738, respectively).

Table 4. Regression relationships between LMs, runoff, and sedimentation.

	LMs	Regression Equation	R^2	Sig.
Runoff	Patch density	-4.457PD $+ 5.010$	0.916	0.003 **
	Contagion	-0.113contagion $+ 8.191$	0.738	0.028 *
	Patch cohesion	-0.717cohesion $+ 71.936$	0.773	0.021 *
	Shannon's diversity	3.312SHDI-3.361	0.930	0.002 **
	Shannon's evenness	12.280SHEI-4.937	0.934	0.002 **
Sedimentation	Contagion	-0.006contagion $+ 0.474$	0.693	0.04 *
	Patch cohesion	-0.043cohesion $+ 4.294$	0.760	0.024 *

* Significant at $p < 0.05$; ** significant at $p < 0.01$.

4. Discussion

The Chinese government initiated the Grain for Green Program (GGP) in 1999 and this nationwide project has gradually changed the national land use structure [41]. The Loess Plateau was particularly affected by the program because it was considered as a priority region [42]. Large areas (Table 2) have been converted to various alternative landscapes in the study watersheds. More check dams were constructed in Tu watershed than in Gu watershed, which play a vital role in intercepting sediment. It was confirmed by the lower coefficient determination in the relationship between runoff and sedimentation in TU watershed (Figure 3). Both watersheds have been subjected to continuous deforestation and conversion of cropland to forest. In the process, patch connectivity developed, which led to species migration and other ecological processes. This was confirmed by the increase in the largest patch index, patch cohesion, and contagion values. The landscape shape index decrease in the TU watershed showed that many patches were affected by anthropogenic activities, which led to a regular and simple patch pattern. This was confirmed by the decrease in the perimeter area fractal dimension values.

In the TU watershed, number of patches decreased with time, but patch cohesion and contagion increased, which indicated that good connectivity was formed by merging a landscape type with species migration and other ecological processes [43]. The variable decreases in landscape shape index and perimeter area fractal dimension illustrated that many patches were being affected by human activities, which also showed that the landscape consisted of regular and simple patches. Large stretches of grassland were recreated in 1996, which led to the lowest value for Shannon's diversity. The lower patch density and area parameters resulted in a complex landscape system in the TU watershed. Therefore, the Shannon's diversity value for the TU watershed was higher than that of the GU watershed. The LS for the urban and rural land in the TU watershed decreased over time due to increased anthropogenic activities (Table 5). More than five programs, including the conversion of cropland to forests program, have been initiated since 1978 in an attempt to control desertification and soil loss. Furthermore, afforestation has also increased in China over the last decade [44]. Therefore, the LS values for grassland and unused land increased after 2000 and interference due to human activities declined across the two landscape types.

Table 5. Landscape stability variation characteristics.

Landscape	Year	Tuweihe Watershed		Gushanchuan Watershed	
		CS	DS	CS	DS
Cropland	1985–1996	0.996	0.995	0.981	0.973
	1996–2000	0.982	0.978	0.989	0.987
	2000–2010	0.952	0.932	0.909	0.881
	1985–2010	0.938	0.914	0.914	0.893
Forest land	1985–1996	0.973	0.954	0.621	0.441
	1996–2000	0.965	0.943	−0.089	−0.845
	2000–2010	0.959	0.956	0.705	0.541
	1985–2010	0.952	0.948	0.482	0.169
Grassland	1985–1996	0.687	0.714	0.885	0.791
	1996–2000	0.956	0.967	0.813	0.648
	2000–2010	0.936	0.896	0.910	0.841
	1985–2010	0.683	0.661	0.871	0.761
Water area	1985–1996	0.951	0.912	0.962	0.959
	1996–2000	0.984	0.969	0.961	0.978
	2000–2010	0.969	0.955	0.963	0.934
	1985–2010	0.932	0.898	0.994	1.000
Urban and rural land	1985–1996	0.995	1.000	0.835	0.934
	1996–2000	0.968	0.985	0.738	0.880
	2000–2010	0.409	0.881	0.652	0.651
	1985–2010	0.362	0.865	0.598	0.589
Unused land	1985–1996	0.591	0.598	0.519	0.506
	1996–2000	0.883	0.945	0.999	1.000
	2000–2010	0.977	0.990	0.892	0.800
	1985–2010	0.676	0.689	0.564	0.608

The cropland landscape was the key factor affecting soil conservation [45] in the study area and there were more check dams in the TU watershed according to the field investigation, which caused the annual sedimentation yield in the TU watershed to be similar to the yield for the GU watershed, even though the annual runoff in the TU watershed was significantly higher ($p < 0.01$, Figure 3). Fragmented natural landscape indicated intensive agricultural activities, which caused more serious erosion and soil nutrient loss [46]. Higher landscape stability of TU watershed further confirmed its controlling function on sedimentation with higher runoff. In addition, the variation coefficient for annual sedimentation was higher than that for annual runoff, which indicated that the sedimentation was more susceptible to environmental effects than runoff.

Runoff and the sediment deposited in water is contained by the spatial pattern of the landscape [47]. The LMs synthesize the retardation capacity and spatial position, and reflect the potential risk of water and soil loss. For example, the Shannon's diversity value is not a biodiversity metric but, rather, focuses on the unbalanced distribution of various patch types in the landscape. In the study area, the diversity of land uses and low degree of landscape fragmentation exerted significant positive influences on runoff ($p < 0.01$, Table 4). The contagion and patch cohesion values had significant negative correlations with annual runoff and sedimentation ($p < 0.05$), which indicated that water and soil loss decreased when external and internal patch connectivity improved. Most LMs were significantly related to annual runoff, which showed that the landscape had a greater effect on runoff than sedimentation. This means that it can be used as an ecological indicator to predict runoff, relative to sedimentation. The LS changes showed that there was an abrupt runoff change in 1981, which the land use data did not show. The sedimentation-to-runoff ratio was lowest in 2001, and the LS values for forest land and grassland were also at their lowest compared to 1985–1996 and 2000–2010 (Table 5). In the GU watershed, the lowest LS for forest land occurred in 1999 (1996–2000), which indicated that there had been a sharp increase in forest land. This could have caused the sharp decrease in runoff and sediment deposition in 1999. It, therefore, appears that a breakpoint usually occurs when the LS for forest land and grassland is small, which suggests that there has been a major expansion in these types of land use.

A Pearson's analysis was conducted to determine the factors that most strongly influence the annual decrease in runoff (DR) and sedimentation (DSe) (Table 6). The results showed that the correlations between DSe and LS, and the different land use types were not significant ($p > 0.05$). However, the DR was positively correlated to the DS for grassland ($P = 0.740$, $p < 0.05$), which meant that annual runoff in the watershed could be significantly reduced if the grassland had a high DSe. When the grass patches were highly connected, runoff could be effectively intercepted [48]. Therefore, the LMs and LS effects on runoff became more significant.

Table 6. Pearson's analysis between the DR, DSe, and LS values for the different land use types.

		Cropland CS	Cropland DS	Forest Land CS	Forest Land DS	Grassland CS	Grassland DS	Water Area CS	Water Area DS	URL CS	URL DS	Unused Land CS	Unused Land DS
DR	Pearson Correlation	0.001	0.016	0.275	0.268	0.586	**0.740** *	0.363	0.145	0.338	0.127	0.347	0.375
	Sig. (2-tailed)	0.998	0.970	0.510	0.520	0.127	**0.036**	0.377	0.732	0.412	0.765	0.400	0.359
DSe	Pearson Correlation	−0.329	−0.369	−0.154	−0.139	−0.180	−0.378	−0.319	−0.162	−0.636	−0.336	0.013	−0.098
	Sig. (2-tailed)	0.427	0.368	0.716	0.742	0.670	0.356	0.442	0.702	0.090	0.417	0.976	0.818

* Correlation is significant at the 0.05 level (two-tailed). DR and DSe mean decrease in annual runoff and sedimentation, CS and DS mean stability of character and density. URL means urban and rural land.

A stepwise regression analysis was used to determine the most influential variables that were not strongly correlated with one another [49]. Every independent variable was subjected to an F test and then deleted if the F-value showed that the variable was not significant. Furthermore, the previous variable was deleted if the F-value was not significant when a new independent variable was added to the set. This algorithm was repeated until no independent variable could be added or deleted. The optimal regression model was then established after applying this method (Table 7). The variance inflation factors (VIF) were 0.446 and 2.244 for the TU and GU watersheds, respectively, which meant that the collinearity hypothesis could be rejected. The significance values were all lower than 0.05. Therefore, the selected LMs (Shannon's evenness and patch cohesion) were the most significant factors affecting annual runoff and sedimentation. When the dominant landscapes had greater ecological benefits, the annual runoff decreased. Furthermore, measures that promote the value of patch cohesion should be taken if the interception function of water and soil loss is to be improved.

This study applied DEM dataset processed by ASTER GDEM for landscape analysis to assess water and soil loss. Although it has been proved to be an appropriate application in the field of erosion estimation [50] and provided scientific basis for soil erosion prevention and land use management, it is still a worthy study to investigate the result difference with finer or coarser resolution. In addition, the relationship between LMs and erosion we established and discussed was based on the dataset collected in the focalized regions, which are typical watershed on the Loess Plateau. Considering the extension of the scientific research, more analysis should be executed with larger scale and different regions. There is, of course, conventional existing research that concluded that LMs, e.g., Shannon's evenness and patch cohesion, were significantly correlated with soil erosion in the whole region of Loess Plateau [51]. In terms of driving factors, vegetation cover and landscape variables are not the only factors that influence the erosion process; soil properties, climatic conditions, etc., also play a vital role in water and soil loss [52]. Therefore, to increase the validity of analysis and deepen the understanding of soil erosion processes, more related variables should be considered in further related research.

Table 7. Optimal regression model for LMs, runoff, and sedimentation.

Dependent	Model		Unstandardized Coefficients		Standardized Coefficients	t	Sig.	Collinearity Statistics	
			B	Std. Error	Beta			Tolerance	VIF
runoff	1	(Constant)	−4.937	0.876		−5.636	0.005		
		Shannon's evenness	12.280	1.630	0.967	7.534	0.002	1.000	1.000
	2	(Constant)	25.492	6.921		3.683	0.035		
		Shannon's evenness	8.895	1.032	0.700	8.618	0.003	0.446	2.244
		Patch cohesion	−0.292	0.066	−0.358	−4.403	0.022	0.446	2.244
sedimentation	1	(Constant)	4.294	1.184		3.627	0.022		
		Patch cohesion	−0.043	0.012	−0.871	−3.554	0.024	1.000	1.000

5. Conclusions

From 1985 to 2010, the landscape of the study area tended to become regular, connected, and aggregated, while the annual runoff and sedimentation values gradually decreased. The LS values for grassland and unused land gradually increased, but they decreased for cropland and urban and rural land due to human activities. Due to larger cropland area and lower landscape stability in the GU watershed than that in the TU watershed, the annual sedimentation for the two watersheds was similar, even though the annual runoff in the TU watershed was greater than that in the GU watershed. The annual runoff was significantly and positively correlated with sedimentation ($p < 0.01$), and the coefficient of determination for the TU watershed (0.48) was substantially lower than that for the GU watershed (0.85). The LMs had more significant influences on runoff than on sedimentation ($p < 0.01$), especially given that density stability for grassland could significantly decrease the runoff in the study area. The Shannon's evenness and patch cohesion were the crucial factors for affecting water and soil loss, and the measures involving landscape and land use could have a greater influence on runoff than on sedimentation.

Author Contributions: Y.Z. and X.L. conceived the main idea of the paper. X.L. designed and performed the experiment. X.L. wrote the manuscript, and all authors contributed in improving the paper. All authors have read and agreed to the published version of the manuscript.

Funding: This research was funded by National Natural Science Foundation of China. Grant number 42107365 and 42107368.

Institutional Review Board Statement: Not applicable.

Informed Consent Statement: Not applicable.

Data Availability Statement: Not applicable.

Conflicts of Interest: The authors declare no conflict of interest.

References

1. Imeson, A.C.; Prinsen, H.A.M. Vegetation patterns as biological indicators for identifyingrunoff and sediment source and sink areas for semi-aridlandscapes in Spain. *Agric. Ecosyst. Environ.* **2021**, *104*, 333–342. [CrossRef]
2. Neris, J.; Tejedor, M.; Rodríguez, M.; Fuentes, J.; Jiménez, C. Effect of forest floor characteristics on water repellency, infiltration, runoff and soil loss in Andisols of Tenerife Canary Islands, Spain. *Catena* **2013**, *108*, 50–57. [CrossRef]

3. Shore, M.; Jordan, P.; Mellander, P.E.; Kelly-Quinnc, M.; Walld, D.P.; Murphye, P.N.C.; Mellandf, A.R. Evaluating the critical source area concept of phosphorus loss from soils to water-bodies in agricultural catchments. *Sci. Total Environ.* **2014**, *490*, 405–415. [CrossRef] [PubMed]

4. Hancock, G.R.; Tony, W.; Martinez, C. Soil erosion and tolerable soil loss: Insights into erosion rates for a well-managed grassland catchment. *Geoderma* **2015**, *237–238*, 256–265. [CrossRef]

5. Zuazo, V.H.D.; Pleguezuelo, C.R.R. Soil-erosion and runoff prevention by plant covers: A review. *Agron. Sustain. Dev.* **2008**, *28*, 65–86. [CrossRef]

6. Li, Z.G.; Cao, W.W.; Liu, B.Z. Present status and dynamic variation of soil and water loss in China. *Soil Water Conserv. China* **2008**, *12*, 7–10. (In Chinese)

7. Ludwig, J.A.; Wilcox, B.P.; Breshears, D.D. Vegetation patches and runoff-erosion as interacting ecohydrological processes in semiarid landscapes. *Ecology* **2005**, *86*, 288–297. [CrossRef]

8. Xiao, B.; Wang, Q.H.; Wang, H.F.; Wu, J.; Yu, D. The effects of grass hedges and micro-basins on reducing soil and water loss in temperate regions: A case study of Northern China. *Soil Tillage Res.* **2012**, *122*, 22–35. [CrossRef]

9. Mohammad, A.G.; Adam, M.A. The impact of vegetative cover type on runoff and soil erosion under different land uses. *Catena* **2010**, *81*, 97–103. [CrossRef]

10. Nunes, A.N.; Almeida, A.C.D.; Coelho, C.O.A. Impacts of land use and cover type on runoff and soil erosion in a marginalarea of Portugal. *Appl. Geogr.* **2011**, *31*, 687–699. [CrossRef]

11. Gyssels, G.; Poesen, J.; Bochet, E.; Li, Y. Impact of plant roots on the resistance of soils to erosion by water: A review. *Prog. Phys. Geogr.* **2005**, *22*, 189–217. [CrossRef]

12. García-Ruiz, J.M. The effects of land uses on soil erosion in Spain: A review. *Catena* **2010**, *81*, 1–11. [CrossRef]

13. Zhang, L.; Wang, J.M.; Bai, Z.K. Effects of vegetation on runoff and soil erosion on reclaimed land in an opencast coal-mine dump in a loess area. *Catena* **2015**, *128*, 44–53. [CrossRef]

14. Nie, X.J.; Wang, X.D.; Liu, S.Z.; Shi, G.X.; Hai, L. 137Cs tracing dynamics of soil erosion, organic carbon and nitrogen in sloping farmland converted from original grassland in Tibetan plateau. *Appl. Radiat. Isot.* **2010**, *68*, 1650–1655.

15. Dotterweich, M.; Ivester, A.H.; Hanson, P.R.; Larsen, D.; Dye, D.H. Natural and human induced prehistoric and historical soil erosion and landscape development in southwestern Tennessee, USA. *Anthropocene* **2015**, *8*, 6–24. [CrossRef]

16. Six, J.; Paustian, K.; Elliott, E.T.; Combrink, C. Soil structure and organicmatter: I. Distribution of aggregate-size classes and aggregate-associated carbon. *Soil Sci. Soc. Am. Journey* **2000**, *64*, 681–689. [CrossRef]

17. Casalí, J.R.; Giménez, J.; Díez, J.; Álvarez-Mozos, J.; Del Valle de Lersundi, J.; GoI, M.; Campo, M.A.; Chahor, Y.; Gastesi, R.; López, J. Sediment production and water quality of watersheds with contrasting land use in Navarre Spain. *Agric. Water Manag.* **2010**, *97*, 1683–1694. [CrossRef]

18. Yan, B.; Fang, N.F.; Zhang, P.C.; Shi, Z.H. Impacts of land use change on watershed streamflow and sediment yield: An assessment using hydrologic modelling and partial least squares regression. *J. Hydrol.* **2013**, *484*, 26–37. [CrossRef]

19. Vera, F.; Thomas, P.; Anda, C. Predicting Soil Erosion After Land Use Changes for Irrigating Agriculture in a Large Reservoir of Southern Portugal. *Agric. Agric. Sci. Procedia* **2015**, *4*, 40–49.

20. Cerdan, O.; Govers, G.; Bissonnais, Y.L. Rates and spatial variations of soil erosion in Europe: A study based on erosion plot data. *Geomorphology* **2010**, *122*, 167–177. [CrossRef]

21. Xu, X.L.; Ma, K.M.; Fu, B.J.; Song, C.J.; Liu, W. Influence of three plant species with different morphologies on water runoff and soil loss in a dry-warm river valley, SW China. *For. Ecol. Manag.* **2008**, *256*, 656–663. [CrossRef]

22. Cao, L.; Zhang, Y.; Lu, H.; Yuan, J.; Zhu, Y.; Liang, Y. Grass hedge effects on controlling soil loss from concentrated flow: A case study in the red soil region of China. *Soil Tillage Res.* **2015**, *148*, 97–105. [CrossRef]

23. Riechers, M.; Balázsi, Á.; Betz, L.; Betz, L.; Jiren, T.S.; Fischer, J. The erosion of relational values resulting from landscape simplification. *Landscape Ecol.* **2020**, *35*, 2601–2612. [CrossRef]

24. Jian, S.Q.; Zhan, C.Y.; Fang, S.M. Effects of different vegetation restoration on soil water storage and water balance in the Chinese Loess Plateau. *Agric. For. Meteorol.* **2015**, *206*, 85–96. [CrossRef]

25. Xu, Y.; Tang, H.P.; Wang, B.J.; Chen, J. Effects of landscape patterns on soil erosion processes in a mountain–basin system in the North China. *Natural Hazards* **2017**, *87*, 1567–1585. [CrossRef]

26. Ouyang, W.; Andrew, K.S.; Hao, F.H.; Wang, T. Soil erosion dynamics response to landscape pattern. *Sci. Total Environ.* **2010**, *408*, 1358–1366. [CrossRef] [PubMed]

27. Durán-Zuazo, V.H.; Francia-Martínez, J.R.; García-Tejero, I. Implications of land-cover types for soil erosion on semiarid mountain slopes: Towards sustainable land use in problematic landscapes. *Acta Ecol. Sin.* **2013**, *33*, 272–281. [CrossRef]

28. Bakker, M.M.; Gerard, G.; Doorn, A.; Doorn, A.V.; Quetier, F.; Chouvardas, D.; Rounsevell, M. The response of soil erosion and sediment export to land-use change in four areas of Europe: The importance of landscape pattern. *Geomorphology* **2008**, *98*, 213–226. [CrossRef]

29. Xu, J.; Jin, G.Q.; Tang, H.W.; Mo, Y.M.; Wang, Y.G.; Li, L. Response of water quality to land use and sewage outfalls in different seasons. *Sci. Total Environ.* **2019**, *696*, 134014. [CrossRef]

30. Xu, J.; Bai, Y.; You, H.; Wang, X.; Ma, Z.; Zhang, H. Water quality assessment and the influence of landscape metrics at multiple scales in Poyang Lake basin. *Ecol. Indic.* **2022**, *141*, 109096. [CrossRef]

31. Silva, A.M.; Huang, C.H.; Francesconi, W.; Thalika, S.; Jazmin, V. Using landscape metrics to analyze micro-scale soil erosion processes. *Ecol. Indic.* **2015**, *56*, 184–193. [CrossRef]

32. Zhang, S.; Fan, W.; Li, Y.; Yi, Y. The influence of changes in land use and landscape patterns on soil erosion in a watershed. *Sci. Total Environ.* **2017**, *574*, 34–45. [CrossRef]

33. Shi, Z.H.; Ai, L.; Li, X.; Huang, X.D.; Wu, G.L.; Liao, W. Partial least-squares regression for linking land-cover patterns to soil erosion and sediment yield in watersheds. *J. Hydrol.* **2013**, *498*, 165–176. [CrossRef]

34. Wu, J.; Lu, J. Spatial scale effects of landscape metrics on stream water quality and their seasonal changes. *Water Res.* **2021**, *191*, 116811. [CrossRef]

35. Gao, H.; Li, Z.; Jia, L.; Li, P.; Xu, G.; Ren, Z.; Pang, G.; Zhao, B. Capacity of soil loss control in the Loess Plateau based on soil erosion control degree. *J. Geogr. Sci.* **2016**, *26*, 457–472. [CrossRef]

36. Yüksek, F.; Yüksek, T. Growth performance of Sainfoin and its effects on the runoff, soil loss and sediment concentration in a semi-arid region of Turkey. *Catena* **2015**, *133*, 309–317. [CrossRef]

37. Song, J.; Yang, Z.; Xia, J.; Cheng, D. The impact of mining-related human activities on runoff in northern Shaanxi, China. *J. Hydrol.* **2021**, *598*, 126235. [CrossRef]

38. Liu, W.; Shi, C.; Ma, Y.; Wang, Y. Evaluating sediment connectivity and its effects on sediment reduction in a catchment on the Loess Plateau, China. *Geoderma* **2022**, *408*, 115566. [CrossRef]

39. Chefaoui, R.M. Landscape metrics as indicators of coastal morphology: A multi-scale approach. *Ecol. Indic.* **2014**, *45*, 139–147. [CrossRef]

40. Lustig, A.; Stouffer, D.B.; Roigé, M.; Worner, S.P. Towards more predictable and consistent landscape metrics across spatial scales. *Ecol. Indic.* **2015**, *57*, 11–21. [CrossRef]

41. State, F.A. *China Forestry Statistical Yearbook*; China Forestry Publishing House: Beijing, China, 2013.

42. Geng, Q.; Ren, Q.; Yan, H.; Li, L.; Zhao, X.; Mu, X.; Wu, P.; Yu, Q. Target areas for harmonizing the Grain for Green Programme in China's Loess Plateau. *Land Degrad. Dev.* **2020**, *31*, 325–333. [CrossRef]

43. Ma, K.; Huang, X.; Liang, C.; Zhao, H.; Zhou, X.; Wei, X. Effect of land use/cover changes on runoff in the Min River watershed. *River Res. Appl.* **2020**, *36*, 749–759. [CrossRef]

44. Xie, X.H.; Liang, S.L.; Yao, Y.J.; Jia, K.; Meng, S.S.; Li, J. Detection and attribution of changes in hydrological cycle over the Three-North region of China: Climate change versus afforestationeffect. *Agric. For. Meteorol.* **2015**, *203*, 74–87. [CrossRef]

45. Hou, J.; Fu, B.J.; Liu, Y.; Lu, N.; Gao, G.Y.; Zhou, J. Ecological and hydrological response of farmlands abandoned for different lengths of time: Evidence from the Loess Hill Slope of China. *Glob. Planet. Chang.* **2014**, *113*, 59–67. [CrossRef]

46. Hao, F.; Zhang, X.; Wang, X.; Ouyang, W. Assessing the Relationship Between Landscape Patterns and Nonpoint-Source Pollution in the Danjiangkou Reservoir Basin in China. *JAWRA J. Am. Water Resour. Assoc.* **2012**, *48*, 1162–1177. [CrossRef]

47. Rao, E.; Ouyang, Z.Y.; Yu, X.X.; Yi, X. Spatial patterns and impacts of soil conservation service in China. *Geomorphology* **2014**, *207*, 64–70. [CrossRef]

48. Morvan, X.; Naisse, C.; Malam, I.O.; Desprats, J.F.; Combaud, A.; Cerdan, O. Effect of ground-cover type on surface runoff and subsequent soil erosion in Champagne vineyards in France. *Soil Use Manag.* **2014**, *30*, 372–381. [CrossRef]

49. Sharma, M.J.; Yu, S.J. Stepwise regression data envelopment analysis for variable reduction. *Appl. Math. Comput.* **2015**, *253*, 126–134. [CrossRef]

50. Liu, X.J.; Zhang, Y.; Li, Z.B.; Li, P.; Xu, G.C.; Cheng, Y.T.; Zhang, T.G. Response of water quality to land use in hydrologic response unit and riparian buffer along Dan River, China. *Environ. Sci. Pollut. Res.* **2021**, *28*, 28251–28262. [CrossRef]

51. Jiang, C.; Zhao, L.; Dai, J.; Liu, H.; Li, Z.; Wang, X.; Yang, Z.; Zhang, H.; Wen, M.; Wang, J. Examining the soil erosion responses to ecological restoration programs and landscape drivers: A spatial econometric perspective. *J. Arid Environ.* **2020**, *183*, 104255. [CrossRef]

52. Zhang, H.Y.; Shi, Z.H.; Fang, N.F.; Guo, M.H. Linking watershed geomorphic characteristics to sediment yield: Evidence from the Loess Plateau of China. *Geomorphology* **2015**, *234*, 19–27. [CrossRef]

 *life*

Article

Foliar Fertilizer Application Alters the Effect of Girdling on the Nutrient Contents and Yield of *Camellia oleifera*

Shuangling Xie [1], Dongmei Li [2], Zhouying Liu [1], Yuman Wang [1], Zhihua Ren [1], Cheng Li [1], Qinhua Cheng [1], Juan Liu [1], Ling Zhang [1], Linping Zhang [1] and Dongnan Hu [1,*]

[1] Jiangxi Provincial Key Laboratory of Silviculture, College of Forestry, Jiangxi Agricultural University, Nanchang 330045, China
[2] Jiangxi Agricultural University Library, Nanchang 330045, China
* Correspondence: dnhu98@163.com

Abstract: Improving the economic benefits of *Camellia oleifera* is a major problem for *C. oleifera* growers, and girdling and foliar fertilizer have significant effects on improving the economic benefits of plants. This study explains the effects of girdling, girdling + foliar fertilizer on nutrient distribution, and the economic benefits of *C. oleifera* at different times. It also explains the N, P, and K contents of roots, leaves, fruits, and flower buds (sampled in March, May, August, and October 2021) and their economic benefits. The results showed girdling promoted the accumulation of N and K in leaves in March 2021 (before spring shoot emergence) but inhibited the accumulation of P, which led to the accumulation of P in roots and that of N in fruits in August 2021 (fruit expansion period). Foliar fertilizer application after girdling replenished the P content of leaves in March 2021, and P continued to accumulate in large quantities at the subsequent sampling time points. The N and P contents of the root system decreased in March. In October (fruit ripening stage), girdled shrubs showed higher contents of N and K in fruits and flower buds, and consequently lower relative contents of N and K in roots and leaves but higher content of P in leaves. Foliar fertilizer application slowed down the effects of girdling on nutrient accumulation in fruits and flower buds. Spraying foliar fertilizer decreased the N:P ratio in the flower buds and fruits of girdled plants. Thus, foliar fertilizer spray weakened the effects of girdling on the nutrient content and economic benefits of *C. oleifera*. In conclusion, girdling changed the nutrient accumulation pattern in various organs of *C. oleifera* at different stages, increased leaf N:K ratio before shoot emergence, reduced root K content at the fruit expansion stage and the N:K ratio of mature fruit, and promoted economic benefits.

Keywords: *Camellia oleifera*; girdling; foliar fertilizer; nutrient content; yield

Citation: Xie, S.; Li, D.; Liu, Z.; Wang, Y.; Ren, Z.; Li, C.; Cheng, Q.; Liu, J.; Zhang, L.; Zhang, L.; et al. Foliar Fertilizer Application Alters the Effect of Girdling on the Nutrient Contents and Yield of *Camellia oleifera*. *Life* **2023**, *13*, 591. https://doi.org/10.3390/life13020591

Academic Editor: Othmane Merah

Received: 19 January 2023
Revised: 9 February 2023
Accepted: 17 February 2023
Published: 20 February 2023

1. Introduction

Camellia oleifera Abel (Theaceae) is an evergreen shrub widely planted in 18 provinces and regions of China, including Hunan, Jiangxi, and Guangxi (in order of planting area) [1,2]. Unlike other fruit trees, *C. oleifera* is characterized by the coexistence of fruit and flowers as well as vegetative growth and reproductive growth. This results in different periods of distribution patterns of *C. oleifera* being different from other fruit trees. Therefore, scientific management of the nutrient content of *C. oleifera* is an important measure for maintaining its yield at high levels [1,3,4]. In recent years, scholars have conducted extensive research on the fertilization of *C. oleifera* forests [5]; however, fertilization in forestland was found to be difficult and expensive. Tree girdling, as a means of nutrient content regulation, has been widely used in citrus [6], grape [7], apple [8], kiwi [9], and other fruit trees [10]. The objective of girdling is to sever the phloem and prevent the flow of carbohydrates to the underground plant parts, thus promoting reproductive organ growth, flowering, and fruit development and quality [11,12]. However, girdling should be carried out when the nutrient content of the tree is sufficient. Foliar fertilizer is applied to plant stems and leaves so that plants can absorb various nutrients through stems and leaves and improve their nutritional status [12]. Meanwhile, with the application of unmanned aerial vehicles (UAVs) for plant protection [13,14], the application of foliar fertilizer on *C. oleifera* trees is in the initial stages. Foliar fertilizer will become another important way of nutrient management in *C. oleifera* [15].

A large number of experiments on fruit trees show that girdling effectively reduces the N(Nitrogen), P(Phosphorus), and K(Potassium) contents of the leaves above the girdle [11,16]. Therefore, if the nutrients are not replenished immediately after the girdling, the tree can become weak and eventually die. Therefore, reasonable foliar fertilizer application can effectively improve the nutrient content of plant organs and promote the growth and development of plants [15]. Urea has a high N content and is often used as a common fertilizer for plant nitrogen supplementation [17]. Potassium dihydrogen phosphate contains P and K, and the plant utilization rate is high, which can promote the absorption of N and P by plants, and it has good water solubility, which is the first choice for foliar fertilizer [18]. Girdling can reduce flower and fruit drop. Furthermore, plants are also sprayed with gibberellin [18], naphthaleneacetic acid [19], or boron [20] to reduce flower and fruit drop. *C. oleifera* has been cultivated for a long time [21]; however, its management is extensive, and there has been no way to report the application of girdling. In a preliminary experiment, we showed that girdling could effectively increase the fruit yield of *C. oleifera*. However, the effect of girdling on nutrient distribution in *C. oleifera*, the need to supplement nutrients after the application of girdling, and the relationship between the nutrient characteristics of *C. oleifera* and yield remain unclear. In this study, 10-year-old *C. oleifera* trees were treated with girdling and foliar fertilizer, and the root, leaf, flower, and fruit samples were collected to determine and analyze the key phenological period of *C. oleifera*. Additionally, ripe *C. oleifera* fruits were picked to determine fruit quality. To explore the effects of girdling and foliar fertilizer application on the nutrient content of *C. oleifera* trees, the distribution regulation and stoichiometric ratio characteristics of these trees were analyzed in different periods. Moreover, to develop recommendations and obtain theoretical support for nutrient management in *C. oleifera*, the relationship between nutrient characteristics and yield was analyzed via path analysis.

2. Materials and Methods

2.1. Plant Material and Study Site

Ten-year-old clones of *C. oleifera*, with an average plant height of 2.70 m, ground diameter of 76.83 mm, east-west crown width of 2.55 m, north-south crown width of 2.62 m, and plant row spacing of 2.0 m × 3.0 m, were used in this study. The experiment was conducted in the Jiu long shan Township of Yushui District, Xinyu City, Jiangxi Province, China (27°40′ N, 114°49′ E). The study site has a subtropical monsoon climate, with an average annual temperature of 15 °C and abundant annual rainfall of 1680 mm (Resources come from Xinyu Meteorological Bureau). The woodland was planted in strips on a gentle slope, and the soil contained 25.12 mg kg^{-1} available nitrogen, 5.73 mg kg^{-1} available phosphorus, and 20.25 mg kg^{-1} available potassium. All *C. oleifera* clones were planted in the same period and grown using the same management practices.

2.2. Study Design

A single factor experimental design was adopted. Girdling was applied at the flowering stage, and foliar fertilizer was sprayed after girdling. The treatment subjected to neither girdling nor spray foliar fertilizer served as the check control (CK). Each treatment contained 15 plants (45 plants total). Additionally, the treatments were implemented on three adjacent strips in the middle of the hillside, with one treatment per strip.

2.3. Experimental Method

2.3.1. Girdling Technique

The experiment began in November 2020 (at the first flowering stage). *C. oleifera* shrubs showing uniform growth were selected and girdled 270° with a girdling cutter, completely severing the phloem but without damaging the xylem. The girdle was 2 mm wide, and was located on first-order branches at 10–20 cm above the main stem [22]. Finally, the shrubs were labeled according to the treatment.

2.3.2. Foliar Fertilizer Spray Technique

A sprayer was used to spray the surface of *C. oleifera* leaves with a foliar fertilizer composed of 0.2% urea, 0.2% potassium dihydrogen phosphate, 0.2% borax, 50 mg L^{-1} gibberellin, and 20 mg L^{-1} naphthalene acetic acid (The concentration selection is determined on the basis of comprehensive consideration of previous studies). A total of 7.5 L of foliar fertilizer was applied to 15 *C. oleifera*, with an average of 0.5 L per tree. The fertilizer was sprayed 1 week after girdling on a day with no rain [23].

2.3.3. Sample Collection

Plant organs were sampled in 2021; roots and leaves were sampled on 10 March (before spring shoot emergence), 25 May (after spring shoot emergence), 5 August (fruit expansion stage), and 15 October (fruit maturity stage); fruits were sampled only on 25 May and 15 October, and flower buds were sampled only on 15 October, because flower buds and fruits were either too small or not present at the other time points. Sampling was performed in triplicate on each date, with each replicate containing samples collected from five plants. After collection, the samples were brought to the lab, washed, fixed, dried to a constant weight, pulverized, and stored for testing.

On 15 October 2021, the yield per plant of *C. oleifera* was determined (all fruits from a single *C. oleifera* plant were picked separately, weighed, and recorded). At the same time, before fruit picking, 24 fruits were randomly picked in the upper, middle, and lower layers in the four directions of southeast, southwest, and northwest, with different treatments (one for every five trees, a total of three parts). After collection, it was brought back to the laboratory to determine the fresh weight of single fruit and single fruit seed kernels, and then the seed kernels were put into the oven and baked to constant weight, crushed with a mortar, and stored to determine the oil content of the seed kernels.

2.3.4. Determination of Nutrient Content and Fruit Economic Characteristics

Before the test treatment, three parts of 0–20 cm rhizosphere soil were collected according to the "S" sampling method, dried naturally, passed through a 2 mm sieve, and sealed and stored.

The available nitrogen content was determined using the alkaline hydrolysis method [24]. The available phosphorus content was determined using the molybdenum blue colorimetric method [21]. The available potassium content was measured using an atomic flame photometer [1].

Weigh 0.1 g samples (leaves, roots, fruits, buds) and put them in a boiling tube, add H_2SO_4-H_2O_2, place a small funnel at the mouth of the tube, put it on the cooking furnace at 420 °C to boil until transparent, cool and set the volume to a 100 mL volumetric flask, let stand for 5–7 min, transfer to a 15 mL centrifuge tube for storage, and determine the concentrations of N, P, and K respectively.

The N content of various plant organs was determined using the automatic discontinuous analyzer (Smartchem 200, AMS, Rome, Italy) [22]; P content was determined using the molybdenum blue colorimetric method (GENESYS 180, Shanghai, China) [21]; and K content was determined with a flame photometer (FP6400, Shanghai, China) [1].

Weigh 0.5 g of the sample (dried seed kernels), put it into a folded filter paper packet, extract it using petroleum ether Soxhlet for 8 h, stand at 75 °C for 0.5 h, weigh the weight, calculate the oil content of seed kernels [1], the oil content of fresh fruits, and the oil yield per plant. The calculation formula is as follows:

$$\text{Oil content of fresh fruit (\%)} = \text{Oil content of seed kernels} \times \text{Single fruit fresh weight} \tag{1}$$

$$\text{Oil production per plant (g plant}^{-1}) = \text{Oil content of fresh fruit} \times \text{Yield per plant} \tag{2}$$

2.4. Data Analysis

The data were analyzed using SPSS 19.0. One-way analysis of variance (ANOVA), followed by Duncan's multiple range test (DMRT), were used to identify significant differences ($p < 0.05$) in various parameters among the different treatments. Path analysis was used to determine the relationship between oil yield per plant and nutrient characteristics. Graphs were prepared in ORIGIN 2021.

3. Results

3.1. Effects of Girdling and Foliar Fertilizer Application on the Nutrient Contents of Various C. oleifera Organs

The N, P, and K contents of *C. oleifera* leaves in each treatment at four time points, 10 March (before spring shoot emergence), 25 May (after spring shoot emergence), 5 August (after fruit expansion), and 15 October (after fruit ripening), are shown in Figure 1. Without any treatment (CK), the N nutrient content of leaves was low before spring shoot emergence; however, the N nutrient content of new leaves gradually increased after spring shoot emergence, reaching a higher level at the later growth stage (Figure 1a). The girdling + foliar fertilizer treatments significantly increased the N content of leaves before spring shoot emergence. Compared with CK, the N content of girdling and girdling + foliar fertilizer treatments were 145.05% and 162.44% higher, respectively. During the period from spring shoot emergence to fruit expansion, no significant difference in leaf N content was detected between each treatment and CK. However, N did not accumulate in leaves at the fruit ripening stage; instead, it was transported to the fruit. At the fruit ripening stage, leaf N content in the girdling treatment was 12.66% lower than that in CK, which was significantly lower than that in the previous period. Spraying foliar fertilizer after girdling reduced the N output of mature leaves, and the gap in the N content between the girdling + foliar fertilizer treatment and CK was relatively smaller. The P content of the leaves in CK reached the highest after spring shoot emergence, and gradually decreased with fruit expansion, oil transformation, and flower bud differentiation at the later stage, indicating that P was transported from leaves to other organs (Figure 1b). Girdling decreased the P content by 8.82% compared with CK before spring shoot emergence. However, the P content of leaves increased during the period from the emergence of spring shoots to the expansion of fruits, and then decreased at the later stage. When the fruits matured, the P content of leaves was significantly higher than that in the CK by 55.54%. Spraying foliar fertilizer after girdling significantly increased the leaf P content before spring shoot emergence but inhibited the accumulation of P in leaves after spring shoot emergence. The leaf P content of the girdling + foliar fertilizer treatment was significantly lower than that of the girdling treatment at three stages after spring shoot emergence, but was not significantly different compared with CK.

The effect of the girdling and foliar fertilizer spray on K accumulation was different from that on the N and P accumulations. The K content of leaves in the CK treatment showed an increasing trend until the end of fruit expansion and decreased at fruit maturity (Figure 1c). Girdling promoted the accumulation of K in leaves before spring shoot emergence. However, during the period from spring shoot emergence to fruit expansion, the leaf K content in the girdling treatment was significantly lower than that in CK, and the K output also decreased at the later stage. At the fruit ripening stage, the K content of leaves in the girdling treatment was similar to that in the CK. Except during fruit expansion, foliar fertilizer spray after girdling reduced the difference in the leaf K content between the girdling and CK treatments at all time points. In other words, foliar fertilizer application weakened the regulation of girdling on the K content of *C. oleifera* leaves.

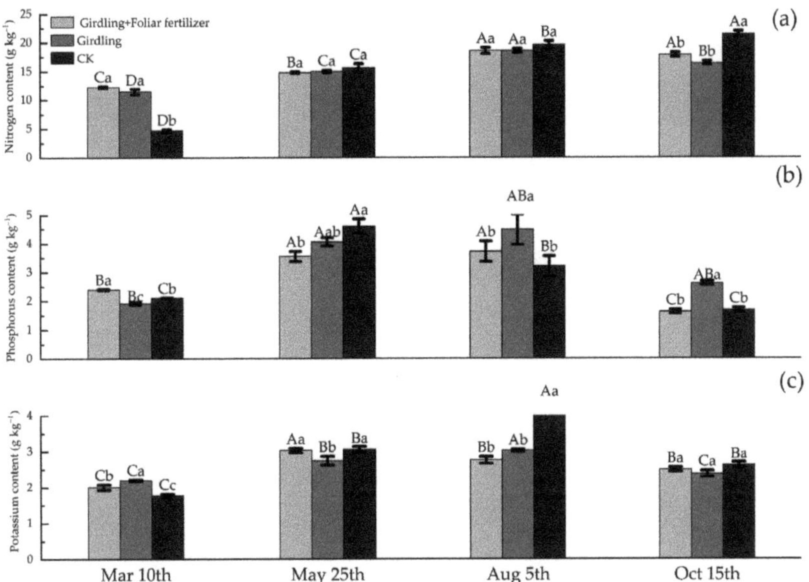

Figure 1. Nutrient content of *C. oleifera* leaves in different stages of each treatment. Data are means $\pm$ SE (n = 3), uppercase letters represent the difference between the same treatment and different periods ($p < 0.05$), lowercase letters represent the difference between different treatments in the same period ($p < 0.05$).

With the change in root growth time, the N, P, and K contents of roots showed different dynamic regulation (Figure 2). Overall, the root N content increased after spring shoot emergence, slightly decreased during fruit expansion, and then increased at fruit maturity; the P content of roots increased at the emergence stage of spring shoot, and then decreased; and the root K content gradually increased with time. However, the change regulation of the different treatments was not completely consistent.

Girdling had no obvious effect on the root N content before and after spring shoot emergence (Figure 2a); however, the root N content was 16.44% after fruit expansion and 15.07% lower after fruit ripening compared with the CK. Therefore, girdling promoted N accumulation in the root system during fruit expansion but decreased N accumulation at fruit maturity. Application of foliar fertilizer after girdling reduced the root N content in each period, indicating that foliar fertilizer supplementation was not conducive to accumulation of N in the roots of girdled *C. oleifera* plants.

Girdling promoted the accumulation and output of root P at spring shoot emergence and fruit maturity (Figure 2b), respectively. The increase in root P was the largest (139.48%) before and after spring shoot emergence, and the decrease in root P was the largest (30.48%) after fruit maturation. Spraying foliar fertilizer after girdling inhibited the accumulation of P in the root system after spring shoot emergence. The root P content was 46.09% lower in the girdling + foliar fertilizer treatment than in the girdling treatment, which was similar to the root P content in CK. However, during other time periods, foliar fertilizer application after girdling had no obvious effect on the root P content.

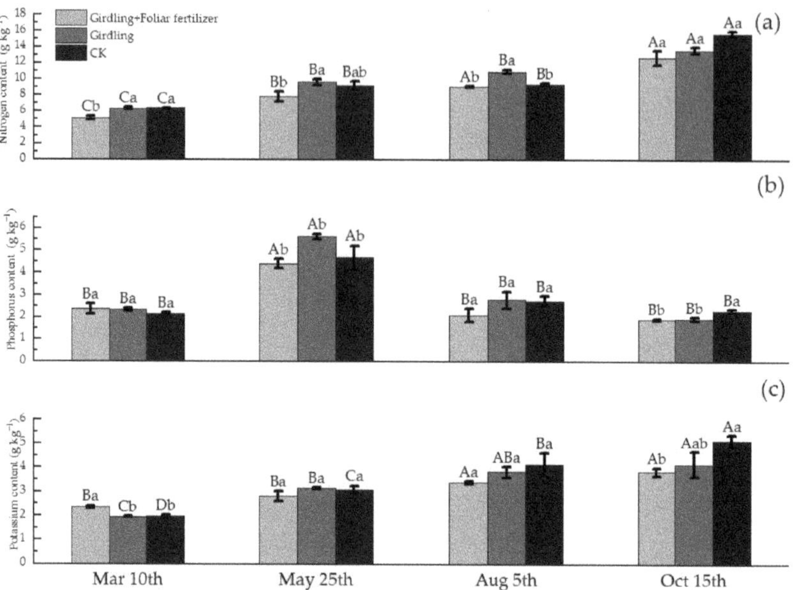

Figure 2. Nutrient content of *C. oleifera* roots in different stages of each treatment. Data are means ± SE (n = 3), uppercase letters represent the difference between the same treatment and different periods ($p < 0.05$), lowercase letters represent the difference between different treatments in the same period ($p < 0.05$).

Girdling had no significant effect of on the root K content in each period (Figure 2c). However, the application of foliar fertilizer after girdling significantly increased the root K content of spring shoots by 20.72% before emergence. Moreover, after the girdling treatment and until fruit ripening, the root K content was reduced by 17.87% and 25.32% after fruit expansion and ripening, respectively, compared with that of CK. The results indicated that spraying foliar fertilizer after girdling was not conducive to the accumulation of K in *C. oleifera* roots after spring shoot emergence.

Girdling had no obvious effect on N and K accumulation in fruits at the fruit expansion stage (Figure 3). However, after fruit ripening, girdling reduced the N and K contents by 28.96% and 10.21%, respectively, compared with CK. The effect of girdling on the fruit P content was observed at the stage of fruit expansion, but little effect was noticed during fruit ripening. Application of foliar fertilizer after girdling reduced the N, P, and K contents of the fruit, and the N, P, and K contents of fruit at the swelling and ripening stages in the girdling + foliar fertilizer treatment showed no significant difference compared with the CK.

Girdling had no obvious effect on the N, P, and K contents of flower buds (Table 1). The N content of flower buds was increased in the girdling + foliar fertilizer treatment, which was 18.36% and 25.21% higher than that in the girdling and CK treatments, respectively. The P and K contents of flower buds were also increased in the girdling + foliar fertilizer treatment, but this increase was not significant compared with the girdling and CK treatments.

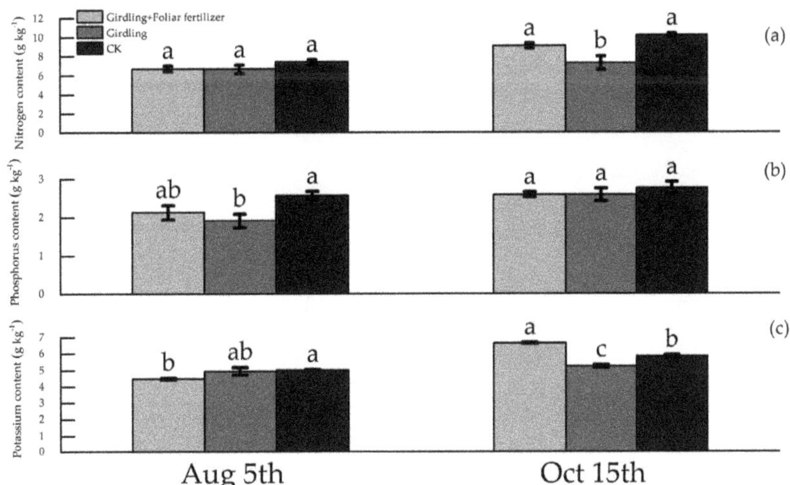

Figure 3. Nutrient contents of *C. oleifera* fruits in each treatment. Data are means ± SE (n = 3), lowercase letters represent the difference between different treatments in the same period (*p* < 0.05).

Table 1. Nutrient contents in flower buds of *C. oleifera* in each treatment.

Treatment	Nutrient		
	N Content (g kg^{-1})	P Content (g kg^{-1})	K Content (g kg^{-1})
Girdling+ foliar fertilizer	14.70 ± 0.73a	2.62 ± 0.14a	4.73 ± 0.15a
Girdling	12.42 ± 0.39b	2.35 ± 0.03a	4.15 ± 0.2a
CK	11.74 ± 0.21b	2.58 ± 0.17a	4.44 ± 0.12a
p value	0.013 *	0.353	0.145

Note Lowercase letters represent the difference between different treatments in the same period (*p* < 0.05); * indicates *p* < 0.05.

3.2. Effects of Girdling and Foliar Fertilizer on Nutrient Distribution in Different *C. oleifera* Organs

Girdling and foliar fertilizer application changed the distribution pattern of N in *C. oleifera* organs (Figure 4). Before spring shoot emergence (10 March 2021), *C. oleifera* fruit had not developed, and flower buds were absent. In the CK treatment, the N nutrient content of roots was significantly higher than that of leaves, and the relative root N content was approximately 60%. Girdling caused more N nutrient accumulation in leaves. The relative N content of leaves in the girdling and girdling + foliar fertilizer treatments was more than 60%, and the relative N content of roots was less than 40%. However, the effect of girdling + foliar fertilizer was more obvious than that of the girdling treatment about the relative leaf N content. After the spring shoot emergence (25 May 2021), the fruit was small, and there were no flower buds. Girdling had little effect on N distribution in roots and leaves at this stage. After fruit expansion (5 August 2021), the root N content was the highest in the girdling treatment among all three treatments, and the relative N content of leaves, roots, and fruits in the girdling + foliar fertilizer treatment were consistent with those in CK. After fruit ripening (15 October 2021), the flower buds were swollen and about to bloom, and the relative N content of flower buds was lowest in CK and highest in the girdling + foliar fertilizer treatment. Additionally, on October 15, the relative N content of leaves was the highest in CK, indicating that girdling and foliar fertilizer application promoted the transport of N from leaves to flowers. Thus, girdling and foliar fertilizer application promoted the transport of N from leaves to other organs.

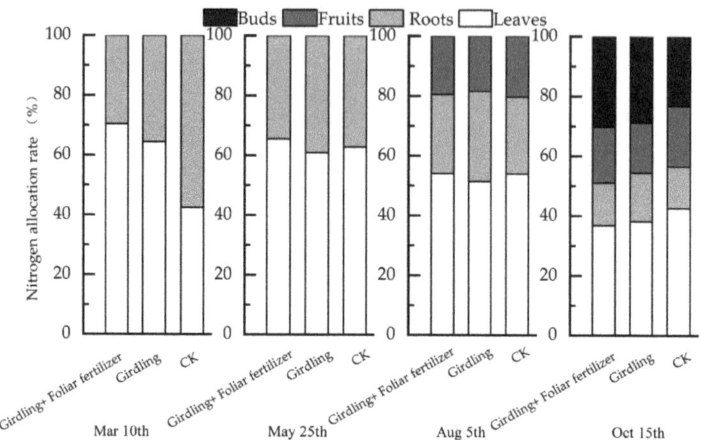

Figure 4. Nitrogen distribution characteristics of *C. oleifera* under different treatments in different periods.

The P distribution patterns in *C. oleifera* organs in different treatments at different stages were shown in Figure 5. At the spring shoot emergence stage, girdling decreased the content of P in leaves and increased P accumulation in the root system. The distribution of P in roots and leaves in the girdling + foliar fertilizer treatment was similar to that in CK. During the period after spring shoot emergence, the relative root P content was the highest in the girdling treatment, indicating that girdling was conducive to the accumulation of P in the root system. However, foliar fertilizer spray on girdled plants slightly decreased the relative root P content. After fruit expansion, the relative P content of fruits was the lowest, while that of leaves was the highest in the girdling treatment; however, the relative P content of roots showed no significant difference among the three treatments. At the fruit maturity stage, the relative P content of leaves was still the highest in the girdling treatment, and the P content in flower buds was relatively less, indicating that the girdling induced the transport of P from roots mainly to leaves. No significant difference was detected in P nutrient allocation between the girdling + foliar fertilizer and CK treatments. In general, the root system of *C. oleifera* accumulated more P in the first half of the year, and more P was transferred to the leaves in the second half of the year when fruits had expanded and ripened. This effect was alleviated by spraying foliar fertilizer after girdling.

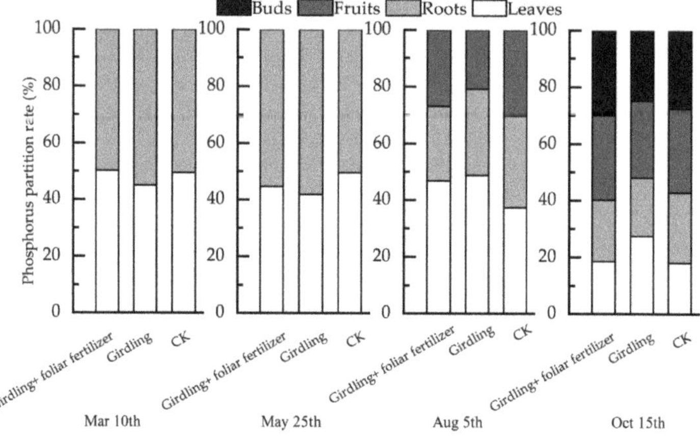

Figure 5. Phosphorus distribution characteristics of *C. oleifera* organs in different treatments in different periods.

Girdling and foliar fertilizer application also affected the distribution of K in various *C. oleifera* organs (Figure 6). Before spring shoot emergence, leaves showed a higher K content in the girdling treatment than in CK; however, spraying foliar fertilizer after girdling had little effect on the distribution nutrients in roots and leaves. After spring shoot emergence, girdling caused more K to accumulate in the root system, reducing the amount of K transported to the leaves. However, spraying foliar fertilizer after girdling further increased the K content of leaves. The relative K content of fruits in the girdling treatment increased with fruit growth and expansion. Foliar fertilizer application after girdling promoted the transport of K from roots and leaves to fruits at the fruit maturity stage, and the relative K content of fruit was highest in the girdling + foliar fertilizer treatment.

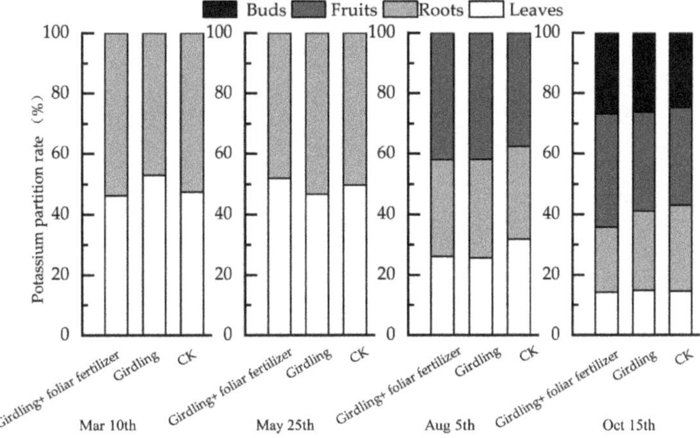

Figure 6. Potassium distribution characteristics of *C. oleifera* organs in different treatments in different periods.

3.3. Effects of Girdling and Foliar Fertilizer Application on N:P and N:K Ratios in Different C. oleifera Organs

Girdling had no significant effect on the leaf N:P ratio from spring shoot emergence to fruit expansion, but significantly increased the leaf N:P ratio before spring shoot emergence and decreased the leaf N:P ratio after fruit ripening (Table 2).

The leaf N:P ratio in the girdling + foliar fertilizer was significantly lower than that in the girdling treatment but significantly higher than that in CK. During other periods, the effect of girdling + foliar fertilizer on the leaf N:P ratio was not obvious. In addition, during the leaf growth period, the N:P ratio in CK gradually increased, reaching a peak at the fruit maturity stage; however, in the other two treatments, the leaf N:P ratio first decreased and then increased at the fruit maturity stage. The root N:P ratio differed among the three treatments only in the first half of the year; after fruit expansion in the second half of the year, girdling had little effect on the root N:P ratio, regardless of the application of foliar fertilizer. The root N:P ratio decreased after girdling; however, after foliar fertilizer spray, the root N:P ratio decreased further, reaching levels significantly lower than those observed in CK, both before and after spring shoot emergence. The effect of girdling on the N:P ratio was not obvious at the fruit expansion stage, but was significant in mature fruits and flowers. Thus, in the CK treatment, the N:P ratio was significantly higher in fruits and significantly lower in flower buds. Application of foliar fertilizer after girdling increased the N:P ratio in mature fruits to different degrees, bringing the N:P ratio in fruits in the girdling + foliar fertilizer treatment closer to that in the CK treatment, although the N:P ratio in flower buds was higher in the girdling + foliar fertilizer treatment than in CK.

Table 2. The ratio of N:P and N:K of different organs in each period.

Date	Treatment	N:P				N:K			
		Leaves	Roots	Fruits	Buds	Leaves	Roots	Fruits	Buds
10 Mar	Girdling+ foliar fertilizer	5.12 ± 0.07b	2.19 ± 0.13b	–	–	6.11 ± 0.12a	2.19 ± 0.09b	–	–
	Girdling	5.98 ± 0.28a	2.69 ± 0.10a	–	–	5.21 ± 0.26b	3.24 ± 0.07a	–	–
	CK	2.22 ± 0.11c	2.96 ± 0.06a	–	–	2.63 ± 0.16c	3.20 ± 0.06a	–	–
25 May	Girdling+ foliar fertilizer	4.19 ± 0.17a	1.52 ± 0.06b	–	–	4.89 ± 0.13a	5.76 ± 0.07ab	–	–
	Girdling	3.71 ± 0.16a	1.28 ± 0.03c	–	–	5.49 ± 0.24a	5.14 ± 0.35b	–	–
	CK	3.41 ± 0.31a	1.79 ± 0.07a	–	–	5.11 ± 0.24a	5.85 ± 0.17a	–	–
5 Aug.	Girdling+ foliar fertilizer	5.11 ± 0.59a	5.75 ± 0.76a	3.19 ± 0.35a	–	6.74 ± 0.30a	5.84 ± 0.36ab	1.50 ± 0.06a	–
	Girdling	4.60 ± 0.91a	6.45 ± 0.96a	3.52 ± 0.36a	–	6.14 ± 0.08a	7.29 ± 1.75a	1.35 ± 0.13a	–
	CK	6.13 ± 0.27a	4.37 ± 0.43a	2.88 ± 0.18a	–	4.59 ± 0.02b	3.74 ± 0.36b	1.47 ± 0.05a	–
15 Oct.	Girdling+ foliar fertilizer	11.06 ± 0.61a	5.78 ± 0.01a	3.50 ± 0.14ab	5.63 ± 0.20a	7.17 ± 0.27b	7.50 ± 0.38a	1.37 ± 0.04a	3.10 ± 0.09a
	Girdling	8.11 ± 2.30b	5.82 ± 0.40a	2.79 ± 0.14b	5.27 ± 0.13b	6.93 ± 0.13b	6.54 ± 0.28a	1.38 ± 0.11a	3.01 ± 0.13ab
	CK	12.79 ± 0.46a	6.11 ± 0.26a	3.71 ± 0.18a	4.59 ± 0.25b	8.18 ± 0.22a	7.36 ± 0.18a	1.75 ± 0.03a	2.65 ± 0.03b

Note: Data are means ± SE (n = 3). Lowercase letters represent the difference between different treatments in the same period ($p < 0.05$).

Compared with CK, the leaf N:K ratio was significantly higher in the girdling treatment before spring shoot emergence and fruit expansion, and lower after fruit ripening; however, the difference in the leaf N:K ratio was not a significant difference between these two treatments during late spring shoot emergence. Application of foliar fertilizer after girdling significantly increased the N:K ratio before spring shoot emergence but had little effect on the N:K ratio at the other stages. In *C. oleifera* roots, girdling had little effect on the N:K ratio before spring shoot emergence and after fruit ripening but significantly reduced the N:K ratio during spring shoot emergence and increased this ratio after fruit expansion. Application of foliar fertilizer after girdling significantly reduced the root N:K ratio before spring shoot emergence but had little effect on the N:K ratio at each stage after spring shoot emergence. Girdling had no obvious effect on the N:K ratio in fruits but increased the N:K ratio in flowers to varying degrees. Girdling + foliar fertilizer treatment showed a significantly higher N:K ratio in flower buds than CK.

3.4. Influence of Girdling and Foliar Fertilizer Application on C. oleifera Yield

Significant differences were observed in the fresh fruit yield, fruit oil content, and per-plant oil yield among the different treatments (Table 3). Girdling significantly increased the fruit yield, while spraying foliar fertilizer reduced the fruit yield increase; nonetheless, compared with CK, the girdling and girdling + foliar fertilizer treatments showed 63.56% and 33.24% higher fruit yield, respectively. However, girdling reduced the oil content of fresh fruit, and the application of foliar fertilizer after girdling further reduced the oil content; compared with CK, the girdling and girdling + foliar fertilizer treatments showed 13.07% and 20.03% lower oil yield, respectively. In addition, the per-plant oil yield in the girdling treatment was significantly increased by 41.91% and 33.65% compared with the girdling + foliar fertilizer and CK treatments, respectively.

3.5. Path Analysis of Nutrient Content, Nutrient Stoichiometric Ratio, and Per-Plant Oil Yield

Only the leaf N:K ratio in March, root K content in August, fruit N:P ratio in August, and fruit N:K ratio in October had significant effects on the per-plant oil yield, and the root P content in March and leaf N content in May had significant effects on per-plant oil production (Table 4). Among these effects, the effect of the N:K ratio on per-plant oil production per plant in March was positive, while those of the root K content in

August, fruit N:P ratio in August, and fruit N:K ratio in October on the per-plant oil yield were negative.

Table 3. Yield of *C. oleifera* in each treatment, oil content of fresh fruit, and oil production per plant.

Treatment	Yield per Plant (kg Plant^{-1})	Oil Content of Fresh Fruit (%)	Oil Production per Plant (g Plant^{-1})
Girdling+ foliar fertilizer	9.58 ± 1.97ab	4.71 ± 0.20b	451.38 ± 85.56b
Girdling	11.76 ± 1.61a	5.12 ± 0.20b	603.29 ± 82.59a
CK	7.19 ± 1.67b	5.89 ± 0.23a	425.13 ± 99.18b
p value	0.173	0.000	0.013

Note: Data are means ± SE (n = 3). Lowercase letters represent the difference between different treatments in the same period (*p* < 0.05).

Table 4. Multiple stepwise regression analysis of oil production per plant and tree nutrients.

Item	Major Affecting Factors	Regression Equation	R^2
Oil production per plant (g·plant^{-1}) Y$_1$	Leaf of N:K ratio in March X$_1$ Root of K content in August X$_2$ Fruit of N:P ratio in August X$_3$ Fruit of N:K ratio in October X$_4$	Y1 = 2164.853 + 175.036X$_1$ − 206.702X$_2$ − 213.266X$_3$ − 320.795X$_4$	0.992

The root K content in August, fruit N:P ratio in August, and fruit N:K ratio in October had direct negative effects on the per-plant oil yield (Table 5). However, the root K content in August, fruit N:P ratio in August, and fruit N:K ratio in October had indirect positive effects on the per-plant oil yield by interacting with each other and the leaf N:K ratio in March, which offsets the direct negative effects among factors. The leaf N:K ratio in March had an indirect negative effect on the per-plant oil yield by interacting with other factors. Among the effects of these factors, the direct positive effect of the leaf N:K ratio in March was greater, and the indirect positive effects of the fruit N:P ratio in August and fruit N:K ratio in October were greater. In addition, the residual path coefficient was 0.089, indicating that the factors in this equation fully explained the variation in oil production per *C. oleifera* plant.

Table 5. Path analysis of tree nutrient factors on oil production per plant of *C. oleifera*.

Independent Variable	Direct Path Coefficient	Indirect Path Coefficient					Residual Path Coefficient
		X1	X2	X3	X4	Total	
X$_1$	1.065	–	−0.524	0.417	−0.959	−1.065	
X$_2$	−0.455	0.224	–	0.122	−0.281	0.065	0.089
X$_3$	−0.436	−0.171	0.117	–	0.262	0.208	
X$_4$	−0.310	0.279	−0.191	0.186	–	0.274	

Note: X1 is N:K ratio of leaves in March, X2 is root of K content in August, X3 is N:P ratio of fruits in August, X4 is N:K ratio of fruits in October.

4. Discussion

4.1. Effect of Girdling on the Nutrient Contents of C. oleifera Plants

Girdling is widely used in fruit trees [6–9] because it causes physical damage to phloem in the trunk, impeding the transport of carbohydrates and inorganic nutrients to the tree roots [25]. Studies show that girdling reduces the N, P, and K contents of leaves, and this effect is related to the healing time of girdling wounds [26]. The growth and development and the nutrient distribution pattern of *C. oleifera* are different from those of most fruit trees. The results of this study showed that girdling of the *C. oleifera* trunk in November significantly increased the N content of leaves before spring shoot emergence in early March of the next year, which was conducive to the accumulation of N in roots at the fruit expansion stage (August), and promoted the transport of N from leaves to fruits at the fruit maturity stage (October). During the period from the girdling of *C. oleifera* to the emergence of spring shoots, the fruits formed in the previous year (2020) had been harvested, and the new fruits (in 2021) had not yet formed; N had been transported to the leaves through the xylem; and competition from other organs was non-existent. Additionally, the girdle in the phloem had not yet begun to heal and could not transport N down to the roots, which significantly increased the N content of leaves during this time. After the girdle healed in mid-May, N was normally transported to above- and below-ground organs. After the girdling, the roots obtained more N and grew better, and the N output of the above-ground leaves increased at the later stage. In the first half of the year, P accumulated more in the roots, so that the P content of leaves before and after spring shoot emergence was significantly lower. However, after the fruit expansion and until the fruit maturity, P was transferred to the leaves in large quantities in the girdling treatment, resulting in a significantly higher leaf P content compared with CK. This result could be attributed to the P uptake characteristics of plants [21]. In *C. oleifera*, the first half of the year is the key period of P uptake and accumulation by roots. Before girdle healing, the normal transport of P is affected, resulting in a greater accumulation of P in the root system. After girdle healing, a large amount of P is transported to the leaves, greatly increasing the leaf P content.

In this study, girdling increased the availability of K to the leaves before spring shoot emergence, for the same reason as that which is responsible for N accumulation in leaves during this period. However, after spring shoot emergence, the K content of *C. oleifera* leaves, roots, fruits, and flowers in the girdling treatment was lower than that in CK, whereas the relative K content of flower buds and fruits was higher in the girdling treatment, so that more K was allocated to fruits and flower buds [27]. These results indicate that girdling promotes the distribution of K to fruits and flower buds, which is conducive to seed setting. However, it should be noted that girdling could consume a large amount of K while promoting seed setting, which may lead to insufficient K availability in the plant, thus requiring supplementation over time.

4.2. Influence of Ring Cutting on the Stoichiometric Ratios of N, P, and K in C. oleifera

N, P, and K are the limiting nutrients affecting plant growth, and the N:P and N:K ratios could be used as the determinants of plant health [28]. When the N:P ratio < 14, plant growth is limited by N; when the N:P ratio > 16, plant growth is limited by P; and when the N:P ratio = 14~16, plan growth is limited by both N and P [29]. The results of this study showed that the N:P of each organ of *C. oleifera* plants was less than 14 in different periods, indicating that the growth of *C. oleifera* was severely limited by N. In the trees girdled in November, the leaves had accumulated a large amount of N in March of the following year, whereas the level of P was low, improving the leaf N:P ratio, which to a certain extent, alleviated the effects of N limitation on *C. oleifera* growth. After the girdle wound healed, phloem returned to its normal transport capacity, gradually reducing the leaf N and P contents to levels consistent with those in the CK. The N:P ratio of the *C. oleifera* roots was relatively low before and after spring shoot emergence, which was severely restricted by N. At the later stage, the root N:P ratio gradually increased. At the fruit maturity stage, large amounts of N and P were accumulated in the vegetative organs before girdle wound healing, which were gradually transferred to the reproductive organs, flower buds, and fruits at the later stage. Therefore, the N:P ratio in *C. oleifera* fruits and flower buds was relatively high at the fruit maturity stage, which reduced the limitation of N in fruits and flower buds to a certain extent. When the N:K ratio is less than 2.1, plant growth is mainly limited by K [30]. The results of this experiment show that the N:K ratio in each organ of *C. oleifera* at different stages was less than 2.1, indicating that the growth of *C. oleifera* plants was limited by K. However, changes in the N:K ratio in different organs at different stages after girdling was the same as that in the N:P ratio, for roughly the same reasons.

4.3. Foliar Fertilizer-Induced Modification of the Effect of Girdling on the Nutrient Content of C. oleifera Organs

Application of foliar fertilizer after girdling affected nutrient accumulation in roots, increased the nutrient contents of fruits and flower buds, brought the N, P, and K contents of leaves, roots, and fruits closer to those in the CK, and weakened the regulatory effect of girdling on the contents of N, P, and K in *C. oleifera*. This result might be related to the vegetative growth of the above-ground parts of *C. oleifera* plants treated with foliar fertilizer [4]; the root system absorbed more nutrients. In addition, foliar fertilizer application after girdling also weakened the effects of girdling on the distribution of N, P, and K in various organs during the period from spring shoot emergence to fruit expansion and reduced the deviation of tree nutrient distribution patterns from the control in this period. Foliar fertilizer increased the relative contents of P and K in ripe fruits and flower buds of girdled *C. oleifera* trees, which may be related to the less fruits treated with foliar fertilizer [31].

4.4. Key Nutrient Characteristics of Girdled C. oleifera Trees for Increasing Yield

Girdling improved the fruit yield and per-plant oil yield of *C. oleifera*, but it did not increase the oil content of fresh fruit. It may be because girdling greatly increased the number of fruits per plant, resulting in reduced seed kernel oil content. However, the substantial increase in yield compensated for the reduction in seed oil content, leading to a substantial increase in oil production per plant.

Path analysis of *C. oleifera* nutrient contents and per-plant oil and fruit yield revealed that the root K content in August, fruit N:P ratio in August, and fruit N:K ratio in October were important factors affecting per-plant oil production. In March, the leaf N:K ratio had direct positive effects on oil production per plant, while the other three factors had indirect negative effects. This indicates that the increase in the leaf N:K ratio in March could directly and effectively promote the increase in oil yield per plant. Girdling promoted the accumulation of N in *C. oleifera* leaves in March, increasing the leaf N:K ratio. At the same time, girdling promoted a reduction in the K content of roots in August, and large amounts of N and P were accumulated in leaves, which were later transported to fruits, but the N and P contents of fruits were relatively low. In October, large amounts of N and K were accumulated in leaves and roots, which were later transported to fruits, and the amount of K transported was greater than that of N, resulting in low N and P contents of fruits.

5. Conclusions

Girdling and foliar fertilizer application affected the nutrient contents of *C. oleifera* organs at the flowering stage. The improvement in the economic benefits of girdling on *C. oleifera* is more obvious. In the girdling treatment, N was accumulated in leaves before girdle wound healing. However, the accumulation of N and K in the roots was limited. With the gradual healing of the girdle wound and the recovery of phloem transport capacity, the underground plant parts began to accumulate N. P accumulated in the root system before girdle wound healing. After wound healing, P began to accumulate in the leaves. At the fruit ripening stage, three nutrients (N, P, and K) previously accumulated in the vegetative organs were gradually transported to the reproductive organs (fruits and flower buds), and the amount of K transported was greater than that of N and P, which promoted flowering and fruiting. Compared with the girdling, the application of foliar fertilizer after girdling weakened the overall effect of girdling on nutrient regulation in *C. oleifera* trees but promoted the accumulation of a large amount of nutrients in roots, flower buds, and fruits.

Author Contributions: Conceptualization, S.X. and D.H.; methodology, S.X. and D.H.; software, S.X. and Y.W.; investigation, Z.L., Z.R., C.L. and Q.C.; resources, D.L.; writing—original draft preparation, S.X.; writing—review and editing, S.X., D.H., L.Z. (Ling Zhang), J.L. and L.Z. (Linping Zhang); visualization, S.X.; project administration, D.H.; funding acquisition, D.H. All authors have read and agreed to the published version of the manuscript.

Funding: This research were funded by National Key R&D Program of China, grant number "2018YFD1000603", and Science and Technology Innovation Project of Forestry Department of Jiangxi Province, grant number Innovation Project [2020] No.2.

Institutional Review Board Statement: Not applicable.

Informed Consent Statement: Not applicable.

Data Availability Statement: The data that support the findings of this study are openly available in PubMed or available in other sources.

Conflicts of Interest: The authors declare no conflict of interest.

References

1.   You, L.; Yu, S.; Liu, H.; Wang, C.; Zhou, Z.; Zhang, L.; Hu, D. Effects of Biogas Slurry Fertilization on Fruit Economic Traits and Soil Nutrients of Camellia Oleifera Abel. *PLoS ONE* **2019**, *14*, e0208289. [CrossRef] [PubMed]
2.   He, Z.; Liu, C.; Zhang, Z.; Wang, R.; Chen, Y. Integration of Mrna and Mirna Analysis Reveals the Differentially Regulatory Network in Two Different Camellia Oleifera Cultivars under Drought Stress. *Front. Plant Sci.* **2022**, *13*, 1001357. [CrossRef] [PubMed]
3.   Ye, H.-L.; Chen, Z.-G.; Jia, T.-T.; Su, Q.-W.; Su, S.-C. Response of Different Organic Mulch Treatments on Yield and Quality of Camellia Oleifera. *Agric. Water Manag.* **2021**, *245*, 1001357. [CrossRef]
4.   Yu, F.; Xin, M.; Yao, Y.; Wang, X.; Liu, K.; Li, Y. Manganese Accumulation and Plant Physiology Behaviour of *Camellia oleifera* in Response to Different Levels of Phosphate Fertilization. *J. Soil Sci. Plant Nutr.* **2021**, *21*, 2562–2572. [CrossRef]
5.   He, X. Effects of Foliar Fertilizers on Fruit Setting in Camellia Oleifera. *Nonwood For. Res.* **2015**, *33*, 18–24.
6.   Lu, Z.-J.; Yu, H.-Z.; Mi, L.-F.; Liu, Y.-X.; Huang, Y.-L.; Xie, Y.-X.; Li, N.-Y.; Zhong, B.-L. The Effects of Inarching *Citrus reticulata* Blanco Var. Tangerine on the Tree Vigor, Nutrient Status and Fruit Quality of *Citrus sinensis* Osbeck 'Newhall' Trees That Have *Poncirus trifoliata* (L.) Raf. As Rootstocks. *Sci. Hortic.* **2019**, *256*, 108600. [CrossRef]
7.   Tyagi, K.; Maoz, I.; Lewinsohn, E.; Lerno, L.; Ebeler, S.E.; Lichter, A. Girdling of Table Grapes at Fruit Set Can Divert the Phenylpropanoid Pathway Towards Accumulation of Proanthocyanidins and Change the Volatile Composition. *Plant Sci.* **2020**, *296*, 110495. [CrossRef]
8.   Baïram, E.; le Morvan, C.; Delaire, M.; Buck-Sorlin, G. Fruit and Leaf Response to Different Source-Sink Ratios in Apple, at the Scale of the Fruit-Bearing Branch. *Front. Plant Sci.* **2019**, *10*, 1039. [CrossRef]
9.   Figiel-Kroczyńska, M.; Ochmian, I.; Lachowicz, S.; Krupa-Małkiewicz, M.; Wróbel, J.; Gamrat, R. Actinidia (Mini Kiwi) Fruit Quality in Relation to Summer Cutting. *Agronomy* **2021**, *11*, 964. [CrossRef]
10.   Christopoulos, M.V.; Kafkaletou, M.; Karantzi, A.D.; Tsantili, E. Girdling Effects on Fruit Maturity, Kernel Quality, and Nutritional Value of Walnut (*Juglans regia* L.) Alongside the Effects on Leaf Physiological Characteristics. *Agronomy* **2021**, *11*, 200. [CrossRef]
11.   Singh, D.; Dhillon, W.S.; Singh, N.P.; Gill, P.P.S. Effect of Girdling on Leaf Nutrient Levels in Pear Cultivars Patharnakh and Punjab Beauty. *Indian J. Hortic.* **2015**, *72*, 319–324. [CrossRef]
12.   de Lange, J.H.; Skarup, O.; Vincent, A.P. The Influence of Cross-Pollination and Girdling on Fruit Set and Seed Content of Citrus 'Ortanique'. *Sci. Hortic.* **1974**, *2*, 285–292. [CrossRef]
13.   Mann, M.S.; Josan, J.S.; Chohan, G.S.; Vij, V.K. Effect of Foliar Application of Micronutrients on Leaf Composition, Fruit Yield and Quality of Sweet Orange (*Citrus sinensis* Osbeck) Cv. Blood Red. *Indian J. Hortic.* **1985**, *42*, 45–49.
14.   Norozi, M.; ValizadehKaji, B.; Karimi, R.; Sedghi, M. Effects of Foliar Application of Potassium and Zinc on Pistachio (*Pistacia vera* L.) Fruit Yield. *J. Hortic. Sci.* **2019**, *6*, 113–123.
15.   Duan, W. Effect of Foliar Fertilizer on Leaf Anatomy Structure and Photosynthetic Characteristics of *Camellia oleifera* Container Seedlings. *J. Northwest A F Univ.* **2015**, *43*, 92–97.
16.   Yilmaz, B.; Çimen, B.; Incesu, M.; Yeşiloğlu, T.; Yilmaz, M.I. Influence of Girdling on the Seasonal Leaf Nutrition Status and Fruit Size of Robinson Mandarin (*Citrus reticulata* Blanco). *Appl. Ecol. Environ. Res.* **2018**, *16*, 6205–6218. [CrossRef]
17.   Li, Y.; Zou, N.; Liang, X.; Zhou, X.; Guo, S.; Wang, Y.; Qin, X.; Tian, Y.; Lin, J. Effects of Nitrogen Input on Soil Bacterial Community Structure and Soil Nitrogen Cycling in the Rhizosphere Soil of *Lycium barbarum* L. *Front. Microbiol.* **2022**, *13*, 1070817. [CrossRef]
18.   Meng, D.; Xu, P.; Dong, Q.; Wang, S.; Wang, Z. Comparison of Foliar and Root Application of Potassium Dihydrogen Phosphate in Regulating Cadmium Translocation and Accumulation in Tall Fescue (*Festuca arundinacea*). *Water Air Soil Pollut.* **2017**, *228*, 118. [CrossRef]

19. Sánchez-Ramos, M.; Berman-Bahena, S.; Alvarez, L.; Sánchez-Carranza, J.N.; Bernabé-Antonio, A.; Román-Guerrero, A.; Marquina-Bahena, S.; Cruz-Sosa, F. Effect of Plant Growth Regulators on Different Explants of *Artemisia ludoviciana* under Photoperiod and Darkness Conditions and Their Influence on Achillin Production. *Processes* **2022**, *10*, 1439. [CrossRef]

20. Thakur, P.; Kumar, P.; Shukla, A.K.; Butail, N.P.; Sharma, M.; Kumar, P.; Sharma, U. Quantitative, Qualitative, and Energy Assessment of Boron Fertilization on Maize Production in North-West Himalayan Region. *Int. J. Plant Prod.* **2023**. [CrossRef]

21. Zeng, J.; Liu, J.; Lian, L.; Xu, A.; Guo, X.; Zhang, L.; Zhang, W.; Hu, D. Effects of Scion Variety on the Phosphorus Efficiency of Grafted *Camellia oleifera* Seedlings. *Forests* **2022**, *13*, 203. [CrossRef]

22. Levin, A.G.; Lavee, S. The Influence of Girdling on Flower Type, Number, Inflorescence Density, Fruit Set, and Yields in Three Different Olive Cultivars (Barnea, Picual, and Souri). *Crop Pasture Sci.* **2005**, *56*, 827–831. [CrossRef]

23. Ma, J.; Zhang, M.; Liu, Z.; Wang, M.; Sun, Y.; Zheng, W.; Lu, H. Copper-Based-Zinc-Boron Foliar Fertilizer Improved Yield, Quality, Physiological Characteristics, and Microelement Concentration of Celery (*Apium graveolens* L.). *Environ. Pollut. Bioavailab.* **2019**, *31*, 261–271. [CrossRef]

24. Wu, F.; Li, J.; Chen, Y.; Zhang, L.; Zhang, Y.; Wang, S.; Shi, X.; Li, L.; Liang, J. Effects of Phosphate Solubilizing Bacteria on the Growth, Photosynthesis, and Nutrient Uptake of Camellia Oleifera Abel. *Forests* **2019**, *10*, 348. [CrossRef]

25. Berüter, J.; Feusi, M.E.S. The Effect of Girdling on Carbohydrate Partitioning in the Growing Apple Fruit. *J. Plant Physiol.* **1997**, *151*, 277–285. [CrossRef]

26. Huang, D.; Liu, S.; Zhou, X.; Qian, D.; Wu, C.; Xu, Y. The Effect of Spiral Girdling on the Fruit-Setting and Development of Young Litchi Trees. *Acta Hortic.* **2012**, 145–150. [CrossRef]

27. Baninasab, B.; Rahemi, M.; Shariatmadari, H. Seasonal Changes in Mineral Content of Different Organs in the Alternate Bearing of Pistachio Trees. *Commun. Soil Sci. Plant Anal.* **2007**, *38*, 241–258. [CrossRef]

28. Zhong, Z.; Zhang, X.; Wang, X.; Dai, Y.; Chen, Z.X.; Han, X.; Yang, G.; Ren, C.; Wang, X. C:N:P Stoichiometries Explain Soil Organic Carbon Accumulation During Afforestation. *Nutr. Cycling Agroecosyst.* **2020**, *117*, 243–259. [CrossRef]

29. Yan, Z.; Kim, N.; Guo, Y.; Han, T.-S.; Du, E.Z.; Han, W.; Fang, J. Effects of Nitrogen and Phosphorus Fertilization on Leaf Carbon, Nitrogen and Phosphorus Stoichiometry of *Arabidopsis thaliana*. *Chin. J. Plant Ecol.* **2013**, *37*, 551–557. [CrossRef]

30. Szczepanek, M.; Siwik-Ziomek, A. P and K Accumulation by Rapeseed as Affected by Biostimulant under Different Npk and S Fertilization Doses. *Agronomy* **2019**, *9*, 477. [CrossRef]

31. González, C.E.A.; Hernández, M.M.; Fernández-Falcón, M. Nutrient Distribution and Stem Length in Flowering Stems of Protea Plants. *J. Plant Nutr.* **2008**, *31*, 1624–1641. [CrossRef]

Article

Dynamic Changes of Endogenous Hormones in Different Seasons of *Idesia polycarpa* Maxim

Song Huang [1], Wei Zheng [1], Yanmei Wang [2], Huiping Yan [2], Chenbo Zhou [1] and Tianxiao Ma [1,*]

1 College of Forestry, Xinyang Agricultural and Forestry University, Xinyang 464000, China
2 College of Forestry, Henan Agricultural University, Zhengzhou 450002, China
* Correspondence: matianxiao1979@163.com

Abstract: *Idesia polycarpa* Maxim is a native dioecious tree from East Asia cultivated for its fruits and as an ornamental plant throughout temperate regions. Given the economic potential, comparative studies on cultivated genotypes are of current interest. This study aims to discover the dynamic changes and potential functions of endogenous hormones in *I. polycarpa*, as well as the differences in endogenous hormone contents in different growth stages among different *I. polycarpa* provenances. We used High-Performance Liquid Chromatography (HPLC) to measure and compare the levels of abscisic acid (ABA), indole-3-acetic acid (IAA), gibberellin A3 (GA3), and trans-Zeatin-riboside (tZR) in the leaves, flowers, and fruits of *I. polycarpa* from various provenances between April and October. Our findings indicated that changes in the ABA and GA3 content of plants from Jiyuan and Tokyo were minimal from April to October. However, the levels of these two hormones in Chengdu plants vary greatly at different stages of development. The peak of IAA content in the three plant materials occurred primarily during the early fruit stage and the fruit expansion stage. The concentration of tZR in the three plant materials varies greatly. Furthermore, we discovered that the contents of endogenous hormones in *I. polycarpa* leaves, flowers, and fruits from Chengdu provenances were slightly higher than those from Tokyo and Jiyuan provenances. The content of IAA was higher in male flowers than in female flowers, and the content of ABA, GA3, and tZR was higher in female flowers than in male flowers. According to the findings, the contents of these four endogenous hormones in *I. polycarpa* are primarily determined by the genetic characteristics of the trees and are less affected by cultivation conditions. The gender of *I. polycarpa* had a great influence on these four endogenous hormones. The findings of this study will provide a theoretical foundation and practical guidance for artificially regulating the flowering and fruiting of *I. polycarpa*.

Keywords: *Idesia polycarpa*; abscisic acid (ABA); indole-3-acetic acid (IAA); gibberellinA3 (GA3); trans-Zeatin-riboside (tZR); High-Performance Liquid Chromatography (HPLC)

Citation: Huang, S.; Zheng, W.; Wang, Y.; Yan, H.; Zhou, C.; Ma, T. Dynamic Changes of Endogenous Hormones in Different Seasons of *Idesia polycarpa* Maxim. *Life* **2023**, *13*, 788. https://doi.org/10.3390/life13030788

Academic Editor: Ling Zhang

Received: 2 January 2023
Revised: 5 March 2023
Accepted: 11 March 2023
Published: 15 March 2023

1. Introduction

Idesia polycarpa Maxim is a dioecious tree in the Salicaceae family, which is economically significant and well-known throughout the temperate region [1,2]. There is only one species of this genus in the world [1]. This species is native to East Asia, including China, Japan, and Korea, and has since spread throughout the region [3]. There are obvious differences in the growth form, development speed, environmental adaptability, and physiological and biochemical characteristics of *I. polycarpa* due to long-term geographical isolation, natural selection, and climate [4].

I. polycarpa is a tall deciduous tree that blooms from April to May. It has yellow-green flowers with a fragrant scent. In the distribution area of *I. polycarpa*, the suitable annual average temperature ranges from 13 °C to 21 °C [5,6]. *I. polycarpa* is also known as the 'Beautiful Tree Oil Depot' because its fruits have a high oil content, in addition to being used as an ornamental and greening tree species [7]. The oil content in the flesh of the fruits ranges between 28.38 and 48.35%. In comparison, the oil content of the seeds is approximately 12.6–28.17%, and the oil extracted from the fruit has high commercial and medicinal value, with various applications, including edible oil, lubricating oil, biodiesel production, and medicines to reduce cholesterol and blood pressure [8]. Because *I. polycarpa* has an appealing color and shape and potential nutritional and medicinal value, more and more areas are attempting to cultivate it artificially. However, most *I. polycarpa* is still in the wild, and cultivation technology is not standardized or mature [9]. There are numerous factors that influence plant growth and development, but endogenous hormones, such as auxin (IAA), gibberellin (GA3), abscisic acid (ABA), and mitogen (tZR), play critical regulatory roles in the growth and development of *I. polycarpa* [10]. Currently, most *I. polycarpa* research focuses on ecology, cultivation, and breeding, with few studies on endogenous hormones. Endogenous hormone changes during flowering and fruiting, as well as their effects on the growth and development of *I. polycarpa*, remain unknown. Therefore, it is intended to investigate the dynamic changes in the endogenous hormones at various stages of the flowering and fruiting processes of *I. polycarpa*.

Plant hormones are a class of organic substances that are synthesized during plant metabolism and are required for various physiological processes in plants [11]. Plant growth regulators have been widely used in various developmental stages to improve plant growth and development, as well as the yield and quality of fruits, due to the importance of plant hormones to plant growth and regulation [12]. Previous research found that plant hormones directly affect fruit formation and development [13]. Therefore, it is essential to study the endogenous hormones of the sprouting, flowering, and fruiting of *I. polycarpa* from different origins and then apply exogenous hormones based on the local climate in the introduction and artificial planting process of this tree species to improve the success rate.

Auxin is a plant hormone that is essential for plant growth and development [14]. Auxin biosynthesis is important in a variety of plant development processes, including root development, embryogenesis, endosperm development, and flower development [15]. Similarly, IAA (indole-3-acetic acid) plays an important role in fruit formation and development [16]. Auxin response factor (ARF) controls the fate of fruit initiation events by controlling the level of gibberellin (GA) [17] and interacts with Aux/IAA proteins [18].

Abscisic acid (ABA) is a plant hormone derived from isoprene that accumulates as a result of a lack of water and is associated with seed dormancy, maturity, and development [19,20]. It also has a significant impact on plant responses to various abiotic stresses, such as drought, high temperature, low temperature, and salt [21]. ABA is also an important ripening control factor, as demonstrated by the following: (1) the content of ABA increased clearly in the early stages of apple fruit ripening [22]; (2) the content in peach and grapefruit fruits increases before ethylene release [23]; (3) exogenous ABA can promote the production of several metabolites related to fruit ripening [24]; (4) the fruits in tomato mutants lacking ABA could not maintain the normal growth pattern [25]; and (5) the maturation stage was delayed in ABA-deficient orange mutants [26].

Gibberellin (GA3) is a plant hormone that promotes cell division and proliferation [27]. GA3 is widely used in plant growth and development for a variety of purposes, including promoting seed germination and coping with abiotic stress [28], enhancing fruit growth [29], accelerating stem elongation [30], flowering [31], and other physiological effects caused by interactions with other plant hormones [32].

Trans zeatin riboside is a type of cytokinin that can promote cell division during tomato fruit development [33]. Liu et al. discovered that tZR has a significant impact on cucumber fruit growth and development [34]. Similarly, Honda et al. discovered that tZR plays an important role in pepper fruit expansion [35]. Furthermore, the content of tZR increased significantly during hop development, indicating that tZR was closely related to hop development [36].

It is well understood that phytohormones play an important role in seed germination. However, there is a lack of studies on how various endogenous hormones play a role in *I. polycarpa* and whether there are differences in the content of endogenous hormones in different growth stages of *I. polycarpa* from different provenances. The current study aims to quantify the changes in IAA, ABA, GA3, and tZR concentrations in the leaves, flowers, and fruits of *I. polycarpa* during different stages of development for three provenances from the natural range of the species. The findings provide a theoretical foundation for understanding the regulation of the flowering and fruiting period and the quantity of *I. polycarpa*, with potential applications in breeding, phenological manipulation, optimized cultivation, adequate harvesting activity planning, and performance evaluation.

2. Materials and Methods

2.1. Plant Materials and Growth Conditions

Experiments were carried out at Henan Agricultural University's Forestry Experiment Station in Zhengzhou City, China (113°38′ E, 34°48′ N). The research site is located in the warm temperate zone and has a temperate monsoon climate with an annual precipitation of ~650 mm. The maximum, minimum, and annual mean temperatures are 43.0 °C, −17.9 °C, and 14.2 °C, respectively, and the average annual sunshine duration is 2400 h (Figure 1, Table 1). The accumulated temperature ≥10 °C is 4717 °C, with a frost-free period of 215 d and a pH of about 7.0 in the experimental fields.

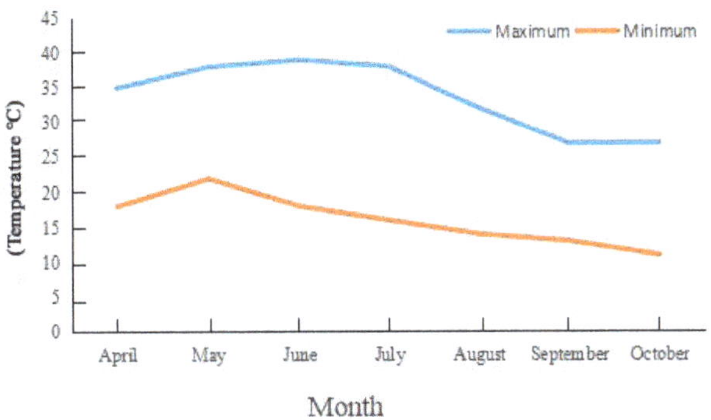

Figure 1. Monthly, minimum, and maximum temperature in Zhengzhou City, China. Note: The data come from https://www.tianqi24.com/zhengzhou/history2021.html (accessed on 2 January 2023).

Table 1. Monthly statistics of air quality in the experimental area.

Month	NO$_2$	CO	SO$_2$	O$_3$
2021–04	31.3 ± 1.53	0.64 ± 0.08	9.17 ± 1.75	113.5 ± 2.36
2021–05	25.9 ± 3.06	0.61 ± 0.13	9.42 ± 2.16	143.8 ± 3.78
2021–06	23.7 ± 3.51	0.71 ± 0.19	6.81 ± 2.39	178.6 ± 4.31
2021–07	16.3 ± 1.27	0.56 ± 0.11	3.40 ± 0.86	134.3 ± 4.54
2021–08	17.6 ± 2.55	0.66 ± 0.24	4.47 ± 1.59	137.9 ± 3.21
2021–09	24.1 ± 4.51	0.75 ± 0.16	5.93 ± 2.13	126.4 ± 7.52
2021–10	42.9 ± 3.72	0.81 ± 0.21	8.74 ± 2.28	84.1 ± 4.14

Note: The data came from the Henan Meteorological Service.

The experimental material was *I. polycarpa* grown at Henan Agricultural University's Forestry Experiment Station. Jiyuan (112°57′ E, 35°08′ N), Tokyo (139°69′ E, 35°68′ N), and Chengdu (104°07′ E, 30°67′ N) provided the plants. Tissue samples were collected from leaves, flowers, and fruits at various stages of development between April and October. The sampling period was divided into seven stages based on the stage of growth and development (Table 2, Figure 2). The plant samples were stored at −80 °C in liquid nitrogen. Furthermore, each sample was divided into three replicates.

Table 2. Sampling Time of *I. polycarpa*.

Period	Stage		Sample Time		
			Jiyuan	Tokyo	Chengdu
Initial Flower	Early April–Late April	I	24 April 2021	26 April 2021	26 April 2021
Full Bloom	Late April–Early May	II	28 April 2021	1 May 2021	1 May 2021
Flower Drop	Early May	III	4 May 2021	6 May 2021	6 May 2021
Initial Fruit	Early May–Late May	IV	13 May 2021	16 May 2021	16 May 2021
Expansion	Late May–Middle August	V	24 May 2021	26 May 2021	26 May 2021
Yellowing	Middle August–Early September	VI	1 September 2021	3 September 2021	3 September 2021
Reddening	Early September–Early October	VII	8 October 2021	10 October 2021	10 October 2021

Figure 2. Morphological changes of *Idesia polycarpa* Maxim. During the flower stage and fruit stages.

2.2. Extraction and Analysis of Phytohormones

The concentrations of IAA, ABA, GA3, and tZR were determined using High-Performance Liquid Chromatography, which was slightly modified from Li [37]. We extracted the samples using a C_{18} solid phase extraction cartridge and analyzed them by HPLC-ESI-MS/MS using 0.05% formic acid in methanol and H_2O as mobile phases for HPLC because the plant samples were different. Lyophilized samples were ground in liquid nitrogen, and 0.5 g powdered samples were extracted with 5 mL acetonitrile extraction solvent containing 30 µg/mL antioxidant. For 12 h, the extracted samples were refrigerated at 4 °C. After centrifuging the extract (10,000 rpm for 20 min at 4 °C), the supernatants were collected and re-extracted with 5 mL of extraction solvent before centrifuging the extraction solution (10,000 rpm for 10 min at 4 °C) again. The supernatants were then combined and distilled in a rotary evaporator at 40 °C before being dissolved in 4 mL of chloroform and 8 mL of phosphate buffer. The mixture was then treated with 150 mg PVPP (polyvinylpyrrolidone) and centrifuged at 8000 rpm for 10 min at 4 °C. The pH was adjusted to 3.0, and 3 mL of the mixture was treated with formic acid and extracted three times with an equal volume of ethyl acetate. The extracted samples were combined and distilled using a rotary evaporator (40 °C), after which they were redissolved in

1 mL of methanol and filtered through a strainer (0.22 µm). Finally, each sample was injected into the HPLC with 10 µL. The following were the mobile phase conditions: water—methanol (52:48) with 0.5‰ formic acid; column temperature—40 °C; detection wavelength—254 nm; injection volume—10 µL; flow rate—0.7 mL/min, constant gradient elution for 23 min.

Calculation Method of Endogenous Hormone Content

The standard curve was based on different peaks, and each peak time corresponds to different concentrations of calibration samples corresponding to plant hormones; the hormone content was calculated using the following method [38]:

Endogenous hormone content ($µg·g^{-1}$) = As·V Css Vss/(Ass Ms·Vs)

As: Peak area of the sample; V: Volume of a final constant volume of sample pre-treatment (mL); Css: Concentration of standard sample ($g·L^{-1}$); Vss: Injection volume of standard sample (µL); Ass: Peak area of the standard sample; Ms: Dry weight of the sample (g); Vs: Injection volume of sample (mL).

2.3. Analytical Methods

Origin 2017 was used to create the graph for each hormone analysis. Using SPSS 24.0 software (IBM, Armonk, NY, USA), analysis of variance (ANOVA) and Duncan's test at $p = 0.05$ or $p = 0.01$ were used to compare the variations of endogenous hormone levels in samples from different provenances and sample time treatments.

3. Results
3.1. Changes of Hormones in Female Leaves of I. polycarpa from Different Provenances

Endogenous concentrations of ABA, GA3, IAA, and tZR in female leaves of *I. polycarpa* from various provenances were examined (Figure 3). There were no discernible differences in the ABA content of plants from Jiyuan and Tokyo. On the contrary, the ABA content of Chengdu plants decreased rapidly, particularly from stage I to stage II. Because ABA biosynthesis is closely related to carotenoid synthesis, we infer that the decrease in ABA is related to the orange deepening of mature fruits. During the developmental stage I, the ABA content of plants from Chengdu was significantly higher than that of plants from the other two provenances ($p < 0.01$).

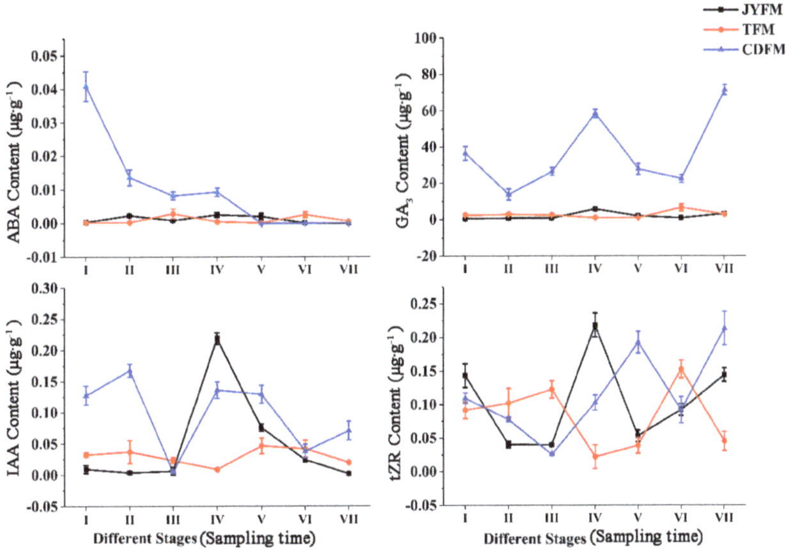

Figure 3. Dynamic changes of hormones in female leaves of *I. polycarpa* from April to October.

The change in trends of GA3 content was comparable to that of ABA within the Jiyuan and Tokyo plants. The difference is that the GA3 content of plants from Chengdu exhibited a "W" trend, which fluctuates significantly more than the other two provenances. GA3 levels were higher in stages IV (58.567 µg·g^{-1}) and VII (71.450 µg·g^{-1}), indicating that GA3 could promote fruit formation and development. In every stage of development, the ABA content of Chengdu plants was significantly higher than that of the other two provenances ($p < 0.05$).

The change in IAA content in Jiyuan provenance plants showed a rising and then falling trend. The content of IAA increased and then decreased with the typical double peak trend in the plants from Tokyo and Chengdu, but it fluctuated more pronouncedly in the Chengdu plants. The plants' IAA content was highest in stages II, IV, and V. IAA was critical in regulating plant growth, particularly during the early stages of flowering and fruiting. Lower IAA concentrations were required for fruit ripening. During stages, I and II, the IAA content of Chengdu plants was significantly higher than that of the other two provenances ($p < 0.05$).

The overall findings revealed that there were significant differences in the content of tZR in plants from all three provenances. The highest concentration of tZR was found in stage IV (0.218 µg·g^{-1}) for Jiyuan plants and stage VII (0.214 µg·g^{-1}) for Chengdu plants. However, the level of tZR in the Tokyo plants was highest in stage VI, with a concentration of 0.153 µg·g^{-1}. The highest concentration of tZR content was found in the first three stages of fruit development, indicating that tZR could promote cell division and was closely linked to fruit development. During stage V, the tZR content of Chengdu plants was higher than that of the other two provenances ($p < 0.05$).

3.2. Changes of Hormones in Male Leaves of I. polycarpa from Different Provenances

Figure 4 depicts the dynamic changes in hormones in male leaves of *I. polycarpa* from April to October. Male leaf ABA content changes were comparable to female leaf ABA content changes in Jiyuan and Tokyo. However, there was an unusual fluctuation in the ABA content of Chengdu plants from stage II to stage IV, as shown in the figure. The ABA content of Chengdu male leaves was significantly higher than that of the other two provenances in stages I, II, and IV ($p < 0.01$).

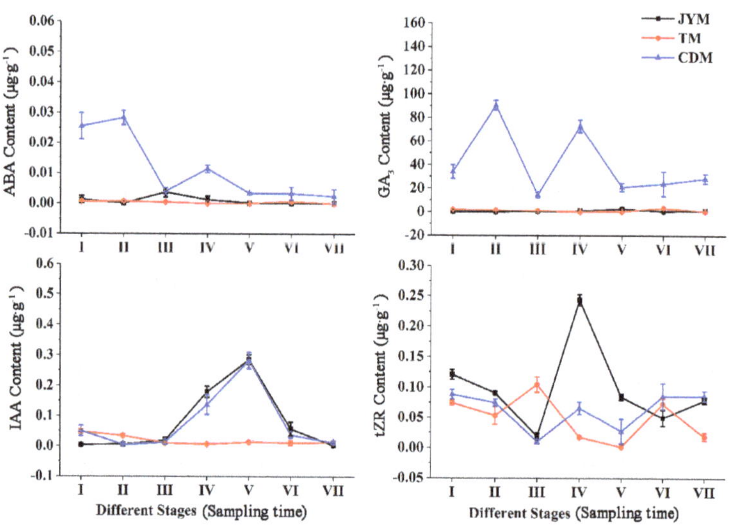

Figure 4. Dynamic changes of hormones in male leaves of *I. polycarpa* from April to October.

Changes in the GA3 content of female leaves were not visible in Jiyuan or Tokyo. The GA3 content in Chengdu leaves was significantly higher at all stages than in the other two provenances ($p < 0.01$), and the changing trend of the M-type was obvious.

The shift in IAA content was not immediately apparent from Tokyo. While the IAA content of male leaves from Jiyuan and Chengdu was highest at stage V, the content of other stages was comparable to Jiyuan. In stages IV and V, the IAA content from Tokyo was significantly lower than that from the other two provenances ($p < 0.01$).

The tZR content in Jiyuan male leaves was highest in stage IV (0.243 $\mu g \cdot g^{-1}$), with the greatest fluctuation. From stage I to stage VII, the content of tZR from Tokyo and Chengdu had a smaller change trend, with a periodic rise and fall. In stage V, the tZR content of male leaves from Jiyuan was significantly higher than that of the other two provenances ($p < 0.01$).

3.3. Changes of Hormones in Female Flowers of I. polycarpa from Different Provenances

The levels of ABA, GA3, IAA, and tZR in female flowers differed statistically between *I. polycarpa* provenances (Figure 5). Changes in the contents of ABA, GA3, IAA, and tZR were not observed in female flowers from Jiyuan and Tokyo, but were observed in Chengdu flowers. The maximum values of ABA, GA3, IAA, and tZR were in stages III (0.111 $\mu g \cdot g^{-1}$), III (37.811 $\mu g \cdot g^{-1}$), I (0.357 $\mu g \cdot g^{-1}$), and III (0.873 $\mu g \cdot g^{-1}$), respectively. This indicated that IAA promotes flower opening and fruit formation.

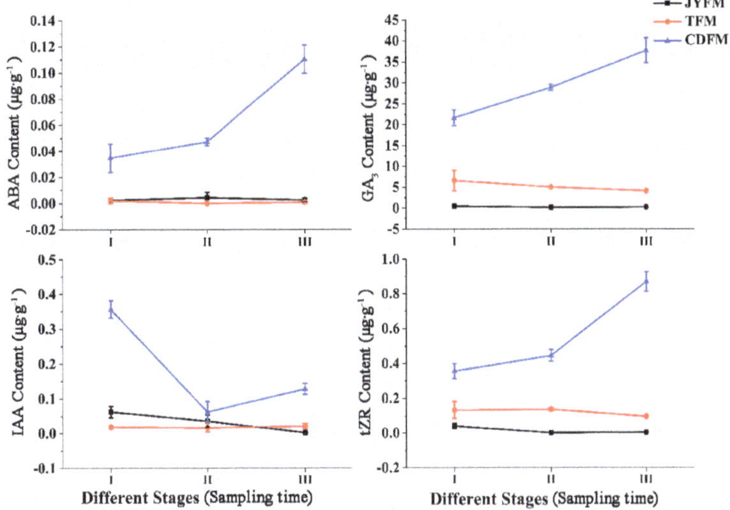

Figure 5. Dynamic changes of hormones in female flowers of *I. polycarpa* from April to May.

The increasing ABA content in Chengdu flowers suggests that ABA is related to female flower abscission in the later period. Similarly, the increasing content of GA3 in Chengdu flowers suggests that GA3 is related to plant fruit setting. The increasing content of tZR in the same Chengdu flowers, on the other hand, suggests that tZR can promote female flower differentiation. In stages II and III, the content of ABA, GA3, and tZR in female flowers from Chengdu was significantly higher than in the other two provenances ($p < 0.05$). Similarly, Chengdu's IAA content in stage I was significantly higher than that of the other two provenances ($p < 0.05$).

3.4. Changes of Hormones in Male Flowers of I. polycarpa from Different Provenances

Changes in the content of ABA, GA3, and IAA were not observed in male flowers from Jiyuan and Tokyo, as shown in Figure 6, and the content of tZR decreased first and then increased. The ABA content of male flowers from Chengdu was higher in stages I (0.114 µg·g^{-1}) and II (0.099 µg·g^{-1}), but lower in stage III (0.020 µg·g^{-1}), possibly due to ABA transfer from flowers to leaves. Chengdu flowers had significantly higher ABA content in stages I and II than in the other two provenances ($p < 0.01$). The primary function of ABA is to promote male flower differentiation and flowering.

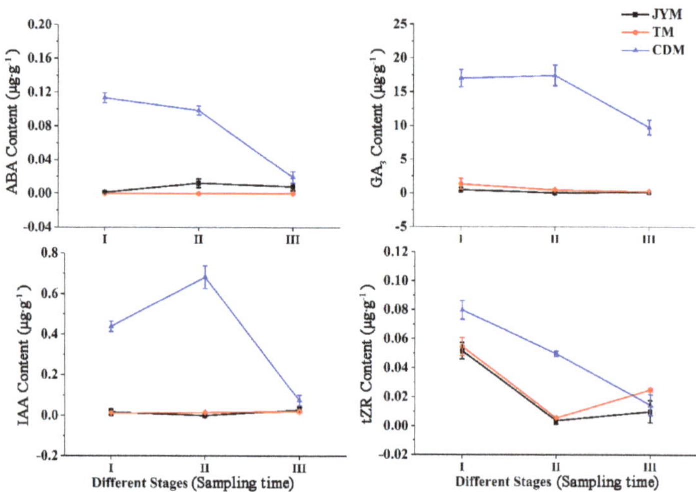

Figure 6. Dynamic changes of hormones in male flowers of *I. polycarpa* from April to May.

GA3 levels in Chengdu male flowers were highest in stages I (17.081 µg·g^{-1}) and II (17.476 µg·g^{-1}). Furthermore, based on our findings, it was clear that the GA3 content in Chengdu flowers at all stages was significantly higher than that of the other two provenances ($p < 0.01$).

The IAA content of male flowers from the Jiyuan and Tokyo provenances was low throughout all stages; however, the IAA content of male flowers from Chengdu was highest in stage II (0.683 µg·g^{-1}), indicating that IAA promotes male flower flowering at this stage of development. Furthermore, we discovered that the IAA content of Chengdu flowers in stages I and II was significantly higher than that of the other two provenances ($p < 0.01$).

The content of tZR revealed that its peak levels of flowers from Jiyuan, Tokyo, and Chengdu occurred in stage I, followed by a drop in stages II and III. This phenomenon demonstrated that tZR was associated with the formation of male flowers. Overall, the tZR content of the Chengdu flowers in stage II was significantly higher than that of the other two provenances ($p < 0.01$).

3.5. Changes of Hormones in Fruits of I. polycarpa from Different Provenances

Figure 7 depicts the changes in hormone concentrations. The concentration of ABA did not change significantly in fruits from Jiyuan and Tokyo, but it did fall and then rise in Chengdu fruits. The ABA content of Chengdu fruits in stages VI and VII was significantly higher than that of the other two provenances ($p < 0.05$). This phenomenon may be associated with the production of stress-resistant proteins in response to external stress.

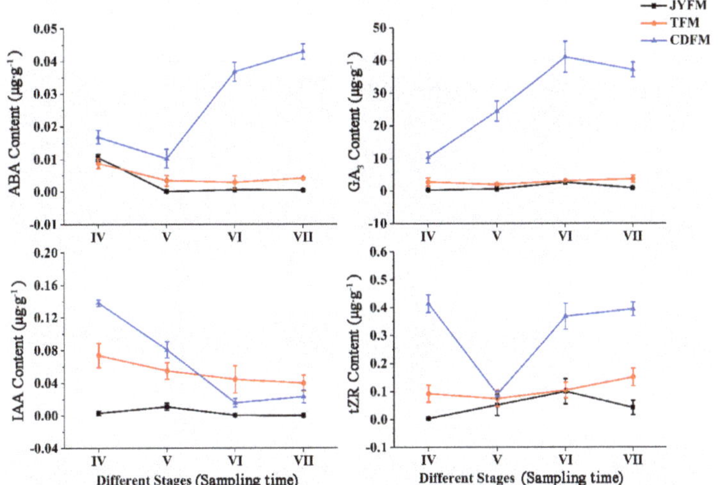

Figure 7. Dynamic changes of hormones in fruits of *I. polycarpa* from May to October.

The content of GA3 was unchanged in the fruits from Jiyuan and Tokyo, but increased gradually and then decreased in the fruits from Chengdu. The highest level of GA3 in the latter case was stage VI ($0.043\ \mu g\cdot g^{-1}$). The GA3 content in Chengdu fruits in stages VI and VII was significantly higher than in the other two provenances ($p < 0.05$).

The content of IAA in the fruits from Jiyuan did not change significantly, while the content of IAA in the fruits from Tokyo and Chengdu had the highest peak values of $0.074\ \mu g\cdot g^{-1}$ and $0.138\ \mu g\cdot g^{-1}$ in stage IV, respectively. The lower IAA content in stage VI suggested that fruit ripening is more sensitive to IAA, whereas higher IAA content could have the opposite effect.

The content of tZR in the fruits from Tokyo and Chengdu had maximum values at stage VII ($0.151\ \mu g\cdot g^{-1}$) and stage IV ($0.415\ \mu g\cdot g^{-1}$), respectively. In the case of the fruits from Jiyuan, the maximum level was in stage VI, with a peak value of $0.100\ \mu g\cdot g^{-1}$. The tZR content of Chengdu fruits in stages IV, VI, and VII was significantly higher than that of the other two provenances ($p < 0.05$). The tZR content in the fruits from the three provenances was similar in stage V, indicating that a certain amount of tZR could promote fruit differentiation.

4. Discussion

The growth and development of *I. polycarpa* were the results of the combined action of many endogenous hormones, including ABA, GA3, IAA, and tZR, each of which had different effects on different organs and stages. The literature currently lacks information on whether cultivation conditions affect the four endogenous hormones in different organs and stages; thus, this study provides a research foundation for its cultivation and development.

Earlier research suggested that ABA could accelerate or delay the flowering time in different plant species and developmental stages. In this study, the ABA content of female flowers from Chengdu increased significantly during the falling flowers period. Many plant hormones, such as abscisic acid, ethylene, and jasmonic acid, are important in regulating organ aging and abscission mechanisms, according to Mohd Gulfishan [39]. In the study of the molecular mechanism of ABA-induced leaf senescence [40], it was also demonstrated that ABA can promote organ dormancy and abscission and has a role in modulating plant stress resistance. Furthermore, the ABA content of Chengdu provenance fruits increased over time. In woodland strawberries, ABA was found to play a coordinating role in fruit growth and ripening [41].

The content of GA3 increased in female flowers from Chengdu, but decreased in male flowers during the falling flowers period, indicating that GA3 was related to fruit setting and male flower differentiation. Watanabe et al. discovered that exogenous GA3 could promote apple fruit formation [42]. Gibberellin could activate and maintain cell division of the ovary wall in citrus, leading to fruit setting [43]. Other research has found that GA3 can promote vegetative growth and flower bud differentiation [44]. From the initial fruit period to the fruit yellowing period, the content of GA3 in Chengdu fruits has increased. Gibberellin was discovered to promote parthenocarpy and fruit expansion in sweet cherries, as evidenced by the qPCR results of related genes, the fruit setting rate, and the parthenocarpy fruit size [45]. More interestingly, high temperature promotes the content of endogenous hormones for seed germination in *I. polycarpa*, causing the balance of endogenous hormone content to break and the ratio of GA3/ABA to increase toward seed germination. Our data showed that the content of GA3 and ABA in the fruit showed an inverse trend and that in the later fruit, both were increased, implying that a balance between them was achieved in the fruit.

The initial flowering stage and full flowering stage of Chengdu flowers had higher IAA content. Physiological and molecular research on *Arabidopsis thaliana* revealed that polar auxin transport was required for flower formation [46]. High-performance liquid chromatography (HPLC) [47] was used to investigate the relationship between IAA and flower formation, and it was discovered that high levels of IAA promoted flower bud induction in apple trees (*Malus pumila*), whereas Liu et al. discovered that low levels of IAA promote flowering in loquat (*Eriobotrya japonica*) [48]. Furthermore, endogenous auxin concentration was the limiting factor in controlling apple fruit size [49]. Almudena Bermejo discovered that IAA can regulate GA metabolism in citrus, resulting in significant changes in the level of active GA1 in the ovule and pericarp and, ultimately, fruit setting [50]. Increased ethylene production, which directly regulates fruit ripening, necessitates higher IAA concentrations for fruit ripening [51]. Our research also revealed that the IAA content in the fruits of the three provenances was higher before the fruit turned yellow, but then dropped until the fruit was fully ripe.

The tZR content of Jiyuan flowers was higher during the early flowering stage. tZR typically promotes female flower development, as evidenced by the higher tZR content in female flowers from the initial flowering stage to the falling flowers stage in a study on *Glycyrrhiza uralensis* [52]. Male flowers had a relatively high tZR content at the start of flower formation, which is supported by the fact that high IAA and tZR contents during the flower bud differentiation stage are conducive to flower bud differentiation [53]. tZR also inhibited many developmental processes, such as bud and root elongation, cell differentiation, bud regeneration, and meristem activity [54,55].

Previous research has shown that high IAA and tZR contents promote flower bud differentiation, whereas high GA and ABA contents promote flower and fruit abscission [52]. High levels of IAA and tZR regulate plant vegetative growth and flower induction [56]. Furthermore, ABA promotes flowering, whereas GA3 and IAA do not [57]. A similar study on the changes in endogenous hormones during the flowering period of *Gnetum parvifolium* found that high levels of GA3 and tZR promote male flower differentiation, while high levels of IAA promote female flower differentiation [58]. The peak contents of different plant organs' endogenous hormones may be related to the transfer of such hormones to different organs at different stages.

Plants can change their growth and development in response to their surroundings, which is controlled by endogenous plant hormones. Several endogenous hormones, including ABA, GA3, IAA, and tZR, act together to promote the growth and development of *I. polycarpa*. Each hormone had a distinct effect on different organs and developmental stages. The fluctuation was obvious, and the concentrations of several endogenous hormones in various organs of the Chengdu provenance were significantly higher than those of the other two provenances. Geographical location and environmental factors, such as light intensity/availability, soil quality, and precipitation, may all have a causal relationship

Life **2023**, *13*, 788

with endogenous hormone concentration variation. Nonetheless, these environmental influences have a minor impact. According to the findings, the contents of these four endogenous hormones in *I. polycarpa* are primarily determined by genetic characteristics of the trees and are less affected by cultivation conditions. The gender in *I. polycarpa* had a great influence on these four endogenous hormones.

5. Conclusions

The materials used in the experiment came from the Forestry Experimental Station of Henan Agricultural University. The differences in hormone contents between provenances could be due to genetic differences in *I. polycarpa*. The contents of various endogenous hormones in different organs of the Chengdu provenance were significantly higher than those of the other two provenances, with a clear fluctuation. This discovery may provide insight into how differences in endogenous hormone concentration can lead to differences in flowering and fruiting time and quantity, providing cultural direction. In the future, we will increase the sample size and lengthen the period of observation to confirm the results.

Author Contributions: Conceptualization, S.H. and W.Z.; methodology, S.H.; software, H.Y.; validation, Y.W., H.Y. and C.Z.; formal analysis, W.Z.; investigation, T.M.; resources, T.M.; data curation, S.H.; writing—original draft preparation, S.H.; writing—review and editing, T.M.; visualization, Y.W.; supervision, T.M.; project administration, T.M.; funding acquisition, T.M. All authors have read and agreed to the published version of the manuscript.

Funding: This research was funded by the Forestry Science and Technology Development Project, the Science and Technology Development Center of State Forestry and Grassland Administration (KJZXSA2019041), and the Science and Technology Revitalization Forestry Project of Henan Provincial Department of Forestry [2018(68)], and the Dabie Mountain Forestry Resources Innovation Theory and Technology innovation team of Xinyang Agriculture and Forestry University (XNKJTD-004).

Institutional Review Board Statement: Not applicable.

Informed Consent Statement: Not applicable.

Data Availability Statement: The data presented in this study are available on request from the corresponding author.

Acknowledgments: The authors wish to express their sincere thanks to the Key Laboratory of the State Forestry Administration for Cultivating Forest Resources in Central China and Henan Agricultural University for their kind assistance in the experiments.

Conflicts of Interest: The authors declare no conflict of interest.

References

1. Wang, C.; Luo, H.; Hu, Z.L.; Diao, Y. Research Progress of *Idesia polycarpa* Maxim. *World J. For.* **2019**, *8*, 40–45. [CrossRef]
2. Wang, H.M.; Rana, S.; Li, Z.; Geng, X.D.; Wang, Y.M.; Cai, Q.F.; Li, S.F.; Sun, J.L.; Liu, Z. Morphological and anatomical changes during floral bud development of the trioecious *Idesia polycarpa* Maxim. *Braz. J. Bot.* **2022**, *45*, 679–688. [CrossRef]
3. Chen, J.S.; Wu, J.; Fang, L.S.; Wang, Y.M.; Gen, X.D.; Liu, Z. Effect of temperature and photoperiod on the branching after sprouted from buds in *Iidesia polycarpa*. *J. Northwest Univ.* **2019**, *34*, 111–117. (In Chinese)
4. Akagi, T.; Henry, I.; Ohtani, H.; Morimoto, T.; Beppu, K.; Kataoka, I.; Tao, R. A Y-encoded suppressor of feminization arose via lineage-specific duplication of a cytokinin response regulator in kiwifruit. *Plant Cell* **2018**, *30*, 780–795. [CrossRef] [PubMed]
5. Su, S.; Li, Z.J.; Ni, J.W.; Gen, Y.H.; Xu, X.Q. Diversity analysis on characters of spikes and fruits of *Idesia Polycarpa*. *J. Plant Resour. Environ.* **2021**, *30*, 35–44. (In Chinese)
6. Zhang, H.Y.; Li, X.L.; Zhang, H.S. Research Progress on Seedling Breeding of *Idesia polycarpa*. *Chin. Wild Plant Resour.* **2022**, *41*, 77–80. (In Chinese)
7. Liu, Y.J.; Zhang, C.S.; Tu, C.H.; Fen, J.; Gen, F.D.; Zheng, Q. Research progress in the nutritive value and refining process technology of *Idesia Polycarpa* Maxim. seed oil. *Farm Prod. Process.* **2020**, *7*, 73–74. [CrossRef]
8. Li, T.T.; Li, F.S.; Mei, L.J.; Li, N.; Yao, M.; Tang, L. Transcriptome analysis of *Idesia polycarpa* Maxim. var vestita Diels fowers during sex diferentiation. *J. For. Res.* **2020**, *31*, 2463–2478. [CrossRef]
9. Yan, H.P. Dynamic changes of endogenous hormones during flowering and fruiting of *Idesia polycarpa* Maxim. *Henan Agric. Univ.* **2019**, 1–59. [CrossRef]

10. Wang, Y.M.; Wang, L.J.; Bing, Y.; Zhen, L.; Fei, L. Changes in ABA, IAA, GA3, and ZR Levels during Seed Dormancy Release in Idesia polycarpa Maxim from Jiyuan. *Pol. J. Environ. Stud.* **2018**, *27*, 1833–1839. [CrossRef]

11. Yang, Y.S.; Chen, Z.Y.; Zou, R.; Tang, J.M.; Jiang, Y.S. Comparative study of endogenous hormone dynamics during the first and second growth cycles in *Sophora japonica* cv. jinhuai harvested twice a year. *J. Biotech Res.* **2023**, *14*, 59–66.

12. Wilson, W.C. The use of exogenous plant growth regulators on citrus. In *Plant Growth Regulating Chemicals*; CRC Press: Boca Raton, FL, USA, 2018; pp. 207–232.

13. Yu, T.; Li, J.G.; Shi, R.; Zhang, J.; Wang, N.; Cao, Y.J. Relationship between endogenous hormone content in Jun jujube and physiological fruit drop. *Non-Wood For. Res.* **2016**, *34*, 45–49. [CrossRef]

14. Zhao, Y. Essential Roles of local auxin biosynthesis in plant development and in adaptation to environmental changes. *Annu. Rev. Plant Biol.* **2018**, *69*, 417–435. [CrossRef] [PubMed]

15. Brumos, J.; Robles, L.M.; Yun, J.; Vu, T.C.; Jackson, S.; Alonso, J.M.; Stepanova, A.N. Local auxin biosynthesis is a key regulator of plant development. *Dev. Cell* **2018**, *47*, 306–318. [CrossRef]

16. Sun, Y.D.; Li, Z.X.; Luo, W.R.; Chang, H.C.; Wang, G.Y. Effects of Indole Acetic Acid on Growth and Endogenous Hormone Levels of Cucumber Fruit. *Int. J. Agric. Biol.* **2018**, *20*, 197–202. [CrossRef]

17. Kosakivska, I.V.; Voytenko, L.V.; Vasyuk, V.A.; Shcherbatiuk, M.M. Effect of pre-sowing priming of seeds with exogenous abscisic acid on endogenous hormonal balance of spelt wheat under heat stress. *Zemdirb. Agric.* **2022**, *109*, 21–26. [CrossRef]

18. Kumar, P.; Sangwan, M.S.; Mehta, N. A protocol for efficient callus induction from hypocotyls explant in chickpea (Cicer arietinum). *Indian J. Agric. Sci.* **2014**, *84*, 167–169.

19. Yoshida, T.; Christmann, A.; Yamaguchi-Shinozaki, K.; Grill, E.; Fernie, A.R. Revisiting the Basal Role of ABA—Roles Outside of Stress. *Trends Plant Sci.* **2019**, *24*, 625–635. [CrossRef]

20. Leng, P.; Yuan, B.; Guo, Y.D. The role of abscisic acid in fruit ripening and responses to abiotic stress. *J. Exp. Bot.* **2014**, *65*, 4577–4588. [CrossRef]

21. Seki, M.; Matsui, A.; Kim, J.M.; Ishida, J.; Nakajima, M.; Morosawa, T.; Kawashima, M.; Satou, M.; To, T.K.; Kurihara, Y. Arabidopsis whole-genome transcriptome analysis under drought, cold, high-salinity, and ABA treatment conditions using tiling array and 454 sequencing technology. *Plant Cell Physiol.* **2007**, *48*, S8.

22. Buesa, C.; Dominguez, M.; Vendrell, M. Abscisic-acid effects on ethylene production and respiration rate in detached apple fruits at different stages of development. *Rev. Esp. De Cienc. Y Tecnol. De Aliment.* **1994**, *34*, 495–506.

23. Zhang, M.; Leng, P.; Zhang, G.L.; Li, X.X. Cloning and functional analysis of 9-cis-epoxycarotenoid dioxygenase (NCED) genes encoding a key enzyme during abscisic acid biosynthesis from peach and grapefruits. *J. Plant Physiol.* **2009**, *166*, 1241–1252. [CrossRef] [PubMed]

24. Giribaldi, M.; Geny, L.; Delrot, S.; Schubert, A. Proteomic analysis of the effects of ABA treatments on ripening Vitis vinifera berries. *J. Exp. Bot.* **2010**, *61*, 2447–2458. [CrossRef] [PubMed]

25. Galpaz, N.; Wang, Q.; Menda, N.; Zamir, D.; Hirschberg, J. Abscisic acid deficiency in the tomato mutant high-pigment 3 leading to increase plastid number and higher fruit lycopene content. *Plant J.* **2008**, *53*, 717–730. [CrossRef]

26. Camara, M.C.; Vandenberghe, L.P.S.; Rodrigues, C.; de Oliveira, J.; Faulds, C.; Bertrand, E.; Soccol, C.R. Current advances in gibberellic acid (GA3) production, patented technologies and potential applications. *Planta* **2018**, *248*, 1049–1062. [CrossRef] [PubMed]

27. Chen, Y.; Jin, M.; Wu, C.Y.; Bao, J.P. Effects of plant growth regulators on the endogenous hormone content of calyx development in 'Korla' Pear. *Hortscience* **2022**, *57*, 497–503. [CrossRef]

28. Colebrook, E.H.; Thomas, S.G.; Phillips, A.L.; Hedden, P. The role of gibberellin signalling in plant responses to abiotic stress. *J. Exp. Biol.* **2014**, *217*, 67–75. [CrossRef]

29. Li, Y.H.; Wu, Y.J.; Wu, B.; Zou, M.H.; Zhang, Z.; Sun, G.M. Exogenous gibberellic acid increases the fruit weight of 'Comte de Paris' pineapple by enlarging flesh cells without negative effects on fruit quality. *Acta Physiol. Plant.* **2011**, *33*, 1715–1722. [CrossRef]

30. Wang, Y.J.; Zhao, J.; Lu, W.J.; Deng, D.X. Gibberellin in plant height control: Old player, new story. *Plant Cell Rep.* **2017**, *36*, 391–398. [CrossRef]

31. Muñoz-Fambuena, N.; Mesejo, C.; González-Mas, M.C.; Iglesias, D.J.; Primo-Millo, E.; Agusti, M. Gibberellic Acid Reduces Flowering Intensity in Sweet Orange [*Citrus sinensis* (L.) Osbeck] by Repressing CiFT Gene Expression. *J. Plant Growth Regul.* **2012**, *31*, 529–536. [CrossRef]

32. Hedden, P.; Sponsel, V. A Century of Gibberellin Research. *J. Plant Growth Regul.* **2015**, *34*, 740–760. [CrossRef]

33. Zaheer, M.S.; Ali, H.H.; Erinle, K.O.; Wani, S.H.; Okon, O.G.; Nadeem, M.A.; Nawaz, M.; Bodlah, M.A.; Waqas, M.M.; Iqbal, J.; et al. Inoculation of *Azospirillum brasilense* and exogenous application of trans-zeatin riboside alleviates arsenic induced physiological damages in wheat (*Triticum aestivum*). *Environ. Sci. Pollut. Res.* **2022**, *29*, 33909–33919. [CrossRef]

34. Liu, X.X.; Pan, Y.P.; Liu, C.; Ding, Y.Y.; Wang, X.; Cheng, Z.H.; Meng, H.W. Cucumber Fruit Size and Shape Variations Explored from the Aspects of Morphology, Histology, and Endogenous Hormones. *Plants* **2020**, *9*, 772. [CrossRef] [PubMed]

35. Honda, I.; Matsunaga, H.; Kikuchi, K.; Matuo, S.; Fukuda, M.; Imanishi, S. Involvement of Cytokinins, 3-Indoleacetic Acid, and Gibberellins in Early Fruit Growth in Pepper (*Capsicum annuum* L.). *Hortic. J.* **2017**, *86*, 52–60. [CrossRef]

36. Villacorta, N.F.; Fernandez, H.; Prinsen, E.; Bernad, P.L.; Revilla, M.A. Endogenous hormonal profiles in hop development. *J. Plant Growth Regul.* **2008**, *27*, 93–98. [CrossRef]

37. Li, Y.H.; Wei, F.; Dong, X.Y.; Peng, J.H.; Liu, S.Y.; Chen, H. Simultaneous analysis of multiple endogenous plant hormones in leaf tissue of oilseed rape by solid-phase extraction coupled with High-performance Liquid Chromatography–Electrospray Ionization Tandem Mass Spectrometry. *Phytochem. Anal.* **2011**, *22*, 442–449. [CrossRef]

38. Yang, L.; Liu, S.; Wang, Z.; Cao, Q.; Wang, J.; Xing, Y.; Qin, L. Determination of IAA and ABA in plant tissue by the GPC-HPLC/MS/MS. *Adv. Anal. Chem.* **2017**, *7*, 131–138. [CrossRef]

39. Gulfishan, M.; Jahan, A.; Bhat, T.A.; Sahab, D. Plant Senescence and Organ Abscission. In *Senescence Signalling and Control in Plants*; Academic Press: Cambridge, MA, USA, 2019; pp. 255–272. [CrossRef]

40. Zhao, Y.; Chan, Z.; Gao, J.; Xing, L.; Cao, M.; Yu, C.; Hu, Y.; You, J.; Shi, H.; Zhu, Y.; et al. ABA receptor PYL9 promotes drought resistance and leaf senescence. *Proc. Natl. Acad. Sci. USA* **2016**, *113*, 1949–1954. [CrossRef] [PubMed]

41. Liao, X.; Li, M.; Liu, B.; Yan, M.L.; Yu, X.M.; Zi, H.L.; Liu, R.Y.; Yamamuro, C. Interlinked regulatory loops of ABA catabolism and biosynthesis coordinate fruit growth and ripening in woodland strawberry. *Proc. Natl. Acad. Sci. USA* **2018**, *115*, E11542–E11550. [CrossRef]

42. Snouffer, A.; Kraus, C.; van der Knaap, E. The shape of things to come: Ovate family proteins regulate plant organ shape. *Curr. Opin. Plant Biol.* **2020**, *53*, 98–105. [CrossRef]

43. Mesejo, C.; Yuste, R.; Reig, C.; Martinez-Fuentes, A.; Iglesias, D.J.; Munoz-Fambuena, N.; Bermejo, A.; Germana, M.A.; Primo-Millo, E.; Agusti, M. Gibberellin reactivates and maintains ovary-wall cell division causing fruit set in parthenocarpic Citrus species. *Plant Sci.* **2016**, *247*, 13–24. [CrossRef]

44. Sha, J.C.; Wang, F.; Xu, X.X.; Chen, Q.; Zhu, Z.L.; Jiang, Y.M.; Ge, S.F. Studies on the translocation characteristics of ^{13}C-photoassimilates to fruit during the fruit development stage in 'Fuji' apple. *Plant Physiol. Biochem.* **2020**, *154*, 636–645. [CrossRef]

45. Wen, B.B.; Song, W.L.; Sun, M.Y.; Chen, M.; Mu, Q.; Zhang, X.H.; Wu, Q.J.; Chen, X.D.; Gao, D.S.; Wu, H.Y. Identification and characterization of cherry (*Cerasus pseudocerasus* G. Don) genes responding to parthenocarpy induced by GA3 through transcriptome analysis. *BMC Genet.* **2019**, *20*, 65. [CrossRef]

46. Li, H.B.; Cheng, Y.; Murphy, A.; Hagen, G.; Guilfoyle, T.J. Constitutive repression and activation of auxin signaling in Arabidopsis. *Plant Physiol.* **2009**, *149*, 1277–1288. [CrossRef] [PubMed]

47. Milyaev, A.; Kofler, J.; Klaiber, I.; Czemmel, S.; Pfannstiel, J.; Flachowsky, H.; Stefanelli, D.; Hanke, M.V.; Wünsche, J.N. Toward Systematic Understanding of Flower Bud Induction in Apple: A Multi-Omics Approach. *Front. Plant Sci.* **2021**, *25*, 604810. [CrossRef]

48. Jinga, D.L.; Chen, W.W.; Xia, Y.; Shi, M.; Wang, P.; Wang, S.M.; Wu, D.; He, Q.; Liang, G.L.; Guo, Q.G. Homeotic transformation from stamen to petal in *Eriobotrya japonica* is associated with hormone signal transduction and reduction of the transcriptional activity of *EjAG*. *Physiol. Plant.* **2020**, *168*, 893–908. [CrossRef]

49. Guan, L.; Li, Y.; Huang, K.; Cheng, Z.M. Auxin regulation and MdPIN expression during adventitious root initiation in apple cuttings. *Hortic. Res.* **2020**, *7*, 143. [CrossRef] [PubMed]

50. Bermejo, A.; Granero, B.; Mesejo, C.; Reig, C.; Tejedo, V.; Agustí, M.; Primo-Millo, E.; Iglesias, D.J. Auxin and gibberellin interact in citrus fruit set. *J. Plant Growth Regul.* **2018**, *37*, 491–501. [CrossRef]

51. Liu, N.N. Effects of IAA and ABA on the immature peach fruit development process. *Hortic. Plant J.* **2019**, *5*, 145–154. [CrossRef]

52. Yan, B.B.; Hou, J.L.; Cui, J.; He, C.; Li, W.B.; Chen, X.Y.; Li, M.; Wang, W.Q. The Effects of endogenous hormones on the flowering and fruiting of *Glycyrrhiza uralensis*. *Plants* **2019**, *8*, 519. [CrossRef]

53. Weiss, D.; Ori, N. Mechanisms of cross talk between gibberellin and other hormones. *Plant Physiol.* **2007**, *144*, 1240–1246. [CrossRef] [PubMed]

54. Greenboim-Wainberg, Y.; Maymon, I.; Borochov, R.; Alvarez, J.; Olszewski, N.; Ori, N.; Eshed, Y.; Weiss, D. Cross talk between gibberellin and cytokinin: The Arabidopsis GA-response inhibitor SPINDLY plays a positive role in cytokinin signaling. *Plant Cell* **2005**, *17*, 92–102. [CrossRef] [PubMed]

55. Jasinski, S.; Piazza, P.; Craft, J.; Hay, A.; Woolley, L.; Rieu, I.; Phillips, A.; Hedden, P.; Tsiantis, M. KNOX action in Arabidopsis is mediated by coordinate regulation of cytokinin and gibberellin activities. *Curr. Biol.* **2005**, *15*, 1560–1565. [CrossRef] [PubMed]

56. Zhang, Y.J.; Li, A.; Liu, X.Q.; Sun, J.X.; Guo, W.J.; Zhang, J.W.; Lyu, Y.M. Changes in the morphology of the bud meristem and the levels of endogenous hormones after low temperature treatment of different Phalaenopsis cultivars. *South Afr. J. Bot.* **2019**, *125*, 499–504. [CrossRef]

57. Chen, C.; Cao, Y.Y.; Wang, X.J.; Wu, Q.K.; Yu, F.Y. Do Stored Reserves and Endogenous Hormones in Overwintering Twigs Determine Flower Bud Differentiation of Summer Blooming Plant-Styrax tonkinensis? *Int. J. Agric. Biol.* **2019**, *22*, 815–820. [CrossRef]

58. Lan, Q.; Liu, J.F.; Shi, S.Q.; Deng, N.; Jiang, Z.P.; Chang, E.M. Anatomy, microstructure, and endogenous hormone changes in *Gnetum parvifolium* during anthesis. *J. Syst. Evol.* **2018**, *56*, 14–24. [CrossRef]

Article

Diversity Analysis of Leaf Nutrient Endophytes and Metabolites in Dioecious *Idesia polycarpa* Maxim Leaves during Reproductive Stages

Jian Feng [1,†], Sohel Rana [1,†], Zhen Liu [1], Yanmei Wang [1], Qifei Cai [1], Xiaodong Geng [1], Huina Zhou [1], Tao Zhang [1], Shasha Wang [1], Xiaoyan Xue [1], Mingwan Li [1], Razia Sultana Jemim [2] and Zhi Li [1,*]

[1] College of Forestry, Henan Agricultural University, Zhengzhou 450046, China
[2] College of Life Sciences, Henan Agricultural University, Zhengzhou 450046, China
* Correspondence: lizhi876@163.com; Tel.: +86-17752559889
† These authors contributed equally to this work.

Abstract: Leaves are essential vegetative organs of plants. Studying the variations in leaf nutrient content and microbial communities of male and female plants at reproductive stages helps us understand allocation and adaptation strategies. This study aimed to determine the nutrient characteristics and microbial differences in the leaves of male and female *Idesia polycarpa* at reproductive stages. Seven-year-old female and male plants were used as test materials in this experiment. The samples were collected at three stages: flowering (May), fruit matter accumulation (July), and fruit ripening (October). The nitrogen (TN), phosphorus (TP), potassium (TK), carbon (TC), and the pH of the female and male leaves were analyzed. In addition, the leaf microbial diversity and differential metabolites were determined using the Illumina high-throughput sequencing method and the ultra-high performance liquid chromatography–tandem mass spectrometry (UPLC–MS/MS) method at the reproductive developmental stages. This study found that male and female plant leaves had different TN and TK contents over time but no difference in TC and TP content. The significant differences in bacterial diversity between male and female plants and the richness of the fungi of male plants at the flowering and fruit maturity stages were observed. Proteobacteria, Pseudomonadaceae, Ascomycota, and Aspergillus were the dominant bacteria and fungi in the *Idesia polycarpa* leaves. The presence of microorganisms differed in the two sexes in different periods. Alphaproteobacteria and Sordariomycetes were the indicator groups for male leaves, and Pseudomonas and Sordariomycetes were the indicator groups for female leaves. Significant differences in phenolic acid were found between male and female leaves. A KEGG enrichment analysis revealed that differential metabolites were enriched in metabolic pathways, amino acid biosynthesis, and the nucleotide metabolism. According to a correlation analysis, leaf TK and TP were strongly correlated with endophytic bacteria abundance and differential metabolite composition. This study revealed the changes in substances and microorganisms in the leaves of male and female plants in their reproductive stages. It provides a theoretical basis for developing and utilizing the leaves of *Idesia polycarpa* and for field management.

Keywords: dioecious plant; reproductive stages; nutrient characteristics; endophytes; metabolite

Citation: Feng, J.; Rana, S.; Liu, Z.; Wang, Y.; Cai, Q.; Geng, X.; Zhou, H.; Zhang, T.; Wang, S.; Xue, X.; et al. Diversity Analysis of Leaf Nutrient Endophytes and Metabolites in Dioecious *Idesia polycarpa* Maxim Leaves during Reproductive Stages. *Life* **2022**, *12*, 2041. https://doi.org/10.3390/life12122041

Received: 17 November 2022
Accepted: 5 December 2022
Published: 6 December 2022

Publisher's Note: MDPI stays neutral with regard to jurisdictional claims in published maps and institutional affiliations.

1. Introduction

Plants, including dioecious plants, gradually evolved various breeding systems to adapt to their natural environments. Dioecious plants show significant differences in their morphological characteristics, physiological and biochemical responses, and gene expression due to their different reproductive costs [1]. The length–width ratio of females in *Rhus typhina* is significantly higher than that of males [2]. Another study showed that the growth of females exceeded that of males in terms of length extension, diameter growth, and leaf production [3]. Additionally, in the leaves of *Excoecaria agallocha*, total carotenoids, phenolic compounds, and protein concentrations were higher in female plants [4]. The oil palm orthologs of acid phosphatase and deficiency showed male-specific expression patterns [5]. Resource and growth microenvironment distribution between male and female plants are significantly different.

Leaves are one of the main organs by which plants obtain and utilize resources. Changes in the microenvironment of leaves can reflect the adaptation strategies of plants to different environments [6]. The nutrient content in leaves can indicate the plant growth status and habitat conditions. For example, nitrogen is an essential component of various key cellular molecules such as proteins, nucleic acids, and secondary metabolites, which are important for maintaining plant metabolism and growth [7]. Previous studies have shown a significant relationship between leaf nutrient content and gender. For example, the effective nitrogen content of female leaves of *Fraxinus mandshurica*, *Populus davidiana* Dode, and *Taxus cuspidata* was higher than that of male leaves [8]. The content of nutrient elements in the leaves of male *Fraxinus velutina* Torr plants was higher than that of female plants at each stage of the growing season [9]. Therefore, it is important to clarify the dynamic changes in leaf nutrients between dioecious plants for forest land management. In addition, many microbial communities in plant leaves interact and coevolve.

Studies have found that some endophytic microorganisms perform various important biological functions. For example, Zuo et al. [10] have confirmed that *Bacillus velezensis* BHZ-29 in Shihezi cotton plants can increase the activity of defense-related enzymes in cotton to varying degrees, as well as reduce the accumulation of MDA and thus enhance the resistance of cotton plants to diseases. Abdul et al. [11] showed that the endophytic strain Bacillus subtilis of *Solanum lycopersicum* could secrete IAA, which can significantly increase the biomass of *Solanum lycopersicum*. Plants provide a stable living environment and sufficient nutrients for endophytes. By measuring the endophytic bacterial community of *Lycium barbarum* leaves in different periods, Gou et al. [12] found that the α diversity and richness were the highest in young leaves, that it decreased with the growth of the leaves, and that it reached its lowest in old leaves. However, the composition and structure of endophytic microbial communities have differed in different periods [13], different organs [14], and different varieties of plants [15]. There is great significance in revealing their biological characteristics and disease control properties by exploring the differences in endophytes between dioecious plants. Various secondary metabolites are produced in leaf growth, and plant leaves become an essential source of plant bioactive substances in interaction with microorganisms. Rattanawiwatpong et al. [16] showed that raspberry leaves have vigorous antioxidant activity, and Staszowska-Karkut et al. [17] found that raspberry leaves are rich in phenolic compounds with high development value. Rong et al. [18] found that the chlorogenic acid in *Arctium lappa* leaves has a pan-antibacterial effect. Analyzing the different metabolites found in dioecious plants can reveal specific metabolites and provide a reference for leaf development and utilization.

Idesia polycarpa Maxim (Flacourtiaceae) is a deciduous broad-leafed tree. This species is excellent for landscaping because of its unique characteristics, i.e., it is tall, has a straight trunk, and produces bright fruit. [19]. This woody oil tree species has a high development value [20] due to the high oil content in its fruit (77% unsaturated fatty acid content and 62.9% linoleic acid content). In addition, it has the potential to be useful in the preparation of biodiesel [21]. The current state of research on *Idesia polycarpa* focuses mainly on seedling breeding, physiological and biochemical studies, and oil extraction [22]. The lack of reports on the growth differences between male and female plants make such studies even more necessary in this dioecious species.

In this study, we attempted to determine the changes in leaf nutrients, leaf metabolite content, and the differences in the endophytic bacteria community composition between male and female plants in different growth periods. We hypothesized that: (1) Females allocate more resources to reproduction than males. Therefore, the leaves of female plants will accumulate more nutrients and carbohydrates. (2) In our daily observations, males tend to be more resilient than females. Therefore, the endophyte community in male leaves will be more abundant, and the accumulation of stress-resistant substances will be greater.

2. Materials and Methods

2.1. Study Area

This study was undertaken at the experimental research station (112°42′114°14′ E, and 34°16′34°58′ N) of the College of Forestry, Henan Agricultural University, Zhengzhou, Henan Province, China, in 2021. The mean annual temperature of this site is 14.2 °C, the frost-free period is 215 days, the mean annual precipitation of 650.1 mm, the annual sunshine hours are about 2400 h, and the soil is slightly alkaline sandy loam.

2.2. Sample Acquisition

In March 2021, six seven-year-old female and male plants with average growth and no apparent diseases and pests were selected. In May (flowering period), July (fruit matter accumulation period), and October (fruit ripening period), 20 intact mature leaves (each period) were collected from the east, west, north, and south side of the crown and brought back to the laboratory in an ice box. They were fixed at 105 °C for 30 min and dried at 80 °C until the sample weight was constant. The samples were then crushed using a pulverizer, passed through a 100-mesh sieve, and sealed for nutrient analysis. A portion of the sample was frozen in liquid nitrogen before being stored in a −80 °C ultra-low temperature refrigerator to determine the endophytic bacteria and metabolites.

2.3. Determination of Leaf Nutrients and Analysis

The leaf nutrients were measured using three samples for each parameter (with three replicates). The leaf pH was measured using an acidometer (LC-PH-3S, Shanghai LiChen Bangxi Instrument Equipment Co., Ltd., Shanghai, China). The total carbon (TC) and total nitrogen (TN) contents were measured using an automatic elemental analyzer (Euro Vector EA3000, Shanghai Wolong Instrument Co., Ltd., Shanghai, China). The total phosphorus (TP) content was determined using the molybdenum antimony anti-colorimetric method, and the total potassium (TK) content was determined using the flame photometer method [23]. Statistical analysis was performed using IBM SPSS v. 26 (IBM Corp., Armonk, NY, USA), and Origin 2017 (www.OriginLab.com (accessed on 30 November 2022)) was used to draw a histogram of nutrient content.

2.4. Determination of Endophytic Bacteria in Leaves

2.4.1. Total DNA Extraction, PCR Amplification, and Sequencing

The total DNA was extracted according to the instructions of the E.Z.N.A.® soil kit (Omega Bio-Tek, Norcross, GA, USA). The DNA concentration and purity were detected using NanoDrop2000, and the DNA extraction quality was detected using 1% agarose gel electrophoresis. The V3–V4 region of the bacterial 16S rRNA gene was amplified by PCR using primers 338F (5′-ACTCCTACGGGAGGCAGCAG-3′) and 806R (5′-GGACTACHVGGGTWTCTAAT-3′). The ITS1-ITS2 region of the fungi was amplified by PCR using primers ITS5-1737F (5′-GGAAGTAAAAGTCGTAACAAGG-3′) and ITS2-2043R (5′-GCTGCGTTCTTCATCGATGC-3′). PCR reaction conditions: 95 °C pre-denaturation 3 min, 27 cycles (95 °C denaturation 30 s, 55 °C annealing 30 s, 72 °C extension 30 s), 72 °C extensions 10 min. The amplified products were purified, and a library was constructed. After the library was qualified, it was sequenced using Illumina's Miseq PE300 platform. The DNA extraction, PCR amplification, and sequencing were entrusted to Wekemo Tech Group Co., Ltd., Shenzhen, China.

2.4.2. Diversity Analysis of Endophytic Bacteria

The sequencing results were subjected to quality control and denoising using the QIIME2 plug-in to obtain valid data. According to the similarity, the sequence was clustered into the operational taxonomic unit (OTU), and species annotation was performed. The alpha diversity index was calculated using QIIME software. LDA (linear discriminant analysis) was performed on each sample using LEfSe software.

2.5. Determination of Metabolites in Leaves

2.5.1. Metabolite Extraction

The leaf samples were placed in a freeze-dryer vacuum, freeze-dried, and ground into a fine powder. Next, 100 mg of the powder was dissolved in 1.2 mL of 70% methanol extract. Vortexing was performed for 30 s every 30 min a total of 6 times, and the material was then placed in a 4 °C refrigerator overnight. The supernatant was centrifuged, filtered, and stored in the sample for UPLC–MS/MS analysis.

2.5.2. The UPLC–MS/MS Analysis

Liquid chromatographic conditions: Agilent SB-C18 1.8 um, 2.1 mm × 100 mm; mobile phase: ultra-pure water (with 0.1% formic acid) for phase A and acetonitrile (with 0.1% formic acid) for phase B; elution gradient: 5% for phase B at 0.00 min, linearly increasing to 95% for phase B at 9.00 min and maintaining at 95% for 1 min. The B-phase ratio decreased to 5% at 10.00–11.10 min and was equilibrated at 5% for 14 min; flow rate: 0.35 mL/min; column temperature: 40 °C; injection: 4 uL.

Mass spectrometry conditions: electrospray ionization (ESI) source for bulk data acquisition with the following operating parameters: the ion source, turbo spray: source temperature 550 °C. Ion spray voltage (IS) positive ion mode 5500 V/negative ion mode −4500 V; ion source gas I (GSI), gas II (GSII), and curtain gas (CUR) were set to 50 psi, 60 psi, and 25 psi, respectively, and collision-induced ionization parameters were set at high. Instrument tuning and mass calibration were performed in triple quadrupole (QQQ) and linear ion trap (LIT) modes with 10 and 100 μmol/L polypropylene glycol solutions. QQQ scans were performed using multiple reaction detection (MRM) mode with collisional nitrogen set to medium. In QQQ, each ion pair was scanned for detection according to the optimized declustering voltage (DP) and collision energy (CE).

2.6. Data Analysis

Based on the self-built database MWDB (metware database) of Maiwei Metabolic Co., Ltd., the secondary spectrum information and the metabolites were quantified using the multi-reaction monitoring mode of triple quadrupole mass spectrometry. The metabolites of the different samples were compared and analyzed, and the identified metabolites were analyzed using a multivariate statistical analysis to explore the metabolic characteristics of the different samples preliminarily. According to the variable importance projection (VIP) score obtained by an orthogonal partial least squares discriminant analysis (OPLS-DA), the metabolites with VIP $\geq$ 1, fold change $\geq$ 2, or fold change $\leq$ 0.5 were defined as significantly changed metabolites (SCMs). In addition, the corresponding differential metabolites were submitted to the KEGG (Kyoto Encyclopedia of Genes and Genomes) database website, and R software (V4.1.0) was used to draw the heat map of the correlation between the differential metabolites and the leaf nutrients.

3. Results

3.1. Leaf Nutrient Characteristics in Different Periods

The characteristics of the nutrient content in the leaves at the flowering, fruit matter accumulation, and fruit repining stages were not significantly different. The pH values in the leaves during the reproductive development stages did not differ, and the pH variation range was 5.61–6.18. In addition, there were no significant differences in TC values and TP contents between the male and female plants. The TC content was higher and TP content was lower in female leaves than in male leaves. The TN and TK contents in the leaves of the female plants were higher than in those of the male plants in different periods, and the TN content at the flowering stage was significantly higher than that of the male plants. At the fruit developing stage, TK content was considerably higher in the female plants than in the male plants (Table 1).

Table 1. The nutrient characteristics of *Idesia polycarpa* leaves in different reproductive stages. The lowercase letters indicate significant differences. Abbreviations: CS, Female; XS, Male.

Period	Sexuality	pH	Total Carbon g·kg^{-1}	Total Nitrogen g·kg^{-1}	Total Phosphorus g·kg^{-1}	Total Potassium g·kg^{-1}
May	CS	5.70 ± 0.48 a	551.17 ± 43.57 a	34.15 ± 0.63 a	4.60 ± 0.17 a	10.78 ± 1.10 a
	XS	5.73 ± 0.53 a	521.20 ± 13.67 a	29.32 ± 1.89 b	5.13 ± 0.1 a	8.80 ± 1.34 a
July	CS	5.67 ± 0.04 a	446.13 ± 3.53 a	18.40 ± 1.98 a	4.37 ± 0.41 a	9.90 ± 2.98 a
	XS	5.72 ± 0.09 a	439.39 ± 4.55 a	17.62 ± 2.14 a	3.17 ± 0.33 a	3.08 ± 0.58 b
October	CS	6.15 ± 0.03 a	489.20 ± 12.53 a	24.84 ± 2.57 a	6.82 ± 0.42 a	5.24 ± 0.03 a
	XS	5.97 ± 0.05 b	492.86 ± 2.89 a	23.83 ± 2.21 a	7.96 ± 0.54 a	4.56 ± 0.39 a

3.2. Differences of Endophytes in Leaves at Different Stages

3.2.1. OTU Distribution and Alpha Diversity

After high-throughput sequencing, at the 97% similarity classification level, 1597 bacterial OTUs were annotated (Figure 1a). There were 25 OTUs in the leaves of the male and female plants at the flowering stage and 24 OTUs at the fruit ripening stage. The number of unique OTUs in the leaves of the male and female plants in each period was XS10 > CS5 > CS10 > XS5.

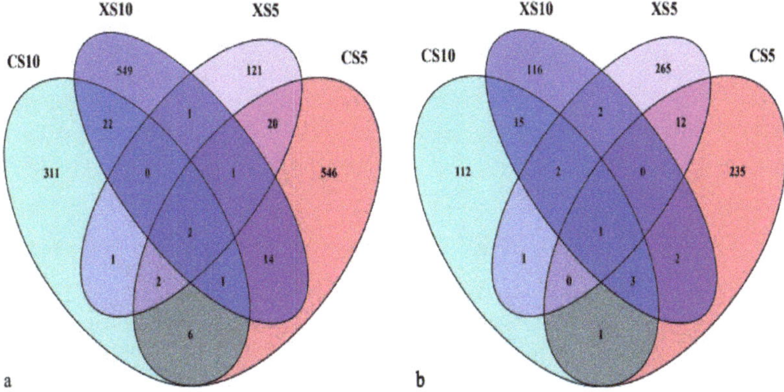

Figure 1. OTU Venn diagram of endophytes in *Idesia polycarpa* leaves. (**a**) Bacteria and (**b**) fungi. Abbreviations: CS10, leaves of the female plants in October; XS10, leaves of the male plants in October; XS5, leaves of the male plants in May; CS5, leaves of the female plants in May.

A total of 767 fungal OTUs were annotated (Figure 1b). There were 13 OTUs in the leaves of the male and female plants at the flowering stage and 21 OTUs at the fruit maturity stage. The number of unique OTUs in the leaves of the male and female plants in each period was XS5 > CS5 > XS10 > CS10. As a result, it was found that the endophytic composition in the leaves of the male and female plants changed significantly over time.

In the alpha diversity comparison, "Goods coverage" refers to the coverage of each sample library; the higher the value, the lower the probability that the sequence in the sample is not measured. "ACE richness" indicates the flora's richness; the smaller the value, the lower the richness. "Shannon and Simpson" indicates the flora's diversity; the smaller the value, the lower the community diversity. The sparse curve of the OTU richness of endophytic bacteria in the leaves of the male and female plants in different periods showed that the number of OTUs of bacteria and fungi in the leaves tended to be stable with the increase in sequencing depth (Figure 2). In addition, the coverage of bacteria and fungi was above 99% (goods coverage), indicating that the sequencing depth of the endophytic bacteria and fungi in the leaves of *Idesia polycarpa* met the requirements of the diversity analysis. There was no significant difference in the ACE index between male and female plants (Table 2). Shannon and Simpson's indices were significantly different at the stage of fruit development, indicating that the bacterial diversity in the leaves of the male and female plants increased significantly with their growth, but that the richness was not changed. There were no significant differences in the Shannon, Simpson, or ACE indices of fungi in the female leaves (Table 3). In contrast, the Shannon and Simpson of the male leaves showed no significant difference, but the ACE showed a significant difference.

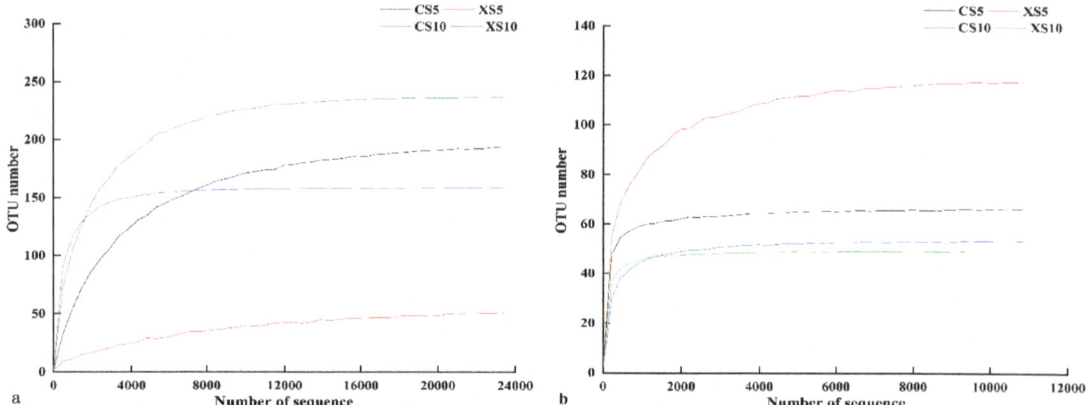

Figure 2. Dilution curve of endophytes in *Idesia polycarpa* leaves. (**a**) Bacteria and (**b**) fungi. Abbreviations: CS5, leaves of the female plants in May; XS5, leaves of the male plants in May; CS10, leaves of the female plants in October; XS10, leaves of the male plants in October.

3.2.2. Phylum Level Analysis of Leaf Endophytic Bacteria Community

A total of 20 phyla of endophytic bacteria were detected in the leaves of the male and female *Idesia polycarpa* plants at two growth stages (flowering and fruit ripening) (Figure 3a). Proteobacteria was the dominant group of endophytic bacteria in the male and female leaves, but the abundance was different in the leaves of each period. It was 95.70–99.76% in male and female leaves at the flowering stage and 68.09–74.87% at the fruit ripening stage. The abundance of Firmicutes (1.90%) and Bacteroidetes (1.37%) in the leaves of the female plants at the flowering stage was higher than 1.00%. The abundance of Actinobacteria (17.71%), Thermi (11.54%), and Firmicutes (1.40%) in the leaves of the female plants at the fruit ripening stage was higher than 1%. The abundance of Firmicutes (14.41%), Bacteroidetes (9.32%), and Actinobacteria (1.01%) was higher than 1.00% in the male leaves. Three phyla of endophytic fungi were detected (Figure 3b), among which Ascomycota was the dominant group, accounting for 26.31–26.97% in the leaves of the female and male plants at the flowering stage and 41.15–53.28% at the fruit ripening stage. Moreover, Basidiomycota was more than 1.00% abundant in the leaves of the male plants at both stages (flowering and fruit ripening), but only during the flowering stage in female plants.

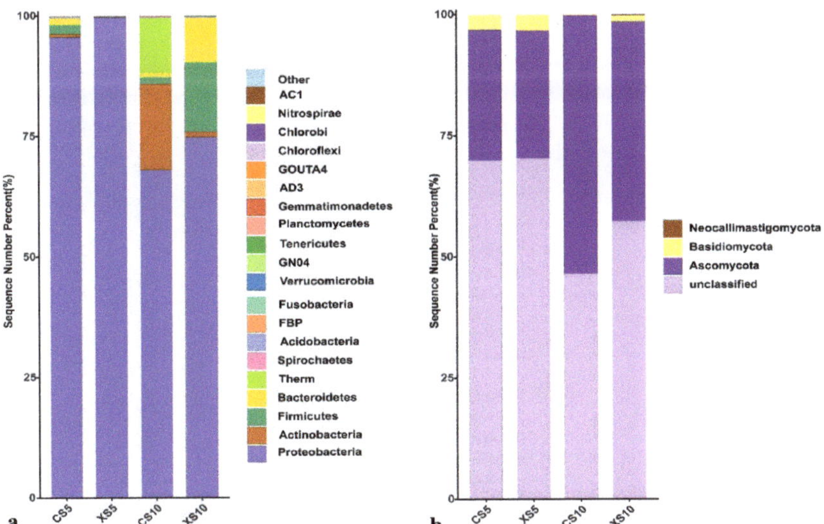

Figure 3. Relative abundance of endophytes in *Idesia polycarpa* leaves at the phylum level. (**a**) Bacteria and (**b**) fungi. Abbreviations: CS5, leaves of the female plants in May; XS5, leaves of the male plants in May; CS10, leaves of the female plants in October; XS10, leaves of the male plants in October.

Table 2. Leaf bacterial α diversity index. Different lowercase letters in the same column indicate significant differences. Abbreviations: CS5, leaves of the female plants in May; XS5, leaves of the male plants in May; CS10, leaves of female plants in October; XS10, leaves of the male plants in October.

Samples	Goods Coverage	Shannon	Simpson	ACE Richness
CS5	1.00 ± 0.00 a	1.44 ± 0.6 bc	0.30 ± 0.06 c	197.38 ± 144.84 a
XS5	1.00 ± 0.00 a	0.93 ± 0.06 c	0.27 ± 0.01 c	55 ± 13.05 a
CS10	1.00 ± 0.00 a	5.18 ± 0.17 a	0.92 ± 0.01 a	157 ± 6.03 a
XS10	1.00 ± 0.00 a	3.51 ± 1.23 ab	0.69 ± 0.12 b	236.67 ± 78.56 a

Table 3. Leaf fungal α diversity index. Different lowercase letters in the same column indicate significant differences. Abbreviations: CS5, leaves of the female plants in May; XS5, leaves of the male plants in May; CS10, leaves of female plants in October; XS10, leaves of the male plants in October.

Samples	Goods Coverage	Shannon	Simpson	ACE Richness
CS5	1.00 ± 0.00 a	5.12 ± 0.25 a	0.95 ± 0.01 a	67.33 ± 1.45 ab
XS5	1.00 ± 0.00 a	5.14 ± 0.57 a	0.95 ± 0.02 a	105.33 ± 25.67 a
CS10	1.00 ± 0.00 a	3.85 ± 0.59 a	0.82 ± 0.09 a	67 ± 13.53 ab
XS10	1.00 ± 0.00 a	4.53 ± 0.14 a	0.92 ± 0.02 a	49.33 ± 3.33 b

3.2.3. Genus Level Analysis of Leaf Endophytic Bacteria Community

At the genus level, a total of 20 genera of endophytic bacteria were detected in the leaves of the male and female plants at the flowering and fruit ripening stages (Figure 4a). During the flowering period, Pseudomonadaceae (58.01–85.86%) was the most abundant endophytic bacteria in the male and female leaves, followed by Ralstonia (6.38–29.91%) and Stenotrophomonas (4.23–6.90%). The dominant endophytic bacteria in the leaves of the female plants at the fruit ripening stage were Methylobacterium (40.66%), followed by Deinococcus (11.52%) and Bradyrhizobium (6.11%). The dominant endophytic bacteria in the leaves of the male plants were Ralstonia (50.67%), followed by Prevotella (3.91%) and Sphingomonas (3.08%).

In addition, a total of 19 genera of endophytic fungi were detected (Figure 4b). Aspergillus (38.75–40.35%), Penicillium (16.50–21.07%), Alternaria (11.39–12.66%), and Candida (8.29–9.34%) were the most abundant endophytic fungi in the male and female leaves at the flowering stage. Colletotrichum (35.67–55.47%), Alternaria (16.53–39.62%), Penicillium (3.08–28.63%), and Discosia (1.36–2.95%) were the most abundant endophytic fungi in the leaves of the male and female *Idesia polycarpa* plants at the fruit ripening stage.

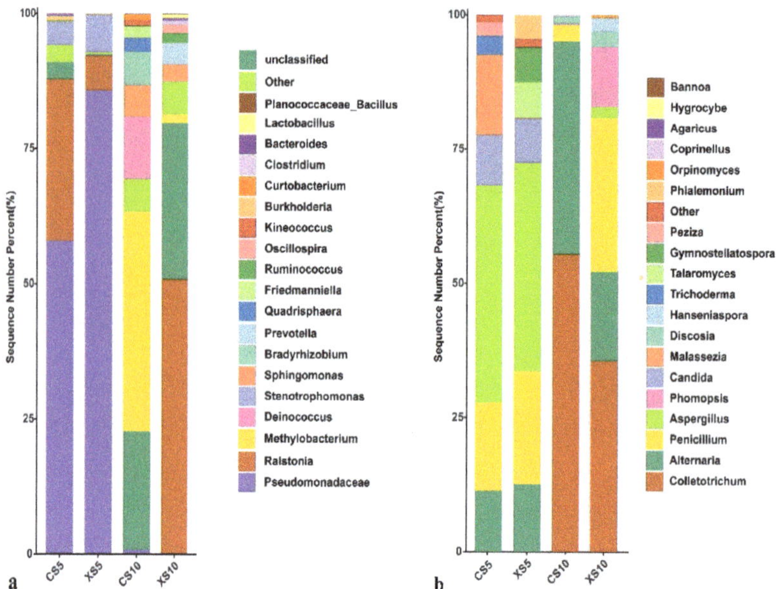

Figure 4. Relative abundance of endophytes in *Idesia polycarpa* leaves at the genus level. (**a**) Bacteria and (**b**) fungi. Abbreviations: CS5, leaves of the female plants in May; XS5, leaves of the male plants in May; CS10, leaves of the female plants in October; XS10, leaves of the male plants in October.

3.2.4. Difference Analysis of Endophytes in Leaves

To screen out the biological indicator species that can represent the microbial community characteristics of the male and female leaves in each period, LEfSe (LDA effect size) was used to analyze the differences in the abundance of endophytes in the leaves of *Idesia polycarpa* at different growth stages (Figure 5). The LEfSe analysis of the endophytic bacteria showed that there were eight differential indicator species in the XS5 samples, among which Gammaproteobacteria and Pseudomonadales containing Pseudomonas were the most significant groups. In the XS10 samples, the most significant taxonomic units were Oxalobacteraceae, Burkholderiales, and Betaproteobacteria. In the CS10 samples, 16 indicator species were identified, among which Alphaproteobacteria, Rhizobiales, and Methylobacterium accounted for the greatest percentage of significant differences. In the CS5 samples, there was no difference in indicator species.

The LEfSe analysis of endophytic fungi revealed four differential indicator species in the CS5 samples, mainly in Malasseziomycetes. In the XS5 samples, Eurotiomycetes and Aspergillus containing Aspergilllaceae were the most significant genera. The most distinct groups of indicator species in the CS10 samples were Sordariomycetes and Glomerellales containing Glomerellaceae. The above results showed significant differences in the composition of endophytic bacteria in the leaves of the male and female plants at different growth stages. Significant differences were found in the relative abundance of some dominant genera in different fields.

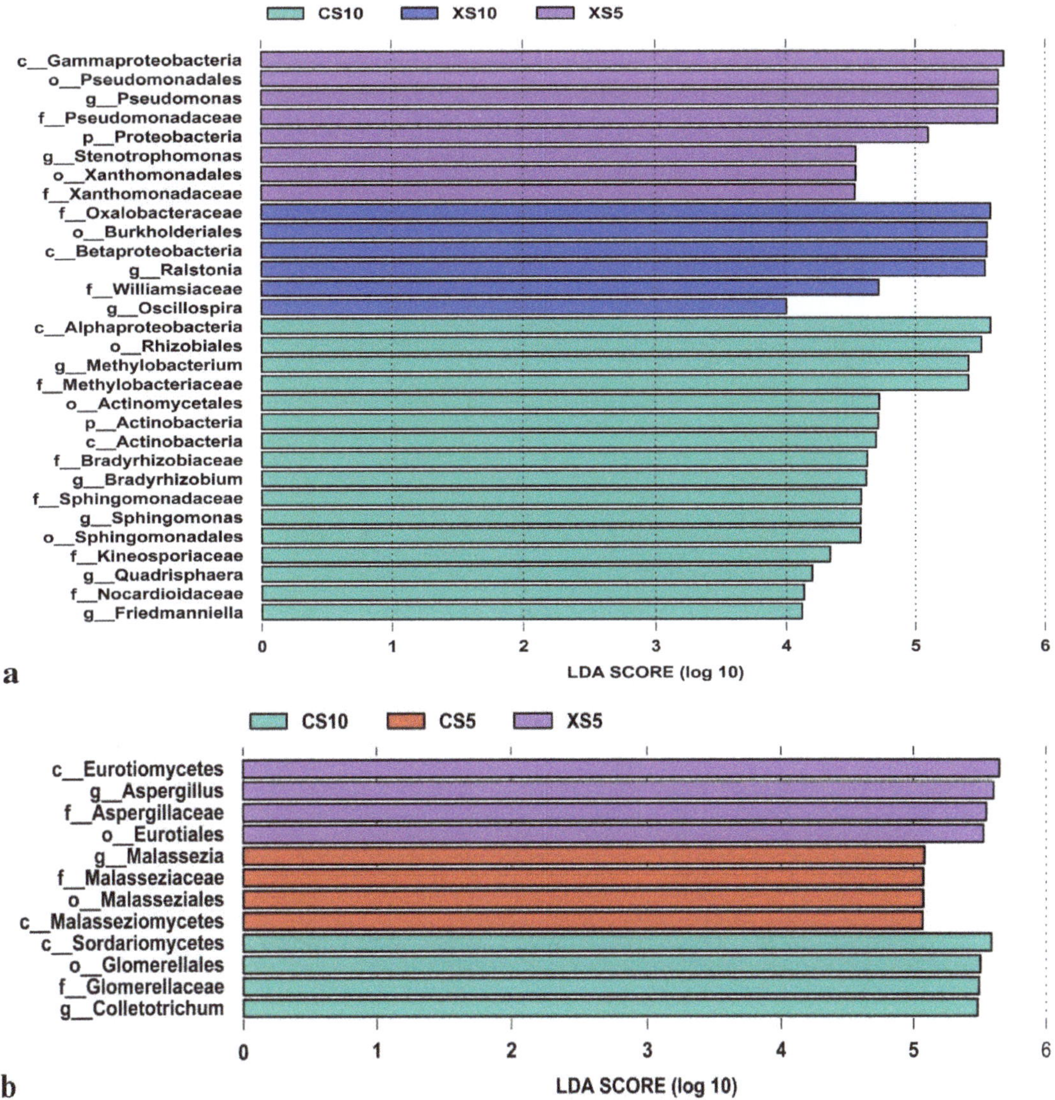

Figure 5. LDA value distribution histogram of endophytes in *Idesia polycarpa* leaves. (**a**) Bacteria and (**b**) fungi. Abbreviations: CS5, leaves of the female plants in May; XS5, leaves of the male plants in May; CS10, leaves of the female plants in October; XS10, leaves of the male plants in October.

3.3. Leaf Metabolite Difference

3.3.1. Sample Principal Component Analysis

A principal component analysis (PCA) was applied to all samples, and the results (Figure 6) showed that the contribution of the first principal component was 49.74%, and that the contribution of the second was 19.39%. Regarding the clustering within the leaf groups of the male and female *Idesia polycarpa* plants at different periods, the distance was small, and the distinction between the sample groups was obvious and distant. On the one hand, the results showed that the repeatability of the determination results was significant. In addition, the leaf metabolites of the male and female plants differed significantly over time.

The OPLS-DA method was used to analyze the differences in metabolites between the male and female leaves in different periods (Figure 7). The results showed that the separation effect between the male and female leaves was significant in two periods. Model reliability verification showed that the four models are stable, and the results are highly reliable.

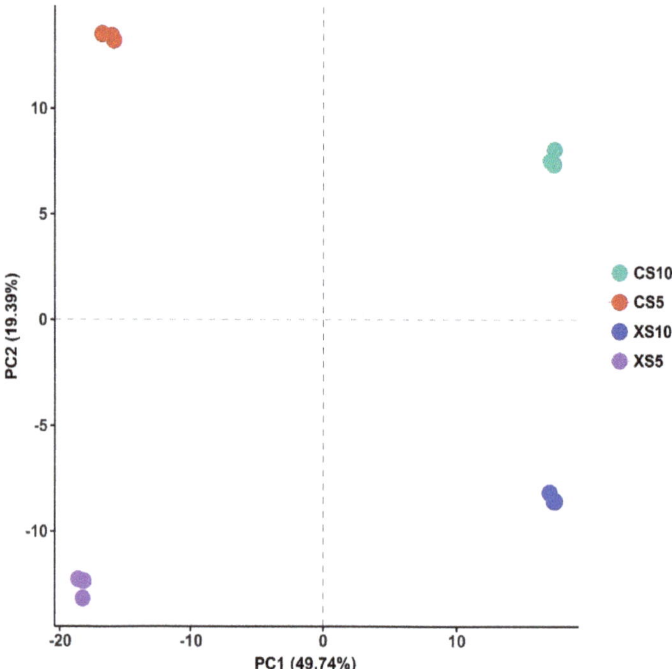

Figure 6. Principal component analysis (PCA) of the leaves. Abbreviations: CS5, leaves of the female plants in May; XS5, leaves of the male plants in May; CS10, leaves of the female plants in October; XS10, leaves of the male plants in October.

3.3.2. Screening of Differential Metabolites in Leaves

A total of 694 metabolites with significant differences were selected from the four comparison groups and were screened based on the p value (p-value < 0.05) of the t-test and the VIP (VIP > 1) of the OPLS-DA model, and most of them were phenolic acids. Each group produced 87 to 254 significantly different metabolites. The top five significantly different metabolites, ranked in descending order by the VIP values of the OPLS-DA model, are shown in Table 4. The abundance of salireposide in the leaves of the male plants was significantly higher than that in the leaves of the female plants at the flowering stage. However, it became significantly lower than that in the leaves of the female plants in October. Stearidonoyl-glycerol was significantly less abundant in the male and female plants during the fruit ripening stage than in the flowering stage.

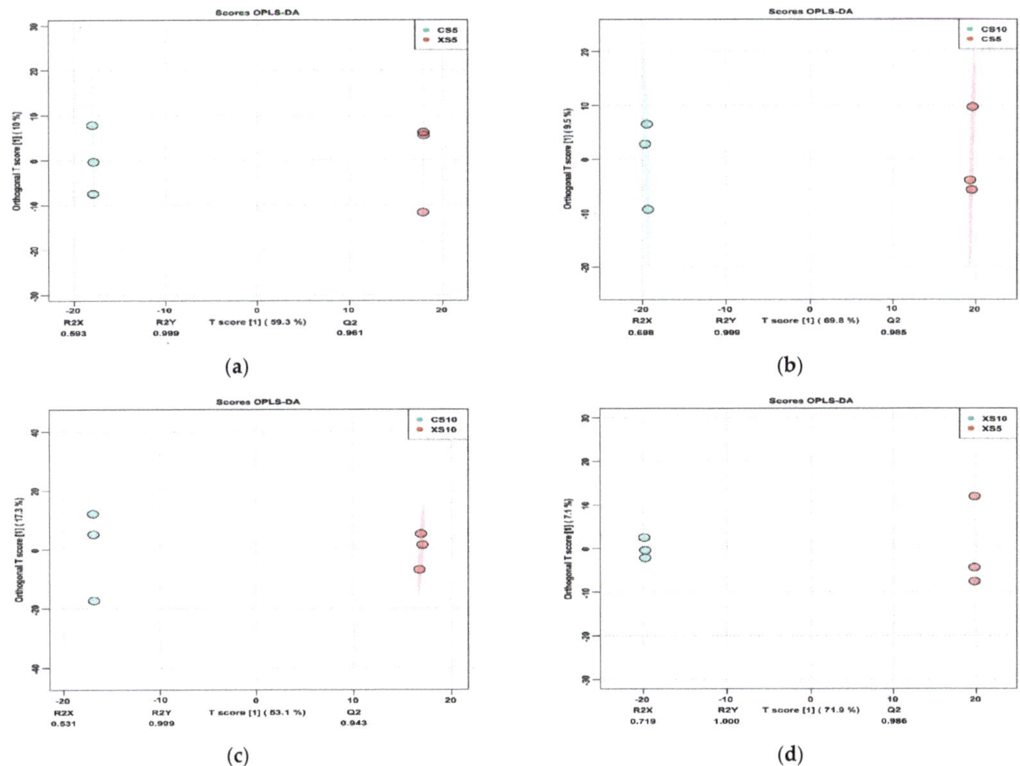

Figure 7. OPLS-DA diagram of differences in metabolites between the two groups of leaves. Abbreviations: (**a**) CS5, leaves of the female plants in May; XS5, leaves of the male plants in May; (**b**) CS10, leaves of the female plants in October; CS5, leaves of the female plants in May; (**c**) CS10, leaves of the female plants in October; XS10, leaves of the male plants in October; (**d**) XS5, leaves of the male plants in May; XS10, leaves of the male plants in October.

Table 4. Top five differential metabolites according to their VIP value in the four comparison groups.

Comparison Group	Total Number of Differential Metabolites	VIP Value Top Five Metabolites	Log2 FC	*p*-Value	VIP	Metabolite Types
CS5 vs. XS5	108	Cyclo (l-phe-l-pro)	9.671	0.000	1.288	Amino acids and derivatives
		2-Phenylethyl beta-D-glucopyranoside	13.186	0.001	1.288	Phenolic acids
		Salireposide	10.385	0.002	1.288	Phenolic acids
		4-O-(6′-O-glucosyl-p-coumaroyl)-4-hydroxybenzyl alcohol	3.356	0.000	1.288	Phenolic acids
		4-Ethoxyphenol	11.342	0.005	1.288	Phenolic acids
CS10 vs. XS10	87	3,4-Dimethoxyphenol	10.558	0.004	1.346	Phenolic acids
		8-Hydroxyguanosine	11.931	0.004	1.346	Nucleotides and derivatives
		Salireposide	−10.336	0.005	1.346	Phenolic acids
		Diethyl phosphate	9.431	0.004	1.346	Organic acids
		Cordycepin (3′-Deoxyadenosine)	10.013	0.005	1.346	Nucleotides and derivatives

Table 4. *Cont.*

Comparison Group	Total Number of Differential Metabolites	VIP Value Top Five Metabolites	Log2 FC	*p*-Value	VIP	Metabolite Types
CS5 vs. CS10	245	1-Stearidonoyl-glycerol	−15.663	0.000	1.175	Lipids
		Cyclo(l-phe-l-pro)	8.051	0.000	1.175	Amino acids and derivatives
		4-Ethoxyphenol	12.138	0.001	1.175	Phenolic acids
		Coniferaldehyde	−12.447	0.001	1.175	Phenolic acids
		Benzoic acid	11.595	0.001	1.175	Phenolic acids
XS5 vs. XS10	254	1-Stearidonoyl-glycerol	−12.498	0.000	1.166	Lipids
		4-(3,4,5-Trihydroxybenzoxy) benzoic acid	3.737	0.000	1.166	Phenolic acids
		1,18-Octadecanediol	−10.018	0.001	1.166	Lipids
		9,12-Octadecadien-6-ynoic acid	−11.759	0.002	1.166	Lipids
		4-Hydroxy-3-methoxymandelate	−14.586	0.003	1.166	Organic acids

3.3.3. Metabolic Pathway Analysis

To determine the mechanism by which the differential metabolites vary from one period to another in the leaves of the male and female *Idesia polycarpa*, we utilized the KEGG database for functional annotation and pathway enrichment analysis. The 108 differential metabolites screened in the CS5 vs. XS5 group were annotated to 34 metabolic pathways (Figure 8a). More metabolites were found in the metabolic pathways, followed by the biosynthesis of amino acids, the nucleotide metabolism, and the 2-oxocarboxylic acid metabolism. In the CS10 vs. XS10 group, the 87 differential metabolites were screened and annotated to 35 metabolic pathways (Figure 8b). More metabolites were concentrated in the metabolic pathways, followed by the ABC transporters and the nucleotide metabolism.

A total of 245 differential metabolites were identified in the CS5 vs. CS10 group and annotated to 128 pathways (Figure 8c). The insulin resistance was significantly enriched ($p < 0.05$). The differential metabolites involved in insulin resistance were D-fuctose6-phosphate*, uridine 5′-diphospho-N-acetylglucosamine, D-glucose 6-phosphate*, D-glucosamine 1-phosphate, and D-glucose*.

A total of 254 differential metabolites screened in the XS5 vs. XS10 group were annotated to 124 pathways (Figure 8d). The alpha-linolenic acid metabolism, phenylalanine metabolism, and linoleic acid metabolism were significantly enriched ($p < 0.01$), with 10, 11, and 13 differential metabolites involved, respectively. The tyrosine metabolism, ubiquinone, and other terpenoid–quinone biosynthesis pathways were significantly enriched ($p < 0.05$), and eight and four differential metabolites were involved, respectively. In addition, a large number of the differential metabolites of the male and female leaves were enriched. The common pathways were the metabolic pathways, the biosynthesis of amino acids, and the nucleotide metabolism.

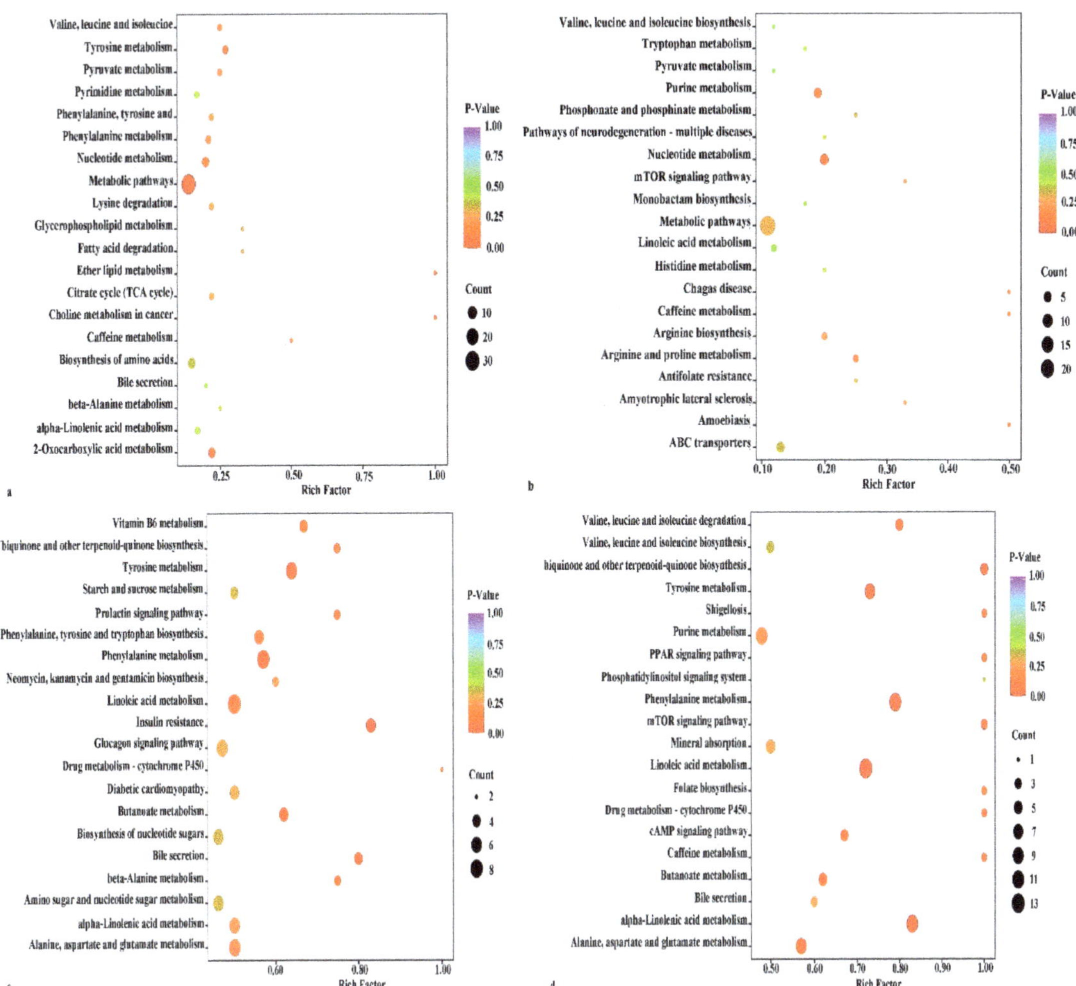

Figure 8. KEGG pathway map for leaf differential metabolites. Abbreviations: (**a**) CS5 vs. XS5, leaves of the female plants in May vs. leaves of the male plants in May; (**b**) CS10 vs. XS10, leaves of the female plants in October vs. leaves of the male plants in October; (**c**) CS5 vs. CS10, leaves of the female plants in May vs. leaves of the female plants in October; (**d**) XS5 vs. XS10, leaves of the male plants in May vs. leaves of the male plants in October.

3.4. Correlation Analysis of Leaf Nutrients with the Microbial Community and Differential Metabolites

Idesia polycarpa leaves exhibit significantly different metabolites and nutrient contents based on the correlation analysis of the top 30 endophytes. The results showed that the TK content in the leaves had the most significant effect on the endophytic bacteria in leaves (Figure 9a), and that these were significantly correlated ($p < 0.05/0.01/0.001$). The pH and total phosphorus content in the leaves had a negative correlation with Proteobacteria and a positive correlation with Actinobacteria. In addition, the total potassium and total nitrogen contents correlated negatively with Proteobacteria and positively with Actinobacteria. There was less correlation between the total carbon content in leaves and the endophytic bacteria. The TK content in the leaves significantly affected the endophytic fungi (Figure 9b) and showed a significant relationship. Colletotrichum was positively correlated with the TC, TN, and TK content in the leaves and negatively correlated with the pH and TP content, while the opposite was true for Aspergillus. The TK content of the leaves significantly affected the content of differential metabolites (Figure 9c). No significant relationship was found between the TC content and the differential metabolites in the leaves. Coniferaldehyde and 1-stearidonoyl-glycerol were positively correlated with TN and TK, and significantly negatively correlated with pH and TP, while the opposite was true for 4-ethoxyphenol. In general, the TK content in the leaves significantly affected the endophytic bacteria and differential metabolites.

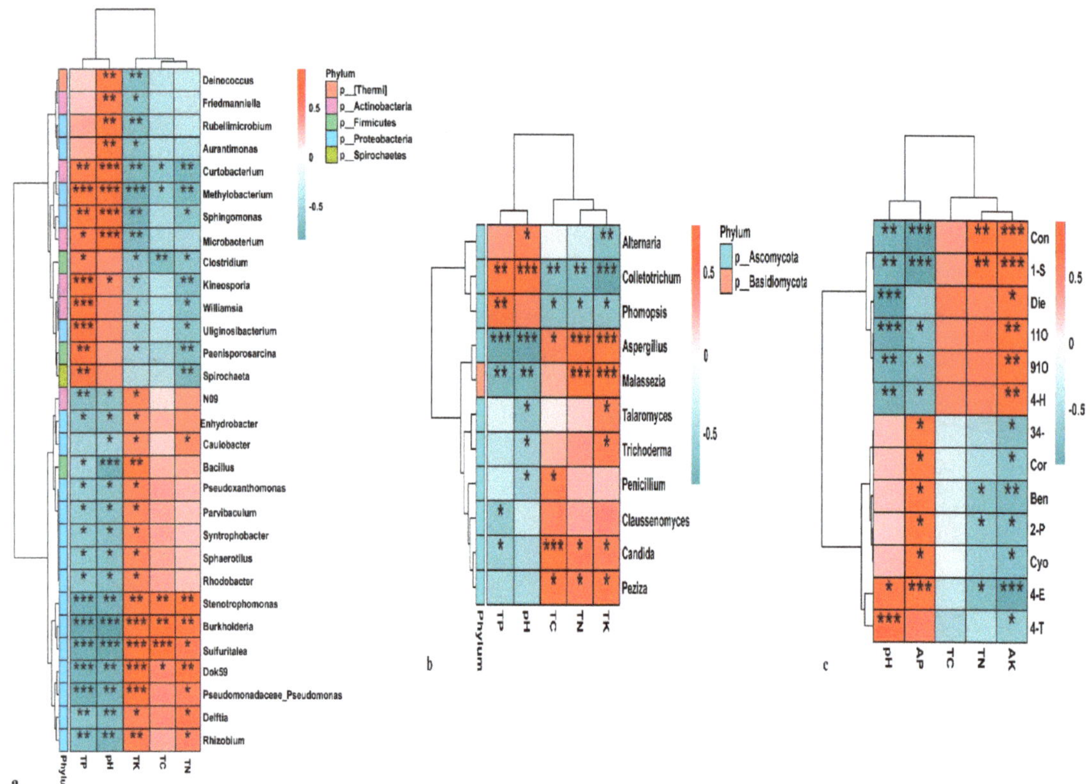

Figure 9. Correlation analysis of leaf nutrients with the microbial community and differential metabolites. (**a**) Bacteria. TP, total phosphorus; TK, total potassium; TC, total carbon, TN, total nitrogen. (**b**) Fungi. TP, total phosphorus; TK, total potassium; TC, total carbon, TN, total nitrogen. (**c**) Metabolites; TP, total phosphorus; TK, total potassium; TC, total carbon, TN, total nitrogen; Cyo, Cyclo(l-phe-l-pro); 2-P, 2-Phenylethyl beta-D-glucopyranoside; Sal, Salireposide; 4-O, 4-*O*-(6′-*O*-glucosyl-p-coumaroyl)-4-hydroxybenzyl alcohol; 4-E, 4-Ethoxyphenol; 3,4-, 3,4-Dimethoxyphenol; 8-H, 8-Hydroxyguanosine; Die, Diethyl phosphate; Cor, Cordycepin (3′-deoxyadenosine); 1-S, 1-Stearidonoyl-glycerol; Con, Coniferaldehyde; Ben, Benzoic acid; 4-T, 4-(3,4,5-Trihydroxybenzoxy)benzoic acid; 11O, 1,18-Octadecanediol; 91O, 9,12-Octadecadien-6-ynoic acid; 4-H, 4-Hydroxy-3-methoxymandelate. The single asterisk (*) mark represents $p < 0.05$ value; double asterisk (**) marks represent $p < 0.01$ value; and triple asterisk (***) marks represent $p < 0.001$ value.

4. Discussion

4.1. Nutrient Contents in Idesia Polycarpa Leaves

The growth and development of plants require nutrients such as carbon, the skeleton element of cells, and nitrogen and phosphorus are essential elements in plant photosynthesis [24]. There is a close correlation between the element characteristics of plant leaves and the basic behavior and functions of plants. The nutrient composition of leaves can also reflect the level of nutrient intake by plants [25]. In this study, there was no significant difference between the male and female plants at different growth stages in terms of TC or TP content. However, the TN content in the leaves of the female plants was significantly higher than that in the leaves of the male plants at the flowering stage. The ratio of carbon to phosphorus and nitrogen to phosphorus in leaves can reflect the ability of a plant to assimilate carbon [26]. The period after flowering is when the plants begin to reproduce. The female plant needs to pay higher reproductive costs than the male plant, which may maintain its consumption by increasing the leaf nitrogen content and enhancing its photosynthesis. Carbon is an important raw material for plant photosynthesis. There was no significant difference in carbon content between the male and female leaves, and the reasons for this need to be explored in more depth. The TK content in the leaves of the female plants was significantly higher than that in the leaves of the male plants at the fruit matter accumulation stage, and similar results were found by Fu Hao et al. [27] in *Zanthoxylum schinifolium*. Potassium can promote the transport of assimilates [28], and thus the accumulation of more potassium in female leaves can promote the accumulation of fruit matter and fruit development.

4.2. Endophytes in Idesia Polycarpa Leaves

Leaf endophytes widely exist in various plants and affect plant growth, development, and metabolic activities [29]. This paper used high-throughput sequencing technology to analyze the community characteristics of endophytic microorganisms in the leaves of male and female plants in different periods. The male and female plants produced different diversity indices in the same period and composed fewer common OTUs in their leaves. There were abundant and diverse microbial communities in the leaves of the male and female plants. Further analysis showed that Proteobacteria, Pseudomonadaceae, Ascomycota, and Aspergillus were the dominant bacteria and fungi in the leaves of *Idesia polycarpa*, and these are similar to the endophytic bacteria found in other plants [30]. Previous studies have shown that microorganisms such as Proteobacteria, Pseudomonadaceae, and Aspergillus play an important role in enhancing plant resistance, promoting plant growth, and preventing disease [31]. According to Liu et al. [32], phyllosphere bacteria differed at the genus level between male and female *Fraxinus chinensis*. We found differences in the composition of endophytes in the leaves of the male and female plants at different stages. At the flowering stage, the relative abundance of Ralstonia was higher in female leaves, while the relative abundance of Pseudomonadaceae was higher in male leaves. It is generally believed that Ralstonia is a pathogen that causes plant diseases [33]. Pseudomonadaceae can increase the diversity of microbial communities and enhance plant resistance by improving nutrient use efficiency [34]. During the fruit accumulation period, the relative abundance of Methylobacterium and Deinococcus in the leaves of the female plants was higher. Methylobacterium can secrete cytokinins to stimulate plant growth [35], and Deinococcus has a strong pollution removal ability [36]. The more beneficial endophytes enriched in the leaves of female plants can enhance plant resistance and ensure the normal development of fruit. The relative abundance of Ralstonia and Prevotella in the leaves of male plants was higher. Prevotella can produce antibiotics and enhance the chemical defense ability of leaves [37]. Plants and these endophytes are mutually beneficial. Plant leaves provide a stable environment and nutrients for endophytes, and endophytes help plants grow and resist adverse conditions. In addition, abundant beneficial endophytic bacteria in the leaves of *Idesia polycarpa* provide a good foundation for excavating functional strains.

4.3. Metabolites in Idesia polycarpa Leaves

The leaf is the main component and a crucial vegetative organ of plants, and its metabolite data can reflect the physiological characteristics of the whole plant [38]. This paper found 694 metabolites with significant differences from the leaves of male and female plants in different periods. There were 195 significantly different metabolites between the male and female leaves, mainly phenolic acids, indicating that the leaves of the male and female plants meet their own needs and adjust phenolic acids accordingly. Compared with the female plants, the phenolic acid content in the leaves of the male plants increased at the flowering and fruit ripening stages. Rabska Mariola et al. [39] conducted a pot experiment on *Juniperus communis* L and found that under abiotic stress, the concentration of phenolic compounds in the leaves of male and female *Juniperus communis* L plants decreased, and that the concentration of phenolic compounds was higher in the female plants than in the male plants. Zhang et al. [40] confirmed that the decrease in the number of phenolic compounds in the leaves of *Populus yunnanensis* under drought stress was greater in female plants than in male plants. The climate and water might be the reason for the difference in phenolic compounds in the leaves of the male and female plants. Phenolic acids have a variety of physiological functions, such as anti-oxidation, regulation of enzyme activity, and resistance to pathogens [41]. This shows that the resistance of male plants is stronger than that of female plants, which may be related to the fact that female plants use more substances for reproduction. Using the leaves at the flowering stage as a control, the lipid content of the leaves at the fruit ripening stage was observed to decreased significantly. Lipids are crucial for plant energy conversion, carbon storage, signal transduction, and stress responses [42]. This is consistent with the growth habit and environment of *Idesia polycarpa* at the reproductive stage.

4.4. Leaf Nutrients with the Microbial Community and Differential Metabolites

It has been found that the dynamic changes in the endophytic bacteria community and metabolite content in plants were affected by many factors, including plant species [43], plant growth, and development stage [44]. Leaves are the environment for the survival of endophytes and the basis for the existence of metabolites. Endophyte composition and metabolite content are closely related to the nutritional content of the plant. Chen et al. [45] found that nitrogen, phosphorus, and potassium in the environment had a significant positive effect on the relative abundance of *Sebacina* sp. Zhang et al. [46] believed that an appropriate amount of phosphate fertilizer could increase the content of chlorogenic acid and total flavonoids in Chrysanthemi indici Flos. As the basic raw material of organic matter, carbon is closely related to the synthesis and metabolism of organic matter. However, this paper has demonstrated no significant correlation between the total carbon content and the differential metabolites in the leaves of *Idesia polycarpa* plants, indicating the efficient utilization of carbon in the leaves of the male and female plants. The TK content in leaves was significantly related to the abundance of most endophytic bacteria and the content of metabolites. This indicated that the potassium content in the leaves played an essential role in constructing the internal environment of leaves, which was also consistent with the view that potassium could improve plant resistance. In addition, the total phosphorus content in the leaves had a positive correlation with Actinobacteria and phenolic acids and a negative correlation with Ascomycota. Phosphate can promote plant growth and an increase in dry-weight biomass [47], and one of the ways it does this might be by increasing the abundance of beneficial endophytes and enhancing the level of metabolism. In production, the richness and diversity of endophytes, the contents of beneficial microorganisms, and the resistant substances in the leaves of *Idesia polycarpa* could be increased by using microbial potassium and phosphorus fertilizers, thus promoting the growth of the *Idesia polycarpa* species.

5. Conclusions

In this study, the nutrient characteristics, endophytes, and metabolites were analyzed in the leaves of male and female Idesia polycarpa at different reproductive stages. The results revealed differences in the absorption of nitrogen and potassium between the male and female plants. We identified significantly different metabolites in the leaves of the female and male plants based on their diversity and endophyte and phenolic acid compositions. Through correlation, it was found that the leaf endophytes and significantly different metabolites were most affected by the content of TK and TP in the leaves, but were less affected by TC and TN content. For future use, this study provides a reference for separating endophytic bacteria, excavating functional strains, developing and utilizing leaves, and managing production.

Author Contributions: Conceptualization, J.F., S.R. and Z.L. (Zhi Li); methodology, J.F., S.R. and Z.L. (Zhi Li); software, J.F.; validation, J.F., Z.L. (Zhen Liu) and Z.L. (Zhi Li); formal analysis, J.F., H.Z., T.Z., S.W., X.X., M.L. and R.S.J.; investigation, J.F.; resources, Z.L. (Zhi Li); data curation, J.F.; writing—original draft preparation, J.F. and Z.L. (Zhi Li); writing—review and editing, S.R. and Z.L. (Zhi Li); visualization, Z.L. (Zhi Li), Z.L. (Zhen Liu), Y.W., Q.C. and X.G.; supervision, Z.L. (Zhi Li), Z.L. (Zhen Liu), Y.W., Q.C. and X.G.; project administration, Z.L. (Zhi Li), Z.L. (Zhen Liu), Y.W., Q.C. and X.G.; funding acquisition, Z.L. (Zhi Li) and Z.L. (Zhen Liu). All authors have read and agreed to the published version of the manuscript.

Funding: This research was funded by the Henan Province Postdoctoral Research Project of China (202002053); the Annual Project of Philosophy and Social Science Planning of Henan Province, China (2021BSH006); the Key Forestry Science and Technology Promotion Project of China Central Government (GTH[2020]17); the National Natural Science Foundation of China (32000267).

Institutional Review Board Statement: Not applicable.

Informed Consent Statement: Not applicable.

Data Availability Statement: The data sets generated and/or analyzed during the current study are available from the corresponding author upon reasonable request.

Acknowledgments: We are grateful to the reviewers for their constructive comments and valuable suggestions which have helped improve the quality of the paper. We also thank all our laboratory members for their help in this experiment.

Conflicts of Interest: The authors declare no conflict of interest.

References

1. Slate, M.L.; Rosenstiel, T.N.; Eppley, S.M. Sex-Specific Morphological and Physiological Differences in the Moss *Ceratodon purpureus* (Dicranales). *Ann. Bot.* **2017**, *120*, 845–854. [CrossRef] [PubMed]
2. Li, G.; Wen, G.; Zhang, M.; Zhou, Z. Study on the Differences of Eco-Physiological Characteristics of Male and Female *Rhus typhina*. *For. Res.* **2013**, *26*, 263–268.
3. Rana, S.; Liu, Z. Study on the Pattern of Vegetative Growth in Young Dioecious Trees of *Idesia polycarpa* Maxim. *Trees—Struct. Funct.* **2021**, *35*, 69–80. [CrossRef]
4. Kader, A.; Sinha, S.N. Sex-Related Differences of *Excoecaria agallocha* L. with a View to Defence and Growth. *Trop. Life Sci. Res.* **2022**, *33*, 55–74. [CrossRef] [PubMed]
5. Ho, H.; Low, J.Z.; Gudimella, R.; Tammi, M.T.; Harikrishna, J.A. Expression Patterns of Inflorescence- and Sex-Specific Transcripts in Male and Female Inflorescences of African Oil Palm (*Elaeis guineensis*). *Ann. Appl. Biol.* **2016**, *168*, 274–289. [CrossRef]
6. Zhang, Y.Q.; Guo, Q.Q.; Luo, S.Q.; Pan, J.W.; Yao, S.; Guo, Y.Y. Stoichiometry Characteristics of Leaves and Soil of Four Shrubs under *Pinus massoniana* Plantations. *Cent. South Univ. For. Technol.* **2022**, *42*, 129–137.
7. Ma, Y.; Zhang, S.; Wu, Z.; Sun, W. Metabolic Variations in Brown Rice Fertilised with Different Levels of Nitrogen. *Foods* **2022**, *11*, 3539. [CrossRef]
8. Sun, X.W.; Wang, X.C.; Quan, X.K.; Yang, Q.J.; Sun, H.Z. Photosynthetic Characteristics of Dioecious *Populus davidiana*, *Fraxinus mandshurica* and *Taxus cuspidata*. *J. Nanjing For. Univ.* **2022**, 1–12. Available online: http://kns.cnki.net/kcms/detail/32.1161.S.20 220409.1715.002.html (accessed on 30 November 2022).
9. Song, A.Y.; Dong, L.S.; Chen, J.X.; Peng, L.; Liu, J.T.; Xia, J.B.; Chen, Y.P. Comparison of Seasonal Dynamics of Mineral Elements Contents in Different Organs of Male and Female Plants of *Fraxinus velutina*. *Sci. Silvae Sin.* **2019**, *55*, 162–170.
10. Zuo, C.G.; Wang, J.Y.; Niu, X.X.; Liu, P.; Guan, L.H.; Dang, W.F.; Yang, H.M.; Chu, M.; Wang, N.; Lin, Q.; et al. Effects of Endophytes and Rhizosphere Bacteria on Cotton Growth Promotion and Disease Resistance Induction. *Southwest China J. Agric. Sci.* **2022**, *35*, 757–763.
11. Abdul, L.K.; Boshra, A.H.; Ali, E.; Sajid, A.; Khadija, A.; Javid, H.; Ahmed, A.; Lee, I. Indole Acetic Acid and ACC Deaminase from Endophytic Bacteria Improves the Growth of *Solanum lycopersicum*. *EJB Electron. J. Biotechnol.* **2016**, *21*, 58–64.

12. Gou, Q.; Lv, Y.; Zhang, T.; Li, J.Y.; Zhao, H.J.; Liu, J.L. Dynamics and Influencing Factors of Endophytic Bacterial Community in Leaves of *Lycium barbarum* during Different Growth Periods. *Chin. J. Ecol.* **2020**, *39*, 2593–2601.

13. Zhang, W.Y.; Yan, L.; Lai, X.J.; Duan, W.J.; Zhang, Y.Z. Analysis of Microbial Diversity and Metabolome of Flue-Cured Tobacco with Different Fermentation Time. *Acta Tab. Sin.* **2022**, *28*, 84–94.

14. Xu, G.Q.; Liu, Y.X.; Cao, P.X.; Ji, Y.L.; Li, J.K.; Li, X.Y.; Liu, X. Endophytes Diversity of *Oxytropis glacialis* at Different Tissues Based on Illumina MiSeq Sequencing. *Acta Ecol. Sin.* **2021**, *41*, 4993–5003.

15. Ren, T.Y.; Han, Y.; Zhen, P.; Li, Y.Q.; Wang, X.Y.; Yan, F.X. Study on the Endophytes of Different Varieties of Barley. *Liquor-Mak. Sci. Technol.* **2022**, *4*, 22–29.

16. Rattanawiwatpong, P.; Wanitphakdeedecha, R.; Bumrungpert, A.; Maiprasert, M. Anti-Aging and Brightening Effects of a Topical Treatment Containing Vitamin C, Vitamin E, and Raspberry Leaf Cell Culture Extract: A Split-Face, Randomized Controlled Trial. *J. Cosmet. Dermatol.* **2020**, *19*, 671–676. [CrossRef]

17. Staszowska-Karkut, M.; Materska, M. Phenolic Composition, Mineral Content, and Beneficial Bioactivities of Leaf Extracts from Black Currant (*Ribes nigrum* L.), Raspberry (*Rubus idaeus*), and Aronia (*Aronia melanocarpa*). *Nutrients* **2020**, *12*, 463. [CrossRef]

18. Rong, B.B.; Xu, G.Q.; Wang, L.H. Rapid Extraction of Active Components from *Arctium lappa* Leaves and Its Anti-Fungal Properties of Wood Rot Fungi. *J. Beijing For. Univ.* **2017**, *39*, 109–116.

19. Wang, H.; Rana, S.; Li, Z.; Geng, X.; Wang, Y.; Cai, Q.; Li, S.; Sun, J.; Liu, Z. Morphological and Anatomical Changes during Floral Bud Development of the Trioecious *Idesia polycarpa* Maxim. *Rev. Bras. Bot.* **2022**, *45*, 679–688. [CrossRef]

20. Wen, L.; Xiang, X.; Wang, Z.; Yang, Q.; Guo, Z.; Huang, P.; Mao, J.; An, X.; Kan, J. Evaluation of Cultivars Diversity and Lipid Composition Properties of *Idesia polycarpa* Var. *vestita* Diels. *J. Food Sci.* **2022**, *87*, 3841–3855. [CrossRef]

21. Yang, F.X.; Su, Y.Q.; Li, X.H.; Zhang, Q.; Sun, R.C. Preparation of Biodiesel from *Idesia polycarpa* Var. *vestita* Fruit Oil. *Ind. Crops Prod.* **2009**, *29*, 622–628. [CrossRef]

22. Rana, S.; Jemim, R.S.; Li, Z.; Geng, X.; Wang, Y.; Cai, Q.; Liu, Z. Study of the Pattern of Reproductive Allocation and Fruit Development in Young Dioecious Trees of *Idesia polycarpa* Maxim. *S. Afr. J. Bot.* **2022**, *146*, 472–480. [CrossRef]

23. Bao, S.D. *Soil Agrochemical Analysis*; China Agriculture Press: Beijing, China, 2000; pp. 263–270.

24. Niklas, K.J.; Cobb, E.D. N, P, and C Stoichiometry of *Eranthis hyemalis* (Ranunculaceae) and the Allometry of Plant Growth. *Am. J. Bot.* **2005**, *92*, 1256–1263. [CrossRef] [PubMed]

25. Manivannan, S.; Chadha, K.L. Standardization of Time of Sampling for Leaf Nutrient Diagnosis on Kinnow Mandarin in North-West India. *J. Plant Nutr.* **2011**, *34*, 1820–1827. [CrossRef]

26. Wang, M.; Murphy, M.T.; Moore, T.R. Nutrient Resorption of Two Evergreen Shrubs in Response to Long term Fertilization in a Bog. *Oecologia* **2014**, *174*, 365–377. [CrossRef]

27. Fu, H.; Ren, Y.; Chen, Z.X.; Shi, Y.L.; Chen, C.; Sun, G.Q.; Zhou, Y.Q.; Li, Q. Nutrient Differences Between Male and Female Zanthoxylum Schinifolium Sieb. et Zucc. and Their Effects on Soil Nutrient. *Non-Wood. For. Res.* **2022**, *40*, 192–199.

28. Mostofa, M.G.; Rahman, M.M.; Ghosh, T.K.; Kabir, A.H.; Abdelrahman, M.; Rahman Khan, M.A.; Mochida, K.; Tran, L.S.P. Potassium in Plant Physiological Adaptation to Abiotic Stresses. *Plant Physiol. Biochem.* **2022**, *186*, 279–289. [CrossRef]

29. Xiong, C.; Singh, B.K.; He, J.Z.; Han, Y.L.; Li, P.P.; Wan, L.H.; Meng, G.Z.; Liu, S.Y.; Wang, J.T.; Wu, C.F.; et al. Plant Developmental Stage Drives the Differentiation in Ecological Role of the Maize Microbiome. *Microbiome* **2021**, *9*, 171. [CrossRef]

30. Trivedi, P.; Leach, J.E.; Tringe, S.G.; Sa, T.; Singh, B.K. Plant–Microbiome Interactions: From Community Assembly to Plant Health. *Nat. Rev. Microbiol.* **2020**, *18*, 607–621. [CrossRef]

31. Sun, Z.B.; Li, H.Y.; Lin, Z.A.; Yuan, L.; Xu, J.K.; Zhang, S.Q.; Li, Y.T.; Zhao, B.Q.; Wen, Y.C. Long-Term Fertilisation Regimes Influence the Diversity and Community of Wheat Leaf Bacterial Endophytes. *Ann. Appl. Biol.* **2021**, *179*, 176–184. [CrossRef]

32. Liu, R.; Yu, W.K.; Hu, S.X.; Shang, S. Microbial Community Structure in Leaf Space of Female and Male Fraxinus Chinensis in Yellow River Delta. *J. Anhui Sci. Technol. Univ.* **2022**, *36*, 35–39.

33. Shi, H.; Xu, P.W.; Yu, W.; Cheng, Y.Z.; Ding, A.M.; Wang, W.F.; Wu, S.X.; Sun, Y.H. Metabolomic and Transcriptomic Analysis of Roots of Tobacco Varieties Resistant and Susceptible to Bacterial Wilt. *Genomics* **2022**, *114*, 110471. [CrossRef] [PubMed]

34. Tsegaye, Z.; Alemu, T.; Desta, F.A.; Assefa, F. Plant Growth-Promoting Rhizobacterial Inoculation to Improve Growth, Yield, and Grain Nutrient Uptake of Teff Varieties. *Front. Microbiol.* **2022**, *13*, 896770. [CrossRef] [PubMed]

35. Vanacore, A.; Forgione, M.C.; Cavasso, D.; Nguyen, H.N.A.; Molinaro, A.; Saenz, J.P.; D'Errico, G.; Paduano, L.; Marchetti, R.; Silipo, A. Role of EPS in Mitigation of Plant Abiotic Stress: The Case of *Methylobacterium extorquens* PA1. *Carbohydr. Polym.* **2022**, *295*, 119863. [CrossRef] [PubMed]

36. Agapov, A.A.; Kulbachinskiy, A.V. Mechanisms of Stress Resistance and Gene Regulation in the Radioresistant Bacterium *Deinococcus radiodurans*. *Biochemistry* **2015**, *80*, 1201–1216. [CrossRef]

37. Zaidi, S.S.E.A.; Mahfouz, M.M.; Mansoor, S. CRISPR-Cpf1: A New Tool for Plant Genome Editing. *Trends Plant Sci.* **2017**, *22*, 550–553. [CrossRef]

38. Wu, Y.; Li, S.; Wu, K.X.; Mu, L.Q. Differences in Leaf Metabolism of Wild *Rosa acicularis* and *Rosa acicularis* 'Luhe' Based on GC-MS. *Bull. Bot. Res.* **2022**, *42*, 1070–1078.

39. Rabska, M.; Pers-Kamczyc, E.; Żytkowiak, R.; Adamczyk, D.; Iszkuło, G. Sexual Dimorphism in the Chemical Composition of Male and Female in the Dioecious Tree, *Juniperus communis* L., Growing under Different Nutritional Conditions. *Int. J. Mol. Sci.* **2020**, *21*, 8094. [CrossRef]

40. Zhang, R.; Liu, J.; Liu, Q.; He, H.; Xu, X.; Dong, T. Sexual differences in growth and defence of *Populus yunnanensis* under drought stress. *Can. J. Res.* **2019**, *49*, 491–499. [CrossRef]

41. Aouadi, M.; Escribano-Bailón, M.T.; Guenni, K.; Salhi Hannachi, A.; Dueñas, M. Qualitative and Quantitative Analyses of Phenolic Compounds by HPLC–DAD–ESI/MS in Tunisian *Pistacia vera* L. Leaves Unveiled a Rich Source of Phenolic Compounds with a Significant Antioxidant Potential. *J. Food Meas. Charact.* **2019**, *13*, 2448–2460. [CrossRef]

42. Zhen, J.B.; Song, S.J.; Liu, L.L.; Liu, D.; Ouyang, Y.F.; Chi, J.N. Comparative Analysis of Differential Metabolites in Flowers of *Gossypium hirsutum* L. and *Aurea helianthus*. *J. China Agric. Univ.* **2021**, *26*, 20–33.

43. Nasslahsen, B.; Prin, Y.; Ferhout, H.; Smouni, A.; Duponnois, R. Management of Plant Beneficial Fungal Endophytes to Improve the Performance of Agroecological Practices. *J. Fungi* **2022**, *8*, 1087. [CrossRef] [PubMed]

44. Wu, J.S.G.L.; Kang, P.; Hu, J.P.; Pan, Y.Q.; Zhou, Y.; Ma, R.; Peng, Y.Y.; Liu, J.Y. Difference in Endophyte Community Structure Between Young and Mature Branches of *Artemisia ordosica*. *Microbiol. China* **2022**, *49*, 569–582.

45. Chen, J.; Wu, Y.; Zhuang, X.; Guo, J.; Hu, X.; Xiao, J. Diversity Analysis of Leaf Endophytic Fungi and Rhizosphere Soil Fungi of Korean Epimedium at Different Growth Stages. *Environ. Microbiomes* **2022**, *17*, 1–14.

46. Zhang, J.H.; Feng, B.B.; Xu, X.Y. Effect of Combined Fertilization of Nitrogen, Phosphorus and Potassium on Intrinsic Quality of *Chrysanthemi indici* Flos. *J. Henan Agric. Sci.* **2013**, *42*, 92–97.

47. Gao, J.N.; Zhang, D.; Yang, M.N.; Chen, Y.L. Effects of Combined Application of Nitrogen, Phosphorus and Potassium on Biomass and Quality Components Content of Tobacco Leaves. *Southwest China J. Agric. Sci.* **2021**, *34*, 1269–1276.

Article

Epidemic of Wheat Stripe Rust Detected by Hyperspectral Remote Sensing and Its Potential Correlation with Soil Nitrogen during Latent Period

Jing Chen [1,2,†] , Ainisai Saimi [1,†], Minghao Zhang [1], Qi Liu [1,2,*] and Zhanhong Ma [2,*]

[1] Department of Plant Pathology, College of Agronomy, Xinjiang Agricultural University, Urumqi 830052, China
[2] Department of Plant Pathology, College of Plant Protection, China Agricultural University, Beijing 100193, China
* Correspondence: liuqi@xjau.edu.cn (Q.L.); mazh@cau.edu.cn (Z.M.)
† These authors contributed equally to this work.

Abstract: Climate change affects crops development, pathogens survival rates and pathogenicity, leading to more severe disease epidemics. There are few reports on early, simple, large-scale quantitative detection technology for wheat diseases against climate change. A new technique for detecting wheat stripe rust (WSR) during the latent period based on hyperspectral technology is proposed. Canopy hyperspectral data of WSR was obtained; meanwhile, duplex PCR was used to measure the content of *Puccinia striiformis* f.sp. *tritici* (*Pst*) in the same canopy section. The content of *Pst* corresponded to its spectrum as the classification label of the model, which is established by discriminant partial least squares (DPLS) and support vector machine (SVM) algorithm. In the spectral region of 325–1075 nm, the model's average recognition accuracy was between 75% and 80%. In the sub-band of 325–1075 nm, the average recognition accuracy of the DPLS was 80% within the 325–474 nm. The average recognition accuracy of the SVM was 83% within the 475–624 nm. Correlation analysis showed that the disease index of WSR was positively correlated with soil nitrogen nutrition, indicating that the soil nitrogen nutrition would affect the severity of WSR during the latent period.

Keywords: climate change; wheat stripe rust; hyperspectral remote sensing; identification model; soil nitrogen

Citation: Chen, J.; Saimi, A.; Zhang, M.; Liu, Q.; Ma, Z. Epidemic of Wheat Stripe Rust Detected by Hyperspectral Remote Sensing and Its Potential Correlation with Soil Nitrogen during Latent Period. *Life* **2022**, *12*, 1377. https://doi.org/10.3390/life12091377

Academic Editors: Othmane Merah and Angel Llamas

Received: 26 July 2022
Accepted: 30 August 2022
Published: 5 September 2022

Publisher's Note: MDPI stays neutral with regard to jurisdictional claims in published maps and institutional affiliations.

1. Introduction

The impact of climate change on agriculture is multi-level and multi-scale. It can affect temperatures, precipitation, climate extremes, atmospheric CO_2, crop yield, products nutritional quality and plant pests and diseases [1]. For example, climate warming not only shortens the growth period of crops and reduces the yield of corn and wheat, it also helps to improve the survival rate, fecundity, and pathogenicity of pathogens [2]. Plant diseases are one of the major threats to agriculture, which impact food production and natural systems, especially under the influence of human activities (agronomic practices, plant material movement) and climate change [3]. Plant diseases are the result of interaction between pathogens, host plants and the environment; this interaction is a continuous process referred to as the disease cycle. The quantification of the relationship between the disease cycle and weather for a given plant disease is the basis for plant disease prediction models that can be used to predict the timing and severity of plant disease in the future [4], that is, disease prediction models have also simulated potential impacts of climate change [5]. Therefore, climate change has a great influence on the plant disease risk.

Pathogens and host plants produce a series of interactions under environments, especially under climate change, and they will produce different response mechanisms. For example, WSR resistance gene *Yr39* is activated by both wheat developmental stages and climate changes [5]; climate warming leads to the expansion of wheat planting areas and the suitable range of rust fungi (*Pucciniales*). Spore germination and colonization depends on the temperature and humidity, the increase of winter temperature make for the spore overwintering, which is in favor of pathogen reproduction and causes the disease spread earlier and more serious yield loss in the following year [2]. At the same time, a suitable climate is more conducive to the pathogens reproduction and spread, but it is difficult to control the pathogens, especially in areas with high levels of inter-annual variability in climate [6]. In addition to climatic factors, soil nutrients are also an important factor affecting the occurrence of diseases. Some studies have pointed out that nitrogen application will aggravate the occurrence of wheat powdery mildew and wheat scab. Excessive nitrogen application leads to an increase in free amino acids, amides and soluble sugars, a decrease in total phenols, flavonoids, and peroxidase activities in plants, and affects the epidermal structure and metabolic activity of the host leaf [7], thereby weakening crop disease resistance and aggravating the crop disease. Meanwhile, excess nitrogen nutrition provides a favorable microclimate for the invasion, development and spread of pathogens [8]. Therefore, climate warming and soil nutrients have created favorable conditions for the outbreak of wheat diseases.

Wheat stripe rust (WSR) is caused by the fungus *Puccinia striiformis* f.sp. *tritici* (*Pst*) [9] that is widely distributed in major wheat-producing regions of the world [10–12]. In China, WSR causes an estimated 13.88 billion kg of loss every year [13]. *Pst* is a living obligate parasite that can spread over long distances with atmospheric pathways [14]. The infection process of *Pst* is divided into a contact period, an invasion period, a latent period, and a symptom period. The latent period is an important stage during which *Pst* multiplies and spreads in the host, and it cannot be directly perceived with the naked eye. Although the wheat does not show symptoms at this time, the parasitic relationship between the pathogen and the wheat will significantly affect the cell internal structure, pigment content, and water content of the wheat [15]. After the latent period, the disease will enter the symptom period in favorable conditions, and the fungus will rapidly expand, causing serious damage for production [16]. The duration of the latent period is one of the most important indicators for evaluating the resistance of crop cultivars [17], and it is also an important parameter in the epidemics of diseases [18]. Early detection and estimation of the dynamic of WSR during the latent period under climate change, could be obtained with more time to take prevention measures and minimize losses before the disease has become widespread.

With the rapid development of information technology, remote sensing has become increasingly used in agriculture [19–22]. Different plants have different spectral features due to their unique morphology and composition. The spectral features of plants comprise comprehensive spectral information generated by their continuous interactions with environmental factors (including biological and non-biological factors) during the growth process, and hyperspectral technology can identify the changes in the characteristic spectrum to determine the corresponding stress factors [23,24]. For example, plant disease caused by the fungal pathogenes destroyed the physical structure and physiology of crops, and manifested typical symptoms, such as foliar chlorosis, wilting or necrosis, and poor growth and development. Hyperspectral technologies have a nanometer-scale spectral resolution, which can respond to the unique disease spectrum, so the hyperspectral remote sensing is gradually applied to detect plant diseases [8,25,26]. However, traditional disease investigation techniques such as manual field scouting and quantitative PCR analysis after manual sampling [27], which has higher accuracy, is labor-intensive, inefficient, expensive and subjective, making it difficult to adapt to large-scale, non-destructive, real-time predictions of disease risk [28].

The objectives of this study were: (1) detecting the field infestation of *Pst* via the wheat canopy hyperspectral during the latent period, and (2) confirming the relationship between soil nitrogen nutrition and the severity of WSR during the latent period and symptom period. The proposed method can accurately, efficiently, and timely monitor the potential spread of WSR during the latent period, and thus is of great significance for forecasting and controlling epidemics of WSR under climate change, while also providing theoretical foundation for the optimization of nitrogen fertilizer application during the epidemic of WSR.

2. Materials and Methods

2.1. Experimental Material

The wheat cultivars used were Mingxian 169, a variety highly susceptible to *Pst*, Beijing 0045, a variety moderately susceptible to *Pst*, and Nongda 195, a variety highly resistant to *Pst*. The test strains were three races of *Pst*, CYR31, CYR32, and CYR33, that were mixed in equal proportions. The gradients of *Pst* spore suspension in 2016–2017 were 2 mg/mL, 1 mg/mL, 0.5 mg/mL, 0.25 mg/mL, and 0.125 mg/mL; in 2017–2018, these were 80 mg/L, 40 mg/L, 20 mg/L, and 10 mg/L. The above materials were provided by the Plant Disease Epidemiology Laboratory of China Agricultural University.

2.2. Experimental Designs

The experiment was conducted at the Kaifeng Experimental Station (34.5° N, 114.2° E) of China Agricultural University during 2016–2018. The Kaifeng Experimental Station is located in Kaifeng, Henan Province, China. There was no source of foreign stripe rust; thus, the location was suitable for artificial inoculation experiments.

A total of 54 plots (3 m × 4 m) were designed for field experiments during 2016–2017 (five inoculation concentration treatments and one healthy controls) and 2017–2018 (four inoculation concentration treatments and two healthy controls). Figure 1 showed the distribution of field plots in this study. The experiment was designed as a complete randomized block with three replicates, and there were protection rows between the plots, with an interval of 1.5 m. The sowing rate was about 225 g/plot, planted in mid-October in 2016 and 2017. Mingxian 169 was planted in the center of the community as an artificial source of induced inoculation. The field inoculation was carried out on 13 March 2017. On the 26th day after inoculation, field investigations revealed that the inoculated wheat had urediniospores, indicating that *Pst* was successfully inoculated. Therefore, the latent period of WSR in this year was 25 days. During the entire latent period, five spectroscopic measurements were performed, one day before inoculation and on the 5th, 10th, 15th, and 20th day after inoculation. The field was treated by spraying a spore suspension, carried out on20 March 2018. On the 21st day after inoculation, the presence of urediniospores was observed, and the latent period of WSR in 2018 was 20 days. During the entire latent period, five spectroscopic measurements were performed, one day before inoculation and on the 1st, 7th, 14th, and 19th day after inoculation. Meanwhile, the 0–20 cm soils of four inoculation concentration treatments with 3 replicates for each treatment were collected to analyze the correlation between the soil nitrogen and the disease index of WSR. The soil total nitrogen was determined by the semi-micro Kjeldahl method. Statistical analysis was performed using SAS v. 9.0 (SAS Institute INC., Cary, NC, USA). The framework of the study is shown in Figure 2.

Figure 1. Distribution of field plots in this study. Note: Numerals, different inoculation concentrations; letters, healthy controls.

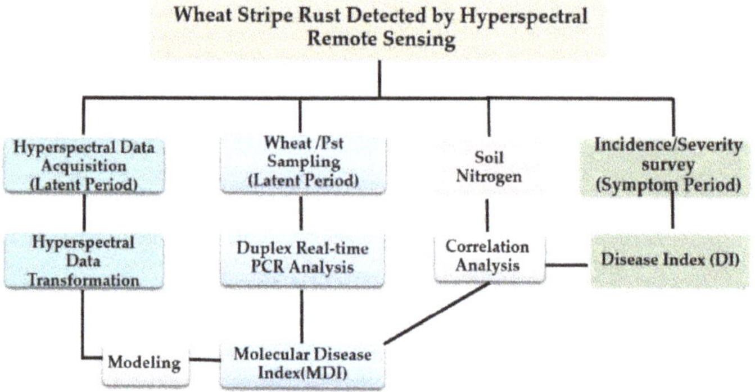

Figure 2. Flowchart.

2.3. Hyperspectral Data Acquisition and Preprocessing

An ASD spectrometer (ASD FieldSpec® HandHeld™ 2, ASD Inc., Boulder, CO, USA) was used to collect wheat canopy hyperspectral data in the wavelength range of 325–1075 nm with a bandwidth of 3 nm. The wavelength accuracy was ±1 nm; the field of view was 25°, and the minimum integration time was 8.5 ms. All hyperspectral data were collected during cloudless weather between 11:30 and 14:30 (Beijing Time). The sampling height was 1.3 m.

In this study, canopy spectrum data were collected three times for each sample, and the average value was used as the canopy spectrum of the sample. During 2016–2017, a total of 4050 canopy spectra were collected in the field experiments, including 810 healthy canopy spectra before inoculation, 540 healthy canopy spectra during the same period (control), and 2700 canopy spectra of different concentration treatments during the latent period. A total of 4320 canopy spectra were obtained in 2017–2018, including 1080 canopy spectra of healthy wheat before inoculation, 1080 canopy spectra of healthy wheat during the same period (control), and 2160 canopy spectra of different concentration treatments during the latent period.

Hyperspectral reflectance is susceptible to environmental and background noise that will affect the accuracy of the signal recognition. Therefore, it was necessary to preprocess the hyperspectral data. The derivative transformation can be used to remove low-frequency background noise spectra [29]. At the same time, the logarithmic transformation of the original spectrum not only reduces the impact of changes in illumination [30] but also enhances the hyperspectral difference in the visible region [31]. In this study, the wheat canopy hyperspectral curve was processed using the ViewSpecPro software and converted into the original hyperspectral reflectance values, and the first derivative and second derivative of the hyperspectral reflectance values were calculated according to Formulas (1) and (2). There were six parameters: reflectance (R), the first derivative of R (R_1st.dv), the second derivative of R (R_2nd.dv), the logarithm of the reciprocal of R ($\log_{10}$ (1/R)), the first derivative of log10 (1/R) ($\log_{10}$(1/R)_1st.dv), and the second derivative of $\log_{10}$ (1/R) ($\log_{10}$(1/R)_2nd.dv), as the first type of hyperspectral feature. The $\log_{10}$ (1/R) is also called the pseudo absorption coefficient, as it can reflect the absorption characteristics of objects [24]. There were 22 vegetation indices, and trilateral variable parameters were used as the second type of spectral feature (refer to references [32]). At the same time, to be able to find the waveband that represented the most effective information, the 325–1075 nm wavelength range was divided into five spectral regions of the same size as the third type of spectral feature. The above three types of spectral features were modeled according to different ratios of the training set to the testing set. The hyperspectral data were statistically analyzed by ViewSpecPro (ASD), SAS 9.0, and Excel 2003.

$$R'(\lambda_i) = \frac{dR(\lambda_i)}{d\lambda} = \frac{R(\lambda_{i+1}) - R(\lambda_{i-1})}{2\Delta\lambda}, \tag{1}$$

$$R''(\lambda_i) = \frac{d^2R(\lambda_i)}{d\lambda^2} = \frac{R'(\lambda_{i+1}) - R'(\lambda_{i-1})}{2\Delta\lambda} = \frac{R(\lambda_{i+2}) - 2R(\lambda_i) + R(\lambda_{i-2})}{4(\Delta\lambda)^2}, \tag{2}$$

2.4. Pst Detection by Duplex Real-Time PCR during Latent Period

DNA of wheat leaves was extracted from 30 leaves per sampling point according to [33], and the DNA of *Pst* was processed as described in [34].

The primers and probes of *Pst* and wheat are refer to references [35,36]. The 20 μL reaction system of duplex PCR comprised 2.0 μL DNA template (500 pg), 2.0 μL Buffer (Mg^{2+} Free), 4.0 μL $MgCl_2$ (25 μM), 2.0 μL dNTP (2500 μM), four primers of *Pst* and wheat 0.4 μL (10 μM) each, two probes 0.3 μL (10 μM) each, 0.4 μL (5 U/μL) Tag enzyme, and 7.4 μL ddH_2O. The reaction conditions were 94 °C for 3 min pre-denaturation; 94 °C for 20 s, 57 °C for 30 s, 72 °C for 20 s, 40 reaction cycles, and fluorescence signal collection at the end of each cycle. The fluorescence intensity thresholds used for threshold cycle (CT) value collection were all set to 100. The equipment used was a $MyiQ^{TM}2$ instrument (Bio-Rad, Hercules, CA, USA). The DNA concentration of *Pst* was calculated according to Equation (3):

$$y = -0.2573x + 5.4837\left(R^2 = 0.9739\right), P < 0.01, \tag{3}$$

The DNA concentration of wheat was calculated according to Equation (4):

$$y = -0.2863x + 8.811(R^2 = 0.9696, P < 0.01), \tag{4}$$

where x is the CT value; y is the logarithm ($\log_{10}C$) value of the DNA concentration. The minimum content detection limits of *Pst* DNA and wheat DNA were 0.4 pg and 0.5 ng, respectively [34].

The molecular disease index (MDI) of *Pst* was calculated according to Equation (5):

$$MDI = PstDNA(pg)/WheatDNA(ng), \tag{5}$$

MDI reflects the DNA content of *Pst* in the latent period. The area under the disease progress curve (AUDPC) reflects the cumulative effect of the development of the disease within a certain period [37]. AUDPC can be obtained with Equation (6).

$$\text{AUDPC} = \sum_{i=1}^{n-1} \left(\frac{Y_i + Y_{i+1}}{2} \right) (t_{i+1} - t_i), \tag{6}$$

where Y_i represents the MDI after inoculation, and t_i represents the inoculation time.

2.5. Field Disease Index Acquisition

After the symptoms of wheat leaves appeared, the five-point sampling method was used to investigate the field diseases, and 30 plants were marked with GPS. For each plant, we surveyed the antepenult leaves, penultimate leaves, and flag leaves for a total of 90 leaves per point. The investigation was performed every seven days until the wheat was mature.

The incidence (I) and severity (S) of disease in the symptomatic period were recorded. Incidence is an indicator reflecting the epidemic degree of a disease and was quantified using Equation (7):

$$I = \frac{n}{N} \times 100, \tag{7}$$

where I is the incidence; n is the number of diseased leaves, and N is the total number of leaves investigated. Severity (S) refers to the degree of damage to plants in the field, and is described in [33]. The severity of WSR was measured every seven days; the average severity was calculated according to Equation (8):

$$\overline{S} = \frac{\sum (S \times n_i)}{n} \times 100, \tag{8}$$

where $\overline{S}$ is the average severity; S is the severity; n_i is the number of diseased leaves corresponding to the severity of the disease; and n is the total number of diseased leaves. Disease Index (DI) is a comprehensive index that considers the incidence and severity given by Equation (9):

$$DI = I \times \overline{S} \times 100, \tag{9}$$

where DI is the disease index; I is the incidence; and $\overline{S}$ is the average severity. AUDPC is quantified using Equation (6). SAS 9.0 software was used to analyze the correlation between MDI-AUDPC and DI-AUDPC to verify whether the MDI during the latent period of WSR could predict the actual disease's occurrence during the symptoms period.

2.6. Recognition Model

On the sampling points marked by GPS, the *Pst* DNA content and the canopy hyperspectral data were obtained at the same time, and the canopy hyperspectral data were matched with MDI point-to-point. The MDI was converted into a classification label of the model. The hyperspectral data were randomly divided into a training set and a testing set. The ratios of the training set to testing set were equal to 1:1, 2:1, 3:1, 4:1, or 5:1 to compare the influence of different ratios on modeling. The DPLS and SVM methods were used to classify healthy and diseased wheat using the three types of spectral features listed above. DPLS is effective in processing data with a small sample size, high dimensionality, and multicollinearity [38] due to its dimension reduction effect [39]. Therefore, the amount of calculation can be reduced, and the calculation efficiency can be improved. SVM can better solve the problems of small samples, over-learning, nonlinearity, high dimensionality, and local minima [40]. These two recognition models were constructed based on MATLAB v.8.2 (R2013b) software (Mathworks, Natick, MA, USA). The model performance was evaluated using the overall identification accuracy.

3. Results

3.1. Wheat Canopy Spectra

During the latent period, after averaging the canopy spectral data of the four sampling times, the spectral curve is shown in Figure 3. The variation trends of the wheat canopy spectral curves at the four sampling times were similar, and there were large differences in the range of 720–1075 nm. In the first 14 days of the latent period, the reflectance increased with the time increase, and reached the maximum on the 14th day; the reflectance values on the 19th day were lower than the 14th day. This phenomenon may be due to the rapid accumulation of *Pst*, breaking through the leaf epidermis to release spores, which changes the physiological structure and biochemical components of wheat leaves, thereby affecting the spectral reflectance. It showed that hyperspectral technology can effectively detect the latent period of WSR.

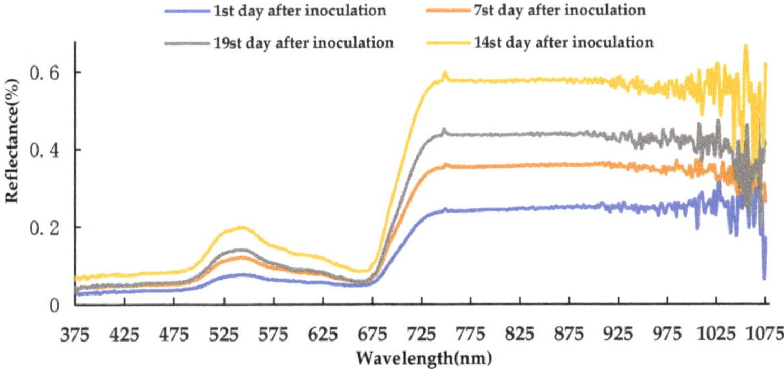

Figure 3. Wheat leaf spectra curves of four sampling times during latent period.

3.2. Correlation between MDI and DI

This study used MDI-AUDPC and DI-AUDPC for correlation analysis. The results are shown in Table 1. There was a significant correlation between MDI-AUDPC and the DI-AUDPC in 2016–2018. This indicated that the MDI of *Pst* in the latent period could predict the DI symptoms period of WSR.

Table 1. Correlation analysis between MDI-AUDPC and DI –AUDPC in different years.

Year	Correlation Coefficient	Significance Level	Regression Equation	R^2	Root Mean Square Error
2016–2017	0.84840	<0.0001	y = 0.0415 + 11.973X	0.7198	0.1221
2017–2018	0.90056	<0.0001	y = 0.6176 + 6.4193X	0.8110	3.1608

3.3. WSR Recognition with Hyperspectral Features in the 325–1075 nm Waveband

The disease recognition results during 2016–2018 are shown in Figures 4 and 5. The average recognition accuracy values of the models built using DPLS in 2016–2017 and 2017–2018 were 78.56% and 74.42%, respectively. The average recognition accuracy values of the models built using SVM in 2016–2017 and 2017–2018 were 79.58% and 77.39%, respectively. The accuracy of the model built by the SVM was superior to that built using DPLS. The average accuracy of the first type of spectral feature and the second type of spectral feature in 2016–2018 was 77.49% and 68.17%, respectively. Therefore, the average accuracy and the stability of the first type of spectral feature were better than the second type of spectral feature.

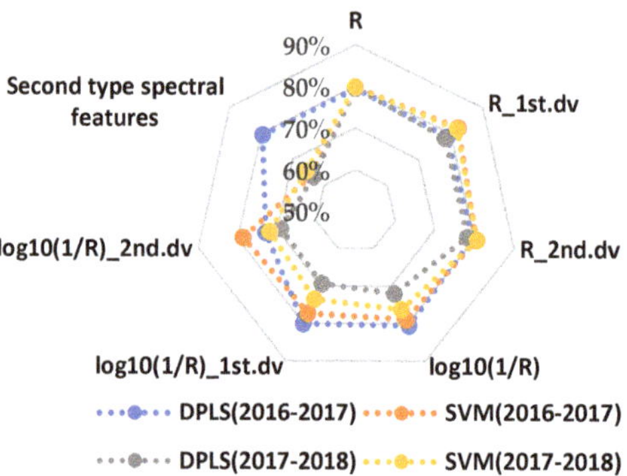

Figure 4. Prediction average accuracy of models resulting from different spectral features and different sampling ratios of the training set to testing set based on DPLS and SVM in all wavebands during 2016–2018.

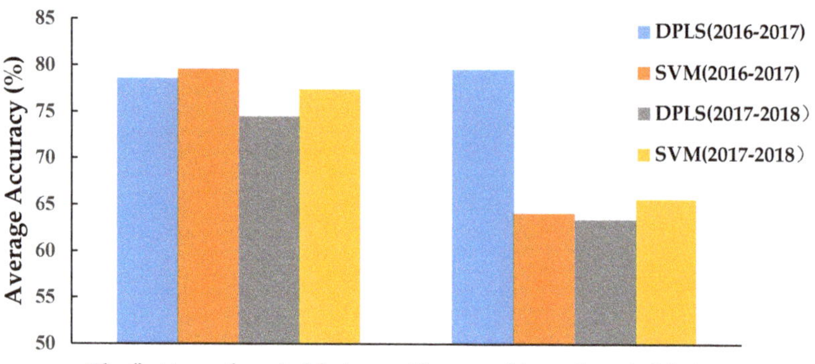

Figure 5. Prediction accuracy of models resulting from different spectral features and modeling methods in the 325–1075 nm waveband.

Tables 2 and 3 show that the recognition accuracy of the best models using DPLS and SVM methods were both in the range of 80–85%. The results demonstrated that it was feasible to use the wheat canopy hyperspectral data and the MDI of *Pst* to establish a mathematical model to detect the occurrence of WSR.

Table 2. Prediction accuracy of the best models based on DPLS in the 325–1075 nm waveband.

Year	Spectral Features	The Ratio of the Training Set to Testing Set	The Principal Component Number	Accuracy	F1 Score	Matthews Correlation Coefficient
2016–2017	$\log_{10}(1/R)$	4: 1	30	84.57	84.21	82.75
2017–2018	R	4: 1	30	82.29	81.82	80.84

Table 3. Prediction accuracy of the best models based on SVM in the 325–1075-nm waveband.

Year	Spectral Features	The Ratio of the Training Set to Testing Set	Optimal Parameter		Accuracy	F1 Score	Matthews Correlation Coefficient
			Best c	Best g			
2016–2017	R_1st.dv	3: 1	6.9644	64	83.17	83.15	82.23
2017–2018	R_1st.dv	4: 1	2.2974	64	84.03	83.65	82.19

3.4. WSR Recognition with Hyperspectral Features in the Sub-Waveband Range

3.4.1. Recognition Results of the DPLS Model in 2016–2017

The 325–1075 nm waveband was divided into five spectral regions (325–474 nm, 475–624 nm, 625–774 nm, 775–924 nm, and 925–1075 nm) of the same size. The DPLS algorithm was used to identify WSR based on different ratios of the training set to testing set and different spectral features during 2016–2017. It can be seen in Figures 6 and 7 that the accuracy of the DPLS model built by the pseudo absorption coefficient was relatively good in all wavebands, and the average accuracy of the testing set was 80.45%.

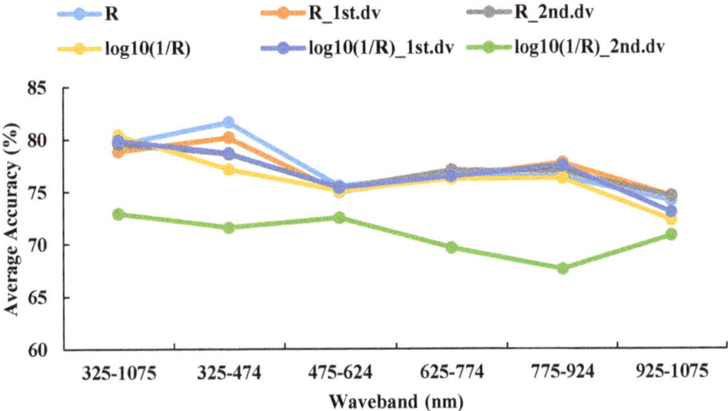

Figure 6. The testing sets average accuracy based on different spectral features and the same waveband using DPLS in 2016–2017.

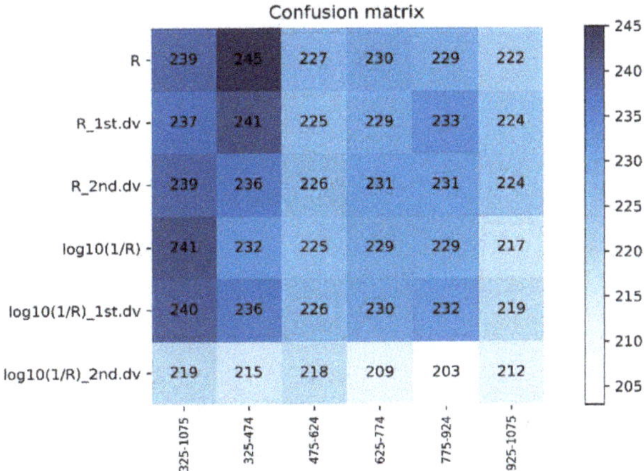

Figure 7. The confusion matrix based on different spectral features and the same waveband using DPLS in 2016–2017.

The recognition accuracy of the DPLS model had the highest values at the waveband range 325–474 nm when using R as the spectral feature, where the average accuracy of the testing set was 81.66%. These values were given priority as the candidate waveband and spectral feature for model establishment. The best model was based on 325–474 nm with R as the spectral feature, with a sampling ratio of 4:1. The accuracy of the testing set was 85.19%.

3.4.2. Recognition Results of SVM Model in 2016–2017

When the R_1st.dv was used as the spectral feature in all wavebands, the average accuracy of the testing set was 82.37%. When $\log_{10}(1/R)$_1st.dv was used as the spectral feature, the model showed a peak within the 475–624 nm range, and the average accuracy of the testing set was 83.32% (Figures 8 and 9). These values were used as the preferred waveband and spectral feature for the model establishment. The best model was characterized by the R_2nd.dv in the range of 325–474 nm. When the sampling ratio was 4:1, the optimal parameters of the model were c = 588.1336; g = 1024, and the accuracy of the testing set was 87.04%.

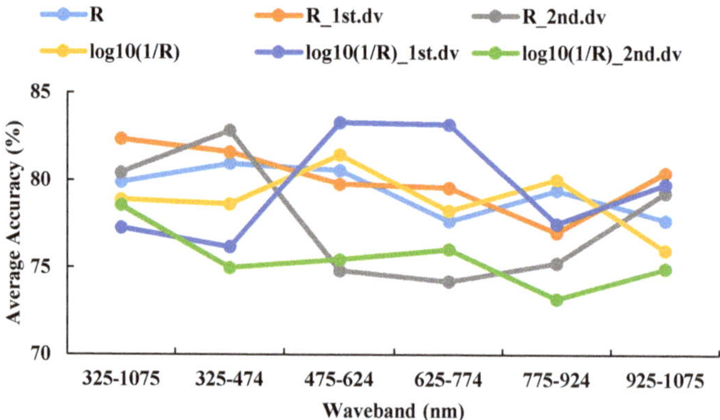

Figure 8. The testing sets average accuracy based on different spectral features and the same waveband using SVM in 2016–2017.

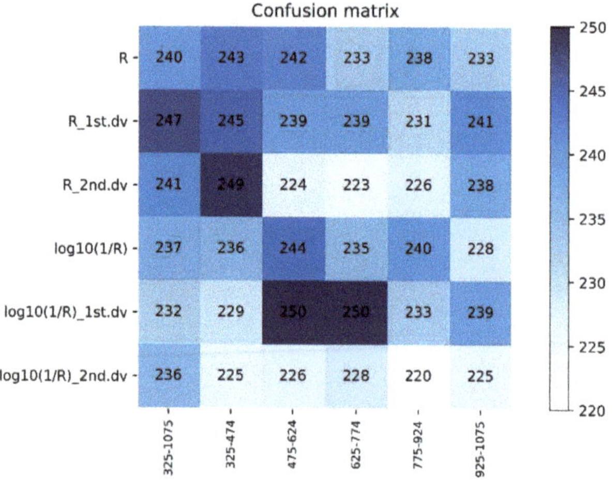

Figure 9. The confusion matrix based on different spectral features and the same waveband using SVM in 2016–2017.

3.4.3. Recognition Results of DPLS Model in 2017–2018

The recognition accuracy of the model built by R was relatively good in all wavebands, and the average accuracy of the testing set was 79.57%. When the waveband range was 325–474 nm and R was the spectral feature, the average accuracy of the model was the best at 80.14% (Figures 10 and 11), and these values can be given priority as the candidate waveband and spectral feature for model establishment. The best model comprised 325–474 nm as the range; the original spectrum was the spectral feature; the sampling ratio was 4:1, the accuracy of the testing set was 81.60%.

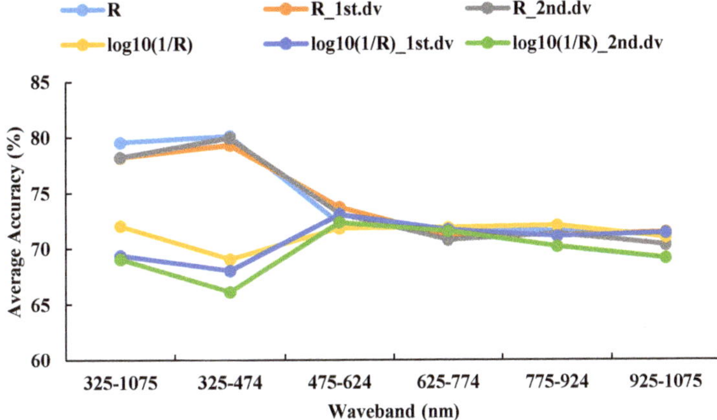

Figure 10. The testing set average accuracy based on different spectral features and the same waveband using DPLS in 2017–2018.

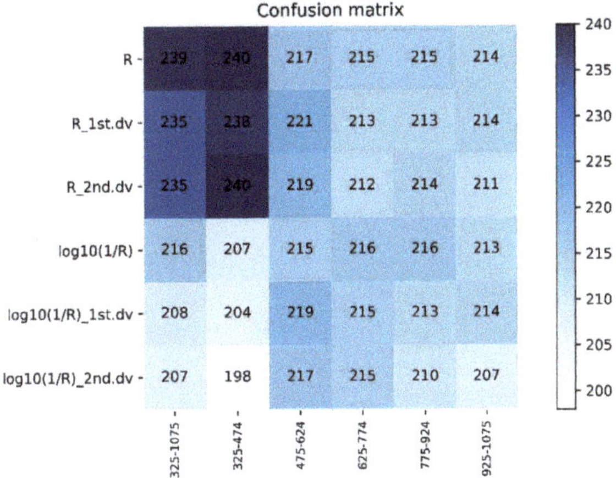

Figure 11. The confusion matrix based on different spectral features and the same waveband using DPLS in 2017–2018.

3.4.4. Recognition Results of SVM Model in 2017–2018

When the R_1st.dv was used as the spectral feature, the recognition accuracy was the best in all wavebands, and the average accuracy of the testing set was 82.05%. When the first derivative of the absorption coefficient was used as the spectral feature within the range 475–624 nm, the average accuracy of the model was the highest at 83.56% (Figures 12 and 13), and thus these values could be prioritized as the candidate waveband and preferred spectral feature for model establishment. The best model used the 325–474 nm range, the original spectrum as the feature, and a sampling ratio of 4:1; the optimal parameter c was 2.2974; g was 337.7940, and the accuracy of the model on the testing set was 85.76%.

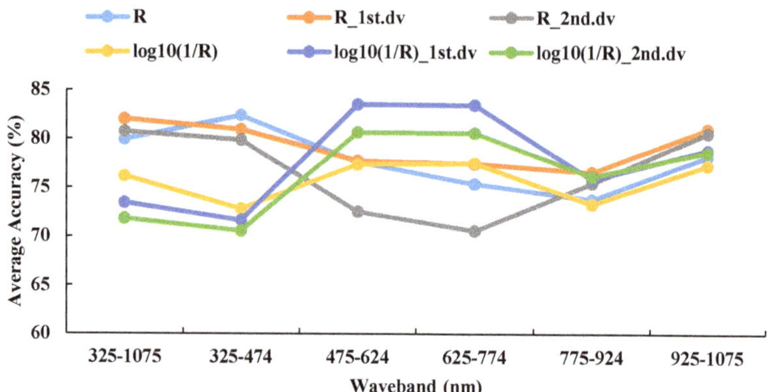

Figure 12. The testing set's average accuracy based on different spectral features and the same waveband using SVM in 2017–2018.

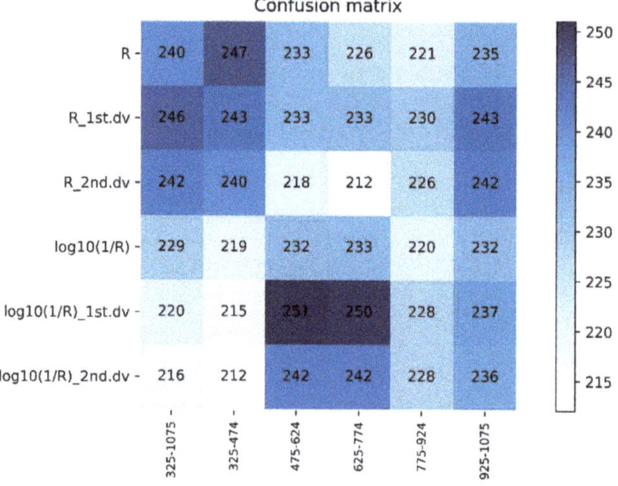

Figure 13. The confusion matrix based on different spectral features and the same waveband using SVM in 2017–2018.

3.5. Correlation between Soil Nitrogen Nutrition and WSR Severity

The correlation between the severity of WSR and soil nitrogen nutrition of different inoculation concentrations of pathogens and different varieties of wheat during the latent period and symptom period were analyzed (Table 4). The results showed that the severity of WSR was positively correlated with soil nitrogen nutrition under artificial inoculation conditions. The correlation between different inoculation concentrations and soil nitrogen nutrition was extremely significant. The correlation between the symptom period and soil nitrogen was slightly higher than the latent period. As the inoculation concentration increased, the severity of the disease gradually increased. With the decrease of inoculation concentration, the correlation between disease index and soil nitrogen nutrition gradually decreased. The disease severity of different wheat resistant varieties under artificial inoculation conditions showed the same regularity as the natural incidence. The susceptible varieties (Mingxian169) had the higher correlation coefficient. Middle-resistant varieties and resistant varieties showed a weaker correlation. Correlation analysis showed that soil nitrogen nutrition was correlated with the occurrence of WSR during the latent and symptom period, but the inoculation concentration and variety had a greater impact on it.

Table 4. Correlation between soil total nitrogen nutrition and disease index of WSR.

Growth Stage	Inoculation Concentration (mg/L)	Mingxian169 Disease Index	Beijing0045 Disease Index	Nongda195 Disease Index
Latent	80	0.916 **	0.574	0.517
Period	40	0.922 **	0.493	0.513
	20	0.801 *	0.354	0.487
	10	0.599	0.277	0.101
Symptom	80	0.982 **	0.673	0.599
Period	40	0.895 **	0.54	0.466
	20	0.838 *	0.13	0.084
	10	0.711	0.063	0.058

Note: *, ** Indicate significant differences at the 5% and 1% levels, respectively.

4. Discussion

4.1. Recognition of Wheat Stripe Rust with Hyperspectral Remote Sensing

Based on the results, it appears feasible to establish models based on the DPLS and SVM methods to identify WSR during the latent period in field settings. Meanwhile, the spectral ranges and spectral characters were good candidates for the fingerprint method for the rapid estimation of *Pst*-infected wheat leaves.

The two-year test results showed that, based on the R spectral feature, the average recognition accuracy of the DPLS algorithm was about 80% in the bands of 325–474 nm. Based on the first derivative of the absorption coefficient as the spectral feature within the 475–624 nm range, the average recognition accuracy of the SVM algorithm was about 83%. In other words, the visible part of the spectrum (400–680 nm) was relevant in distinguishing between the latent period and the symptomatic period of WSR, suggesting that *Pst* infection alters the spectral properties of wheat leaves. For fresh plants, leaf reflectance in the visible spectrum is usually low due to the absorption of pigments (e.g., chlorophyll and anthocyanin) in leaves. However, in the latent period of *Pst* infected wheat leaves, although the external morphology has not changed, internal changes have occurred that destroy leaf pigments, leading to changes in the color of plant leaves and increases in spectral reflectance [41]. Each host–pathogen interaction is unique, and the resulting spectral changes are also unique. By monitoring these spectral changes it is possible to analyze the severity and spread of diseases. This analysis has further confirmed that the canopy spectrum of WSR has a very significant correlation with the disease index.

The highest recognition accuracy rate was 87.04% in 2016–2017, and the highest rate was 85.76% in 2017–2018. The recognition accuracy of the model built in the field experiments was generally lower than for the indoor experiment model [42]. The reason may be due to the complex field environment, the many interference factors, such as illumination intensity, and soil nutrients. Therefore, continuing to optimize the model parameters, wavebands, and spectral characters can improve the model's recognition accuracy. Meanwhile, the detection limit of *Pst* by measuring the hyperspectral data of the wheat canopy with a spectrometer will be the focus of our further studies.

4.2. Correlation between Soil Nitrogen Nutrition and WSR Severity

Climatic factors are the direct factors leading to the WSR, because of its airborne. Additionally, a warmer climate will lead to more days for sporulation, shorter latent period and higher spore reproductive rate will lead to more spores produced in the suitable temperatures [2]. Therefore, it is particularly important to be able to perform a high-throughput screening WSR during the latent period. However, there are many factors affecting the severity of the disease, such as varieties, fertilization, and soil environment. When the crop reaches the optimal "nutrient balance" for growth, the disease resistance is the strongest, but it will change as the nutrient status deviates from the optimal growth degree [43]. Reasonable fertilization promotes the balance of various nutrients in wheat, which is conducive to the control of WSR. It has been reported that high nitrogen application results in increased wheat rust severity [2], and wheat rust disease and N deficiency both cause changes in foliar pigments that result in chlorosis [44]. Therefore, it is crucial for hyperspectral remote sensing to distinguish disease infection from nitrogen nutrient effects during the latent period. We will continue to verify the influence of different nitrogen application levels on disease severity, spectral diagnosis, and the effects of WSR and soil nitrogen interaction on spectral characteristics of wheat during the latent period.

5. Conclusions

In this study, we developed a high degree of accuracy approach for high-throughput detection WSR directly during the latent period.

1. In the 325–1075 nm waveband, the average recognition accuracy of the model built by SVM was better than that using DPLS. The average accuracy of the model built by the first type of spectral feature was better than the model using the second type of spectral feature. The average accuracy values of the DPLS and SVM methods were 75–80%, and the accuracy of the best-performing model was between 80–85%.
2. In the sub-wavebands, the models built based on the DPLS method with the best accuracy in two years were all concentrated in the 325–474 nm range using the original spectrum (R) as the spectral character. The models built based on the SVM method with the best recognition accuracy in the two years were concentrated in the 475–624 nm range using the first derivative of the pseudo absorption coefficient $(\log_{10}(1/R)\_1st.dv)$ as the spectral feature.
3. There was a significant positive correlation between wheat stripe rust and soil nitrogen nutrients during latent period and symptom period, which also provided the theoretical basis for more accurate remote sensing monitoring on the wheat stripe rust.

Author Contributions: Conceptualization, Q.L. and Z.M.; software, M.Z.; investigation, A.S.; writing—original draft preparation, J.C.; writing—review and editing, Q.L.; project administration, Z.M. and Q.L.; funding acquisition, Z.M. and Q.L. All authors have read and agreed to the published version of the manuscript.

Funding: This research was funded by the National Key Research and Development Program of China, grant number 2021YFD1401000; the National Natural Science Foundation of China, grant number 31860477.

Institutional Review Board Statement: Not applicable.

Informed Consent Statement: Not applicable.

Data Availability Statement: The data sets generated during and/or analyzed during the current study are available from the corresponding author upon reasonable request.

Conflicts of Interest: The authors declare no conflict of interest.

References

1. Porter, J.R.; Xie, L.; Challinor, A.J.; Cochrane, K.; Howden, S.M.; Iqbal, M.M.; Lobell, D.B.; Travasso, M.I. Chapter 7. Food Security and Food Production Systems. Climate Change 2014: Impacts, Adaptation and Vulnerability. In *Working Group II Contribution to the IPCC 5th Assessment Report*; IPCC Secretariat: Geneva, Switzerland, 2014; pp. 488–489.
2. Prank, M.; Kenaley, S.C.; Bergstrom, G.C.; Acevedo, M.; Mahowald, N.M. Climate change impacts the spread potential of wheat stem rust, a significant crop disease. *Environ. Res. Lett.* **2019**, *14*, 124053. [CrossRef]
3. Juroszek, P.; Racca, P.; Link, S.; Farhumand, J.; Kleinhenz, B. Overview on the review articles published during the past 30 years relating to the potential climate change effects on plant pathogens and crop disease risks. *Plant Pathol.* **2020**, *69*, 179–193. [CrossRef]
4. De, W.E.; Isard, S.A. Disease Cycle Approach to Plant Disease Prediction. *Annu. Rev. Phytopathol.* **2007**, *45*, 203–220.
5. Garrett, K.A.; Nita, M.; Wolf, E.; Gomez, L.; Sparks, A.H. Plant pathogens as indicators of climate change. In *Climate Change*; Elsevier: Amsterdam, The Netherlands, 2009; pp. 425–437.
6. Luo, C.; Ma, L.; Zhu, J.; Guo, Z.; Dong, K.; Dong, Y. Effects of Nitrogen and Intercropping on the Occurrence of Wheat Powdery Mildew and Stripe Rust and the Relationship With Crop Yield. *Front. Plant Sci.* **2021**, *12*, 637393. [CrossRef]

7. Snoeijers, S.S.; Pérez, G.A.; Joosten, M.; De Wit, P.J.G.M. The Effect of Nitrogen on Disease Development and Gene Expression in Bacterial and Fungal Plant Pathogens. *Eur. J. Plant Pathol.* **2000**, *106*, 493–506. [CrossRef]
8. Devadas, R.; Simpfendorfer, S.; Backhouse, D.; Lamb, D.W. Effect of stripe rust on the yield response of wheat to nitrogen. *Crop J.* **2014**, *2*, 201–206. [CrossRef]
9. Hovmøller, M.S.; Walter, S.; Justesen, A.F. Escalating threat of wheat rusts. *Science* **2010**, *329*, 329–369. [CrossRef]
10. Enjalbert, J.; Duan, X.; Giraud, T.; Vautrin, D.; De Vallavieille-Pope, C.; Solignac, M. Isolation of twelve microsatellite loci, using an enrichment protocol, in the phytopathogenic fungus *Puccinia striiformis* f. sp. *tritici*. *Mol. Ecol. Notes* **2002**, *2*, 563–565. [CrossRef]
11. Boyd, L.A. Can Robigus defeat an old enemy? Yellow rust of wheat. *J. Agric. Sci.* **2005**, *143*, 233–243. [CrossRef]
12. Chen, X.M. Epidemiology and control of stripe rust (*Puccinia striiformis* f.sp. *tritici*) on wheat. *J. Plant Pathol.* **2005**, *27*, 314–337.
13. Hu, X.; Cao, S.; Xu, X. Predicting overwintering of wheat stripe rust in central and north-western China. *Plant Dis.* **2020**, *104*, 44–51. [CrossRef]
14. Zeng, S.M.; Luo, Y. Long-distance spread and interregional epidemics of wheat stripe rust in China. *Plant Dis.* **2006**, *90*, 980–988. [CrossRef] [PubMed]
15. Wang, H.Z.; Li, S.J.; Huo, Z.R.; Pang, J.A. Physiological changes of cucumber after being infected by *Sphaerotheca fuliginea*. *Acta Agric. Boreali-Sin.* **2006**, *21*, 105–109.
16. Wan, A.M.; Chen, X.M.; He, Z.H. Wheat stripe rust in China. *Aust. J. Agric. Res.* **2007**, *58*, 605–619. [CrossRef]
17. Fan, H.Z.; Pei, W.Y. A preliminary investigation on the affecting factors and disease management of rice yellow stunt disease in Kwangtung. *J. South China Agric. Coll.* **1980**, *1*, 1–15.
18. Segarra, J.; Jeger, M.J. Epidemic dynamics and patterns of plant diseases. *Phytopathology* **2001**, *91*, 1001–1010. [CrossRef]
19. Guo, A.; Huang, W.; Ye, H.; Dong, Y.; Ma, H.; Ren, Y.; Ruan, C. Identification of Wheat Yellow Rust using Spectral and Texture Features of Hyperspectral Images. *Remote Sens.* **2020**, *12*, 1419. [CrossRef]
20. Dehkordi, R.H.; Jarroudi, M.E.; Kouadio, L.; Meersmans, J.; Beyer, M. Monitoring Wheat Leaf Rust and Stripe Rust in Winter Wheat Using High-Resolution UAV-Based Red-Green-Blue Imagery. *Remote Sens.* **2020**, *12*, 3696. [CrossRef]
21. Wu, Q.; Wang, J.H.; Wang, C.; Ma, Z.H. Recognition of wheat pre-harvest sprouting based on hyperspectral imaging. *Opt. Eng.* **2012**, *51*, 111710. [CrossRef]
22. Liu, H.J.; Brooke, B.; Trevor, G.; Bettina, B. The Performances of Hyperspectral Sensors for Proximal Sensing of Nitrogen Levels in Wheat. *Sensors* **2020**, *20*, 4550. [CrossRef]
23. Liu, Q.; Li, W.; Wang, C.; Gu, Y.; Wang, R.; Ma, Z. Qualitative identification of canopy spectra in wheat stripe rust based on Logistic, IBk and Random committee methods. *J. Plant Prot.* **2018**, *45*, 146–152.
24. Liu, Q.; Gu, Y.; Wang, C.; Wang, R.; Li, W.; Ma, Z. Canopy hyperspectral features analysis of latent period wheat stripe rust based on discriminant partial least squares. *J. Plant Prot.* **2018**, *45*, 138–145.
25. Golhani, K.; Balasundram, S.K.; Vadamalai, G.; Pradhan, B. A review of neural networks in plant disease detection using hyperspectral data. *Inf. Processing Agric.* **2018**, *5*, 354–371. [CrossRef]
26. Zhang, N.; Yang, G.; Pan, Y.; Yang, X.; Chen, L.; Zhao, C. A review of advanced technologies and development for hyperspectral-based plant disease detection in the past three decades. *Remote Sens.* **2020**, *12*, 3188. [CrossRef]
27. Yan, J.H.; Luo, Y.; Pan, J.J.; Wang, H. Quantification of latent infection of wheat stripe rust in the fields using real-time PCR. *Acta Phytopathol. Sin.* **2011**, *41*, 618–625.
28. Bock, C.H.; Poole, G.H.; Parker, P.E.; Gottwald, T.R. Plant disease severity estimated visually, by digital photography and image analysis, and by hyperspectral imaging. *Crit. Rev. Plant Sci.* **2010**, *29*, 59–107. [CrossRef]
29. Demetriades Shah, T.H.; Steven, M.D.; Clark, J.A. High resolution derivative spectra in remote sensing. *Remote Sens. Environ.* **1990**, *33*, 55–64. [CrossRef]
30. Serrano, L.; Penuelas, J.; Ustin, S.L. Remote sensing of nitrogen and lignin in Mediterranean vegetation from AVIRWAS data: Decomposing biochemical from structural signals. *Remote Sens. Environ.* **2002**, *81*, 355–364. [CrossRef]
31. Wang, J.H.; Zhao, C.J.; Huang, W.J. *The Application of Agricultural Quantitative Remote Sensing*; Science Press: Beijing, China, 2008.
32. Wang, H.; Qin, F.; Ruan, L.; Wang, R.; Liu, Q.; Ma, Z.; Li, X.; Cheng, P.; Wang, H. Identification and Severity Determination of Wheat Stripe Rust and Wheat Leaf Rust Based on Hyperspectral Data Acquired Using a Black-Paper-Based Measuring Method. *PLoS ONE* **2016**, *11*, e0154648. [CrossRef]
33. Rogers, S.O.; Bendich, A.J. Extraction of DNA from milligram amount of fresh herbarium and mummified plant tissues. *Plant Mol. Biol.* **1985**, *5*, 69–76. [CrossRef]
34. Pan, Y.; Gu, Y.L.; Luo, Y.; Ma, Z. Establishment and application of duplex real-time PCR quantitative determination method on latent infection of wheat stripe rust. *Acta Phytopathol. Sin.* **2016**, *46*, 485–491.
35. Zhao, J.; Wang, X.J.; Chen, C.Q.; Huang, L.L.; Kang, Z.S. A PCR-based assay for detection of *Puccinia striiformis* f. sp. *Tritici* in wheat. *Plant Dis.* **2007**, *91*, 1669–1674. [CrossRef] [PubMed]
36. Sandberg, M.; Lundberg, L.; Ferm, M.; Yman, I.M. Real time PCR for the detection and discrimination of cereal contamination in gluten free foods. *Eur. Food Res. Technol.* **2003**, *217*, 344–349. [CrossRef]
37. Simón, M.R.; Cordo, C.A.; Perelló, A.E.; Struik, P.C. Influence of nitrogen supply on the susceptibility of wheat to *Septoria tritici*. *J. Phytopathol.* **2003**, *151*, 283–289. [CrossRef]
38. Penuelas, J.; Bweret, F.; Filella, I. Semiempirical indexes to assess carotenoids chlorophyll-a ratio from leaf spectral reflectance. *Photosynthetica* **1995**, *31*, 221–230.

39. Qin, H.; Wang, H.R.; Li, W.J.; Jin, X.X. Application of DPLS-based LDA in corn qualitative near infrared spectroscopy analysis. *Spectrosc. Spectr. Anal.* **2011**, *31*, 1777–1781.
40. Vapnik, V.N. An overview of statistical learning theory. *IEEE Trans. Neural Netw.* **1999**, *10*, 988–999. [CrossRef]
41. Li, J.; Chen, Y.; Jiang, J. Using Hyperspectral Derivative Index to Identify Winter Wheat Stripe Rust Disease. *Sci. Technol. Rev.* **2007**, *25*, 23–26.
42. Liu, Q.; Gu, Y.; Wang, S.; Wang, C.; Ma, Z. Canopy Spectral Characterization of Wheat Stripe Rust in Latent Period. *J. Spectrosc.* **2015**, *2015*, 1–11. [CrossRef]
43. Xiao, J.X.; Zheng, Y. Nutrients uptake and pests and diseases control of crops in intercropping system. *Chin. Agric. Sci. Bull.* **2005**, *21*, 150–154.
44. Devadas, R.; Lamb, D.W.; Backhouse, D.; Simpfendorfer, S. Sequential application of hyperspectral indices for delineation of stripe rust infection and nitrogen deficiency in wheat. *Precis. Agric.* **2015**, *16*, 477–491. [CrossRef]

Article

Soil Quality Assessment in Tourism-Disturbed Subtropical Mountain Meadow Areas of Wugong Mountain, Central Southeast China

Sohel Rana [1,†] , Ziheng Xu [2,†] , Razia Sultana Jemim [3] , Zhen Liu [1] , Yanmei Wang [1] , Xiaodong Geng [1] , Qifei Cai [1] , Jian Feng [1] , Huina Zhou [1] , Tao Zhang [1] , Mingwan Li [1] , Xiaomin Guo [4] and Zhi Li [1,*]

1 College of Forestry, Henan Agricultural University, Zhengzhou 450002, China; sohel.sam@live.com (S.R.); liuzh20@163.com (Z.L.); wym3554710@163.com (Y.W.); xiaodonggeng@163.com (X.G.); caiqifei0416@163.com (Q.C.); 15713679647@163.com (J.F.); zhn100585@163.com (H.Z.); a17538392107@163.com (T.Z.); limingwan3@126.com (M.L.)
2 Palm Eco-Town Development Co., Ltd., Building 3A, Haihui Center, Zhengdong New Area, Zhengzhou 450000, China; xuzh1013@163.com
3 College of Life Sciences, Henan Agricultural University, Zhengzhou 450002, China; jemim.sam@outlook.com
4 College of Forestry, Jiangxi Agricultural University, Nanchang 330045, China; gxmjxau@163.com
* Correspondence: lizhi876@163.com; Tel.: +86-177-5255-9889
† These authors contributed equally to this work.

Abstract: Meadow soil is a vital ecosystem component and can be influenced by meadow vegetation. Evaluating soil quality in mountain meadows subjected to different levels of tourism disturbance is essential for scientific research, ecological restoration, and sustainable management. This study aimed to evaluate meadow soil quality at different tourism-disturbance levels and attempted to establish a minimum data set (MDS) with compatible indicators for soil quality assessment of subtropical mountain meadows. We analyzed fifteen soil physical, chemical, and biological indicators in control check (CK), light disturbance (LD), medium disturbance (MD), and severe disturbance (SD) meadow areas in Wugong Mountain, west of Jiangxi, China. In addition, a soil quality index (SQI) was determined using the established MDS based on the integrated soil quality index. Average soil permeability, soil pH, available nitrogen (AN), available phosphorus (AP), and number of fungal OTUs were finally introduced into the MDS to evaluate meadow soil quality at different tourism-disturbance levels. The study found that the soil of the Wugong Mountain meadow was acidic, the bulk density was loose, and the nutrient content was rich. Additionally, SQI decreased with increase in tourism-disturbance level. The mean SQI values of the Wugong Mountain meadow areas were: CK, 0.612; LD, 0.493; MD, 0.448; and SD, 0.416. Our results demonstrate that the SQI based on the MDS method could be a valuable tool with which to indicate the soil quality of mountain meadow areas, and the SQI can be regarded as a primary indicator of ecological restoration and sustainable management.

Keywords: minimum data set; mountain meadow; soil quality index; Wugong Mountain; tourism disturbance

Citation: Rana, S.; Xu, Z.; Jemim, R.S.; Liu, Z.; Wang, Y.; Geng, X.; Cai, Q.; Feng, J.; Zhou, H.; Zhang, T.; et al. Soil Quality Assessment in Tourism-Disturbed Subtropical Mountain Meadow Areas of Wugong Mountain, Central Southeast China. *Life* **2022**, *12*, 1136. https://doi.org/10.3390/life12081136

Academic Editors: Yoh Iwasa and Ling Zhang

Received: 14 July 2022
Accepted: 26 July 2022
Published: 28 July 2022

Publisher's Note: MDPI stays neutral with regard to jurisdictional claims in published maps and institutional affiliations.

1. Introduction

Soil is the basis for the survival of humans, animals, and plants [1], as well as the living space for numerous microorganisms [2]. Soil quality is the comprehensive expression of soil's physical, chemical, biological, and other properties. If only analyzed from a single aspect, the differences in soil quality under the action of different environments or external factors cannot be effectively represented [3]. Domestic and foreign scholars have conducted studies on soil quality evaluation using different methods [4–8]. The comprehensive quality evaluation model method has been used to analyze the soil quality of an abandoned mine residue area, and it was considered that pH value, organic carbon, total phosphorus, calcium, and sulfur were the key factors [9]. Geographic information system (GIS) technology was used to analyze the soil quality of Sopron Town in Hungary and it was reckoned that heavy metal pollution was the key factor affecting the soil quality of the town; soil quality risk was determined for an urban park area, which laid a foundation for a follow-up study [3].

The soil quality of a *Phyllostachys heterocycle* (Carr.) Mitford cv. *Pubescens* forest, characterized by different woodland densities, was evaluated using the multi-index method and a reasonable density of Phyllostachys heterocycla (Carr.) Mitford cv. Pubescens forest was the key measure in controlling soil quality [10]. Currently, there is no unified method to evaluate the consistency of soil quality. Most calculation functions are determined according to the different index profiles of a given study area, and various scientific research evaluations are carried out according to different regional locations, environmental conditions, and evaluation purposes. The concept is to determine specific evaluation indicators and then conduct comprehensive screening, using different statistical methods for comprehensive calculation. A minimum data set (MDS) can reflect soil quality with a small number of indicators, which is useful for soil quality evaluation and detection [11] and has the advantage of reducing data redundancy and subjective human factors [12]. The MDS method was used to analyze 41 soil physicochemical and biological indicators of cold waterlogged paddy fields in Fujian Province and six factors were selected, including carbon and nitrogen ratio, bacteria, microbial biomass nitrogen, total reducible matter, physical sand, and total phosphorus, to form a MDS for use in analyzing soil quality in different regions [13]. In addition, soil quality was analyzed in Irish grassland under different management intensities; soil organic carbon, total nitrogen, soil particle density, bulk density, available potassium, and carbon-nitrogen ratio constituted the MDS for soil quality evaluation and it was determined that high-intensity artificial intervention would have adverse effects on the soil quality of the grassland [14].

The Wugong Mountain meadow in Jiangxi is a typical representative of subtropical mountain meadows. It has typicality and particularity in the vertical belt spectrum of vegetation in East China due to its vast area and low distribution datum altitude [15]. In recent years, mountain landscapes, such as southern mountain meadows, have been widely developed, with sharp increases in the numbers of tourists. The meadow ecosystem has been degraded to varying levels due to human trampling, and the grassland plants appear to be dwarf, poor, and sparse. Some even become bare surfaces, resulting in weakened ecosystem functions and reduced resilience [16,17]. In addition, with regard to the hyperspectral characteristics of several kinds of vegetation in this region, it was found that the spectral reflectance of vegetation exhibited the following order: *Carex chinensis > Arundinella anomala > Miscanthus sinensis > Sinarundinaria nitida > Fimbristylis wukunnshanensis* [15]. Studies have reported that under high-temperature conditions, the CO_2 and N_2O emission rates of *Miscanthus sinensis* soil in this area were lower than those of *Carex chinensis* and *Fimbristylis wukunnshanensis* [18,19]. Furthermore, the total amount of inorganic phosphorus in meadow soil was found to increase significantly with increasing elevation [20]. The content of alkali-hydrolyzable nitrogen in this region's surface layer of meadow soil was greater than that in deep soil [21].

It should be noted that there are abundant mountain meadow resources in subtropical areas. While studying this type of ecosystem, previous scholars have focused on the development strategies and suggestions of the animal husbandry industry. In addition, southern meadows are characterized by poor palatability to livestock and are prone to ecological degradation due to thin turf. As a typical representative of the southern meadow ecosystem, Wugong Mountain has been the subject of relevant reports on the impact of tourism development, utilization, and flora research in recent years. However, the research on the particular mountain meadowland of Wugong Mountain has not attracted enough attention. We have made relevant analyses with regard to different aspects, such as vegetation characteristics, individual physical and chemical characteristics of soil, and tourism marketing strategies, but there is still a lack of more in-depth and systematic research.

To our knowledge, systematic studies on the soil quality evaluation of subtropical mountain meadows have been less well documented. For this study, we selected as the scope of our research the core tourist areas, and fifteen soil physical, chemical, and biological indicators were determined. Our research attempted to establish an MDS with compatible indicators for soil quality assessment of subtropical mountain meadows. This study is a new attempt to demonstrate the variation in soil properties under different tourism-disturbance levels and verify the effectiveness of the MDS in this study area. The primary objectives of this study were: (1) to identify the variation in soil properties under three different tourism-disturbance levels, (2) to establish an MDS with the proper indicators for soil quality assessment, and (3) to evaluate the soil quality of different tourism-disturbance levels in the Wugong Mountain region using the SQI method and determine the controlling indicators in order to identify whether the MDS is useful for soil quality evaluation in meadow ecosystems and as a theoretical basis for practical applications related to sustainable ecological restoration and management.

2. Materials and Methods

2.1. Study Site

Wugong Mountain (114°10′–114°17′ E, 27°25′–27°35′ N) is at the junction of three administrative regions (Jian, Pingxiang, and Yichun City) of Jiangxi Province, China. It is the watershed of the Xiangjiang and Ganjiang river systems and stretches for about 120 km, with a total area of about 970 km². The annual average temperature is 14–16 °C, and the highest temperature in summer is 23 °C. The average annual sunshine duration is 1580–1700 h, the average annual evaporation is 1360–1700 mm, the average annual humidity is 70–80%, and the average annual rainfall is 1350–1570 mm. Wugong Mountain rock types are mainly granite and gneiss, and the peak Baihefeng (Jinding) is about 1918.3 m above sea level [22]. Mountain meadows are distributed at an altitude of 1600 m to the top of the mountain range. The soil is subtropical mountain meadow soil, the vegetation mainly Miscanthus sinensis, Arundinella anomala, Perotis indica, etc., with a small number of Polygonaceae, Rosaceae, Labiatae, and Cruciferae plants. One of the most widespread species in the region is *Miscanthus sinensis* [23,24].

2.2. Experimental Design and Sample Collection

2.2.1. Plot Setting

The Jinding (main peak) area of Wugong Mountain is one of the typical tourism-disturbance areas. In the meadow area, the vegetation grows well in the absence of tourists, and there is no other disturbance behavior except tourism activities. Therefore, tourism activities directly lead to the degradation of meadow vegetation coverage in the study area. In October 2019, the altitude (1900 m) range was selected under the condition of excluding differences in altitude, terrain, and other natural factors, with reference to the national standard (GB 19377—2003) of "grading index of natural grassland degradation, desertification and salinization" issued by the Administration of Quality Supervision, Inspection and Quarantine (AQSIQ) in 2004 [25] and the research results of relevant scholars on the grading standard of degraded grassland [26–28]. Based on tourism disturbance, the vegetation coverage rate (CR) decreases the relative percentage (%). A total of four samples were set up in this study; the samples were: control check (CK, CR ≥ 90%), light disturbance (LD, 60% ≤ CR < 90%), medium disturbance (MD, 30% ≤ CR < 60%), and severe disturbance (SD, CR < 30%). The three 10 m × 10 m repeated plots were randomly set for each sample to assess the soil quality of mountain meadows with different disturbance levels. A basic overview of the different research treatments and sample plots of mountain meadows is shown in Table 1.

Table 1. The geographic positions of the mountain meadow areas with different treatments.

Experimental Treatments	Elevation (m)	Slope Degree (°)	Slope Direction	Longitude (E)	Latitude (N)	Main Vegetation Type	Vegetation Coverage Rate (%)
	1907	7	NE25°	114°10′26.09	27°27′16.76	*Miscanthus sinensis*	98
CK	1904	9	NE27°	114°10′25.16	27°27′20.22	*Miscanthus sinensis*	97
	1903	6	NE29°	114°10′24.79	27°27′20.60	*Miscanthus sinensis*	100
	1914	5	NE23°	114°10′24.18	27°27′16.22	*Miscanthus sinensis*	76
LD	1917	8	NE27°	114°10′23.29	27°27′19.03	*Miscanthus sinensis*	82
	1901	7	NE26°	114°10′24.36	27°27′20.92	*Miscanthus sinensis*	73
	1912	6	NE24°	114°10′25.24	27°27′16.24	*Miscanthus sinensis*	47
MD	1910	8	NE28°	114°10′23.59	27°27′16.74	*Miscanthus sinensis*	55
	1911	<5	NE25°	114°10′23.41	27°27′17.16	*Miscanthus sinensis*	39
	1912	7	NE20°	114°10′24.87	27°27′16.18	*Miscanthus sinensis*	21
SD	1918	6	NE23°	114°10′23.04	27°27′17.00	*Miscanthus sinensis*	15
	1917	<5	NE25°	114°10′21.79	27°27′14.76	*Miscanthus sinensis*	19

CK: control check (no disturbance); LD: light disturbance; MD: medium disturbance; SD: severe disturbance; NE: north of due east.

2.2.2. Sample Collection and Determination

(1) The methods for the collection and determination of soil chemical properties, soil enzymes, and microorganisms

In each 10 m × 10 m quadrat, five sampling points were carried out along two diagonal lines and their intersection points. The samples were collected from each point at a soil depth of 0–20 cm, and the samples were mixed. About 500 g of soil was removed by the quartering method (a 100 g soil sample from each sampling point) and put into fresh-keeping bags. Two circular knives were used for sampling (the circular knives were stainless, the upper and lower covers were aluminum, the specification was 50.46 mm × 50 mm, and the cubage was 100 cm^3). The study was conducted according to the standard list of experiments and calculation methods [29,30]. The soil samples were returned to the laboratory for natural air-drying, and plants, animal residues, and stones were removed. The soil was carefully crushed, and samples were prepared for chemical indicator and soil enzyme analysis. Soil pH, organic matter (OM), total nitrogen (TN), total phosphorus (TP), total potassium (TK), available nitrogen (AN), available phosphorus (AP), available potassium (AK), and other chemical indicators were determined by conventional analysis methods [31]. Soil enzymes were determined by the Guansongmeng method [32], sucrase by invertase 3,5-Dinitrosalicylic acid colorimetry, soil catalase by the volumetric method, and urease by indophenol blue colorimetry. In addition, about 50 g of soil was taken, and the samples were immediately put into a dry ice low-temperature box. Afterward, the samples were entrusted to the Beijing Nohe Zhiyuan Biological Information Technology Co., Ltd. for high-throughput sequencing of microbial diversity. Soil bacteria were analyzed using 16S rDNA amplicon sequencing technology, with the V3 and V4 areas selected for amplification, and in fungal 18S rDNA sequences were analyzed. The sample attribution was first determined at higher levels, followed by lower-level attribution analysis based on ITS1 sequences. Bacteria and fungi were all sequenced on a Illumina HiSeq2500 sequencing platform using the paired-end sequencing (paired-end) method to construct small fragment libraries for double-end sequencing, filtered by splicing on reads, OTU (operational taxonomic units) clustering, and, later, species and diversity analysis [33]. Since some OTU results could not be annotated when species interpretation was conducted (to avoid information loss), the diversities of bacteria and fungi were represented by their respective OTU numbers.

(2) Methods for the collection of soil samples and the determination of soil physical properties

In each 10 m × 10 m sample plot, three sampling points were selected according to the shape of the "pin" or along the diagonal line. The spacing of each point was about 5 m. Sampling was conducted with two ring knives in a 0–20 cm soil layer (the ring cutter body was made of stainless steel, and the upper and lower covers were made of aluminum). The specification was 50.46 mm (diameter) × 50 mm (height), and the volume was 100 cm^3. This was in accordance with the experimental operation and calculation methods listed in the forestry industry standards of the People's Republic of China, "Determination of forest soil water-physical properties [30]" and "Determination of forest soil percolation rate [29]", combined with the research results of relevant scholars [34]. Drying and infiltration methods were used to measure sample volume weight and average infiltration rate. The following formula was used:

$$\text{The average infiltration rate} = \frac{\text{The total amount of seepage at the time of steady infiltration}}{\text{The time when the steady infiltration reached}}$$

Since the permeability rate of all soil samples reached a stable level before 60 min, for the convenience of comparisons, the total amount of infiltration was the same as that in the previous 60 min.

2.3. Calculation Method for the Soil Quality Comprehensive Index

2.3.1. Collation of Basic Data Sets

The results of the fifteen indicators in meadow soil were summarized using Microsoft Excel v. 2016 (Microsoft Corp., Redmond, WA, USA). In addition, the basic data set for soil quality evaluation, the SPSS v. 26 (IBM Corp., Armonk, NY, USA) program used for descriptive statistics, principal component analysis, and the functional model were used to determine the overall soil quality.

2.3.2. Construction of the Minimum Data Set

The SPSS program was used to analyze the principal components of fifteen indicators in the basic data set and calculate the principal components whose characteristic roots were greater than 1. The indicators with a principal component factor load greater than or equal to 0.5 in each column were divided into groups. If an indicator load was greater than or equal to 0.5 in two groups of principal components, the index was merged into the group with a lower correlation with other indicators. We calculated each group's norm value, selected the index whose norm value was less than 10% of the highest score, and analyzed the correlation of the selected indicators in each group. If a high correlation (r > 0.5) was found, the index with a high score was determined to enter the MDS to obtain the final MDS. The norm value represents the ability to interpret comprehensive information, and the calculation formula used was as follows:

$$N_{ik} = \sqrt{\sum_{i=1}^{k} (u_{ik}^2 \lambda_k)} \tag{1}$$

In the formula, Nik is the comprehensive load of the i-th variable on the first k principal components whose eigenvalue is greater than 1; u_{ik} is the load of the i-th variable on the k-th principal component; and λ_k is the characteristic root of the k-th principal component.

2.3.3. Comprehensive Evaluation Index of Soil Quality

The formula used for the soil quality comprehensive evaluation index was as follows [10,11]:

$$SQI = \sum_{i=1}^{n} W_i \times N_i \tag{2}$$

In the formula, the soil quality index (SQI) is the comprehensive evaluation index of soil quality; W_i is the index weight coefficient; and the Person correlation analysis in SPSS 21.0 was used to calculate the correlation coefficient of each index. The ratio of the average value of the correlation coefficient between an indicator and other indicators to the average value of the correlation coefficient of all evaluation indicators is the weight coefficient of the index; N_i is the membership degree, and n is the number of indicators.

Since changes in soil indicators are continuous, the continuous membership function was used to standardize the indicators, and the ascending and descending properties of the membership functions were determined by using the positive and negative characteristics of the load of the principal component factors.

The formula of the "S" ascending membership function is:

$$F(X) = \begin{cases} 1 & (X \geq X_{max}) \\ 0.9 \times \frac{X - X_{min}}{X_{max} - X_{min}} + 0.1 & (X_{max} \geq X \geq X_{min}) \\ 0.1 & (X \leq X_{min}) \end{cases} \tag{3}$$

The formula of the "S" descending membership function is:

$$F(X) = \begin{cases} 1 & (X \geq X_{max}) \\ 0.9 \times \frac{X_{max}-X}{X_{max}-X_{min}} + 0.1 & (X_{max} \geq X \geq X_{min}) \\ 0.1 & (X \leq X_{min}) \end{cases} \qquad (4)$$

In the formula, X_{min} and X_{max} are the minimum and maximum values of soil evaluation indicators.

3. Results

3.1. Descriptive Statistics of Soil Physicochemical Properties, Microorganisms, and Enzyme Activities in the Mountain Meadow

The physical properties of mountain meadow soil with different tourism-disturbance levels in the Wugong Mountain region are shown in the descriptive statistical results presented in Table 2. The mean volume of the bulk density increased with the disturbance level, while the average permeability shows that the disturbance meadow area was reduced compared to CK. Regarding chemical properties, the mean value of soil pH decreased with the increase in disturbance level. The mean values of organic matter, total nitrogen, total phosphorus, and available nitrogen were slightly higher than CK in the disturbance area. The total potassium levels in the MD and SD regions were lower than the corresponding CK and LD levels. Available phosphorus in the disturbance areas increased with the increase in disturbance level, but the average values for the LD and MD regions were lower than the value for CK. Available potassium decreased with the increase in disturbance level, but the LD area value was slightly higher than that of the CK area. The individual contributions of the various meadow soil properties in areas of different tourism-disturbance levels are shown in Figure 1.

Table 2. Descriptive statistics for meadow soil characteristics in areas with different tourism-disturbance levels.

Index	CK	LD	MD	SD
Soil bulk density (g·cm^{-3})	0.6 ± 0.03 [a]	0.81 ± 0.02 [b]	0.83 ± 0.05 [b]	0.89 ± 0.03 [b]
Average infiltration rate (mm·min^{-1})	10.64 ± 8.46 [a]	4 ± 0.43 [ab]	4.39 ± 1.04 [ab]	2.25 ± 1.59 [b]
Soil pH	4.74 ± 0.01 [a]	4.67 ± 0.05 [a]	4.54 ± 0.1 [a]	4.47 ± 0.1 [b]
Organic matter (g·kg^{-1})	90.52 ± 13.15 [ab]	87.96 ± 7.58 [abc]	119.03 ± 19.26 [bc]	115.16 ± 20.13 [c]
Total N (g·kg^{-1})	3.84 ± 0.22 [ab]	4.24 ± 0.41 [abc]	5.18 ± 0.56 [c]	4.9 ± 0.64 [bc]
Total P (g·kg^{-1})	1.02 ± 0.11 [ab]	1.01 ± 0.1 [ab]	1.4 ± 0.18 [ab]	1.53 ± 0.22 [a]
Total K (g·kg^{-1})	47.67 ± 5.85 [a]	48.25 ± 1.5 [a]	38.75 ± 1.53 [ab]	41.42 ± 1.8 [ab]
Available nitrogen (mg·kg^{-1})	282.24 ± 82.13 [abc]	337.92 ± 144.18 [ab]	382.02 ± 93.13 [a]	295.47 ± 74.19 [abc]
Available phosphorus (mg·kg^{-1})	27.42 ± 9.84 [ab]	19.71 ± 1.18 [b]	25.59 ± 4.46 [ab]	39.23 ± 15.82 [c]
Available potassium (mg·kg^{-1})	82.67 ± 17.13 [a]	89.95 ± 22.68 [a]	76.06 ± 27.09 [a]	70.2 ± 29.91 [a]
Sucrase (mg·g^{-1})	80.42 ± 4.45 [a]	59.74 ± 15.43 [a]	67.47 ± 22.72 [a]	54.8 ± 16.54 [a]
Catalase (mg·g^{-1})	0.07 ± 0.01 [b]	0.07 ± 00.01 [b]	0.07 ± 0.01 [b]	0.05 ± 0.02 [a]
Urease (mg·g^{-1})	0.55 ± 0.27 [a]	0.37 ± 0.08 [a]	0.37 ± 0.16 [a]	0.41 ± 0.04 [a]
Bacterial OTU number	1635 ± 73 [a]	1620 ± 37 [a]	1544 ± 21 [a]	1722 ± 93 [b]
Fungal OTU number	1112 ± 97 [a]	1177 ± 352 [a]	1174 ± 71 [a]	1252 ± 118 [a]

Notes: CK: control check (no disturbance); LD: light disturbance; MD: medium disturbance; SD: severe disturbance. Data are mean values ± SE. Different lowercase letters with in the same raw for each index indicate a significance differences at $p < 0.05$, respectively.

Regarding soil biological characteristics, the activities of soil invertase in the disturbance areas were significantly lower than in the CK area. The catalase activities in the LD and MD areas were equal to that in the CK area, but the activity was lower in the SD region. Soil urease activity in the disturbance areas was significantly lower than in the CK region. The number of soil bacterial OTUs decreased with the disturbance level, but in the SD, the value was increased. The number of soil fungal OTUs increased with the disturbance levels. The coefficients of variation for each index showed weak variation (CV $\leq$ 10%) or moderate variation (10% < CV ≤ 100%). On the whole, the soil in the study area was acidic and bulk density was loose, while nutrient contents and bacterial and fungal presence were relatively rich. The effect of different tourism-disturbance levels on the soil properties was different. It was necessary to take the data for each index as a basis for comprehensively evaluating the quality of meadow soil in different disturbance areas.

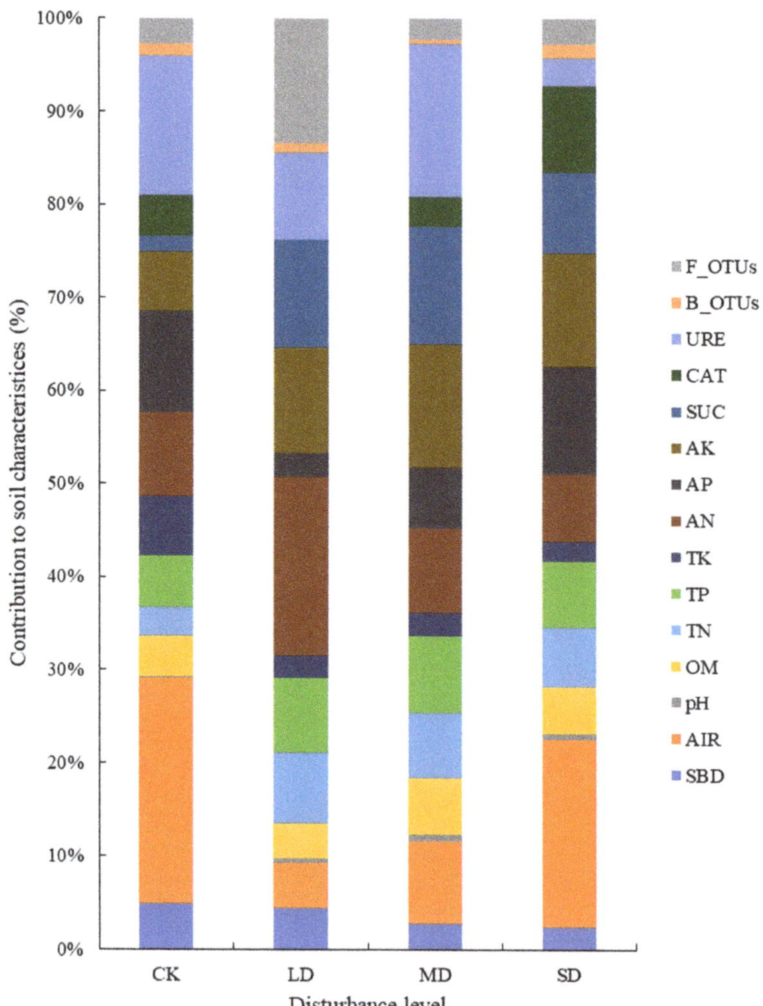

Figure 1. The individual contributions of meadow soil properties in areas of different tourism-disturbance levels. Abbreviations: SBD, soil bulk density; AIR, average infiltration rate; pH, soil pH; OR, organic matter; TN, total nitrogen; TP, total phosphorus; TK, total potassium; AN, available nitrogen; AP, available phosphorus; AK, available potassium; SUC, sucrase; CAT, catalase; URE, urease; B_OTUs, bacterial OTU number; F_OUTs, fungal OTU number.

3.2. Determination of the MDS for Soil Quality Evaluation of Mountain Meadows

The results of the principal component analysis showed that the eigenvalues of the first five principal components were 5.809, 2.474, 2.025, 1.474, and 1.117, respectively (Table 3). The variance contribution rates were 38.728%, 16,491%, 13,499%, 9.825%, and 7.446%. The total cumulative contribution rate was 85,989%, while the cumulative contribution rate of the first four principal components reached 78,542%, which is greater than 70% and meets the requirements for explaining system variation information.

Table 3. Calculation results for the principal component load matrix and norm values.

Index	Principal Component Load Matrix					Grouping	Norm Value
	PC1	PC2	PC3	PC4	PC5		
A1	0.652	−0.227	−0.135	0.442	0.345	1	1.749
A3	−0.932	−0.051	0.136	−0.031	−0.200	1	2.266
A4	0.803	0.134	0.247	−0.426	0.229	1	2.060
A5	0.882	−0.074	0.416	−0.107	0.026	1	2.215
A6	0.874	0.128	0.133	−0.109	0.067	1	2.131
A11	−0.682	0.075	−0.307	−0.152	0.340	1	1.751
A12	−0.731	−0.168	0.286	0.076	0.445	1	1.889
A7	−0.482	0.523	0.388	0.304	−0.333	2	1.610
A10	−0.160	−0.587	0.400	0.112	−0.490	2	1.269
A13	−0.335	0.775	0.333	0.148	0.127	2	1.553
A14	0.264	0.747	−0.237	0.418	−0.017	2	1.469
A9	0.620	0.627	−0.110	−0.171	−0.315	2	1.840
A2	−0.493	0.318	0.587	−0.093	0.330	3	1.580
A8	0.417	−0.194	0.843	0.216	0.043	3	1.616
A15	0.295	−0.192	−0.122	0.811	0.097	4	1.268
Characteristic root	5.809	2.474	2.025	1.474	1.117		
Variance contribution rate (%)	38.728	16.491	13.499	9.825	7.446		
Cumulative contribution rate (%)	38.728	55.218	68.717	78.542	85.989		

According to the load data for the principal component factors, the factors with an absolute value greater than 0.5 were selected and grouped. The indicators entered into the first group were bulk density (A1), pH (A3), organic matter (A4), total nitrogen (A5), total phosphorus (A6), sucrase (A11), and catalase (A12). The indicators entered into the second group were total potassium (A7), available potassium (A10), urease (A13), and bacterial OTU number (A14). The third group's indicators were average infiltration rate (A2) and available nitrogen (A8). Finally, the fourth group was the number of fungal OTUs (A15). Since, for the first four principal components, the load of each factor belonging to the group was greater than the value of the fifth principal component, the basic data set was divided into four groups.

Combining the feature of each indicator factor loading and characteristic root, the norm value for each variable was calculated. It can be seen that the highest norm value for the first group was 2.266 (A3), while that for the second group was 1.84 (A9), that for the third group was 1.616 (A8), and that for the fourth group was 1.268 (A15). In each group, the indicator which was less than 10% of the highest norm value of the group was taken and combined with the correlation indicator (Table 4). If the absolute value of the correlation coefficient between the indicators in the same group was greater than 0.5, the indicator with the higher norm value was retained. Finally, five indicators, such as average soil permeability (A2), pH (A3), available nitrogen (A8), available phosphorus (A9), and fungal OTU quantity (A15), were determined to be entered in the MDS for soil quality evaluation of mountain meadows.

Table 4. Correlation coefficient matrix of soil indicators.

Index	A1	A2	A3	A4	A5	A6	A7	A8	A9	A10	A11	A12	A13	A14	A15
A1	1.00														
A2	−0.45	1.00													
A3	−0.67 *	0.49	1.00												
A4	0.37	−0.08	−0.72 **	1.00											
A5	0.47	−0.22	−0.76 **	0.83 **	1.00										
A6	0.47	−0.22	−0.84 **	0.79 **	0.84 **	1.00									
A7	−0.34	0.39	0.51	−0.45	−0.33	−0.36	1.00								
A8	0.30	0.17	−0.29	0.42	0.73 **	0.40	0.09	1.00							
A9	0.04	−0.28	−0.58 *	0.53	0.49	0.70 *	0.04	0.01	1.00						
A10	−0.19	0.07	0.31	−0.26	0.03	−0.03	0.07	0.33	−0.31	1.00					
A11	−0.44	0.34	0.51	−0.49	−0.68 *	−0.37	0.08	−0.60 *	−0.28	−0.05	1.00				
A12	−0.30	0.53	0.56	−0.52	−0.48	−0.54	0.33	0.02	−0.69 *	0.14	0.60 *	1.00			
A13	−0.34	0.57	0.31	−0.07	−0.25	−0.22	0.67 *	0.04	0.17	−0.35	0.18	0.21	1.00		
A14	0.20	0.04	−0.34	0.05	0.03	0.31	0.28	−0.18	0.58	−0.44	−0.12	−0.33	0.36	1.00	
A15	0.49	−0.26	−0.31	−0.12	0.15	0.19	−0.21	0.23	−0.03	0.14	−0.14	−0.17	−0.09	0.27	1.00

Note: * and ** represent significance differences at $p < 0.05$ and $p < 0.01$, respectively.

3.3. The Comprehensive Evaluation of the Soil Quality of Mountain Meadows at Different Tourism-Disturbance Levels

The soil in the study area was mountainous meadow soil; good soil permeability represents a better water conservation function. The data analysis results showed that soil pH and average permeability are the core factors in the soil quality evaluation of mountain meadows. While the soil was generally acidic, the increase in pH indicated a benign trend in soil quality. The amounts of available nitrogen, available phosphorus, and fungal OTUs in the soil were all positive indicators of soil fertility. Therefore, the membership value for mountain meadow soil quality evaluation had an "S" ascending function, and, according to the correlation coefficients between each indicator in the MDS, the weight coefficient was calculated by referring to the following method (Table 5). From the data analysis results, it can be seen that soil pH and average permeability are the core factors in the soil quality evaluation of mountain meadows.

Table 5. Weight coefficient of soil quality evaluation index.

Index	Weight Coefficient	Subordinate Function
A2	0.228	$F(X) = \begin{cases} 1 & (X \geq X_{20.35}) \\ 0.9 \times \frac{X - X_{0.99}}{X_{20.35} - X_{0.99}} + 0.1 & (X_{20.35} \geq X \geq X_{0.99}) \\ 0.1 & (X \leq X_{0.99}) \end{cases}$
A3	0.317	$F(X) = \begin{cases} 1 & (X \geq X_{4.75}) \\ 0.9 \times \frac{X - X_{4.36}}{X_{4.75} - X_{4.36}} + 0.1 & (X_{max} \geq X \geq X_{4.36}) \\ 0.1 & (X \leq X_{4.36}) \end{cases}$
A8	0.132	$F(X) = \begin{cases} 1 & (X \geq X_{504.39}) \\ 0.9 \times \frac{X - X_{233.18}}{X_{504.39} - X_{233.18}} + 0.1 & (X_{max} \geq X \geq X_{233.18}) \\ 0.1 & (X \leq X_{233.18}) \end{cases}$
A9	0.167	$F(X) = \begin{cases} 1 & (X \geq X_{50.85}) \\ 0.9 \times \frac{X - X_{18.39}}{X_{50.85} - X_{18.39}} + 0.1 & (X_{max} \geq X \geq X_{18.39}) \\ 0.1 & (X \leq X_{18.39}) \end{cases}$
A15	0.156	$F(X) = \begin{cases} 1 & (X \geq X_{1441}) \\ 0.9 \times \frac{X - X_{777}}{X_{1441} - X_{777}} + 0.1 & (X_{max} \geq X \geq X_{777}) \\ 0.1 & (X \leq X_{777}) \end{cases}$

According to the membership value and weight coefficient of the MDS index for soil quality evaluation, the soil quality indexes of mountain meadows subjected to different levels of tourism disturbance were calculated according to Formula (2) (Figure 2). The results showed that the ranking of soil quality for mountain meadow areas subjected to different levels of tourism disturbance was CK > LD > MD > SD, and the soil quality indexes were 0.612, 0.493, 0.448, and 0.416, respectively, indicating that soil quality decreased with the increase in disturbance level and that only the soil quality of the CK area was in the middle-to-high level. The soil quality at each disturbance level decreased; the quality index of LD was 19.45%, that of MD was 26.80%, and that of SD was 32.00%—all lower than that of CK.

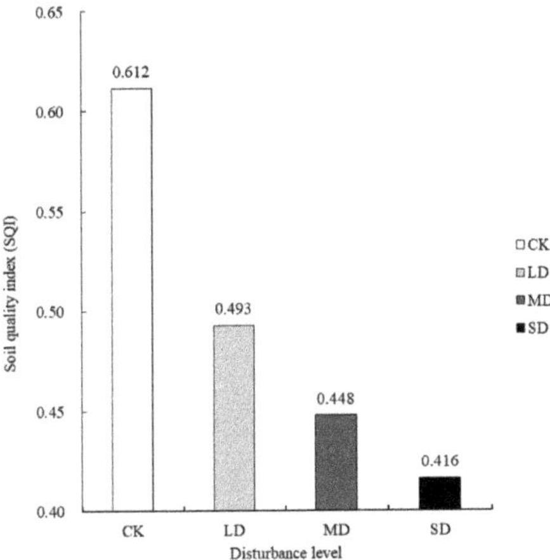

Figure 2. Comprehensive soil quality indexes of mountain meadow areas with different tourism-disturbance levels. Abbreviations: CK, control check; LD, light disturbance; MD, medium disturbance; SD, severe disturbance.

4. Discussion

4.1. Soil Quality Evaluation of Mountain Meadows

This study considered fifteen indicators of soil physical, chemical, and biological properties in typical subtropical mountain meadow areas of Wugong Mountain. On the basis of mathematical statistics and analysis, five indicators (average soil permeability, pH, available nitrogen, available phosphorus, and fungal OTUs) were selected for a minimum data set (MDS) to obtain a comprehensive index of meadow soil quality given different tourism-disturbance levels in the Wugong Mountain region. The results showed that the meadow soil in the study area was acidic, that bulk density was loose, and that the soil organic matter, total nitrogen, available nitrogen, and other nutrient contents and microbial presences were rich. If the comprehensive index of soil quality is greater than 0.5, this indicates that the soil quality is good [35]. The SQI of the meadow in the study area without tourism disturbance was greater than 0.5, while it was less than 0.5 in the tourism-disturbed areas. The comprehensive index of soil quality decreased with the increase in disturbance levels. As a result, the soil quality of meadows in tourism-disturbed areas is worse than in meadows without tourism disturbance.

Studies have reported that comprehensive soil quality was significantly decreased with increase in tourism disturbance [36,37], which was consistent with the results of this study. The mountain meadow was rich in terms of the root system and there was a large amount of humus in the soil. Due to the low temperature, slow microbial decomposition, and high organic matter content, the soil bulk density was loose, but the nutrient content was rich [38]. The source of soil nutrients is mainly the return of nutrients from surface vegetation and underground roots. A previous study [20] has shown that the distribution of soil nutrients in Wugong Mountain meadowland shows strong surface aggregation.

However, the disturbance behavior of tourists has reduced the soil surface vegetation of mountain meadows in Wugong Mountain, affected the source of soil nutrient return, and destroyed the soil structure, which has had a negative impact on soil bulk density, porosity, and permeability. The changes in soil quality with different disturbance levels are comprehensively reflected through the five indicators included in the MDS. In disturbed areas with low comprehensive indexes of soil quality, appropriate methods should be selected for vegetation restoration to prevent further degradation of soil quality, which results in the loss of the survival basis of vegetation and the deterioration of regional ecologies.

4.2. Soil Quality Evaluation Method Based on the MDS and the SQI Model

The combination of the MDS and the SQI can enable the effective evaluation of soil quality under different environmental or external factors. However, there is no unified standard for the determination of a minimum data set [35], including the membership function and weight value in the process of calculating the comprehensive evaluation index of soil quality, and there is also a lack of a unified calculation process [10]. In practice, the calculation function is usually determined according to different indicator profiles of the study area. However, the calculation function has been based on different calculation methods for soil quality evaluation [39,40], such as fuzzy mathematics, artificial neural networks, grey system theory, principal component analysis, etc. Different evaluations of the soil quality of a certain region may obtain different data, but the overall results should be similar [12,35].

In the Wugong Mountain meadow distribution area, previous studies have analyzed different characteristics of or indicators in the soil [41,42], and conclusions have also been based on certain aspects of research [43]. There is a lack of a systematic and representative evaluation metric, but the minimum data set (MDS) can be used to reflect soil quality statistically, ensuring a more accurate evaluation of soil quality [12].

5. Conclusions

The meadow soil in Wugong Mountain was found to be acidic, loose in terms of bulk density, and rich in nutrients. Five indicators, including average soil permeability, pH, available nitrogen, available phosphorus, and number of fungal OTUs, can be used as a minimum data set (MDS) to obtain a comprehensive index of soil quality for areas subjected to different levels of tourism disturbance in the Wugong Mountain region. Among the indicators, soil average permeability and pH are the core factors in soil quality evaluation. Comprehensive soil quality indexes decreased with increase in tourism-disturbance level. The soil quality index ranking with respect to different tourism-disturbance levels was CK > LD > MD > SD, and the soil quality indexes were 0.612, 0.493, 0.448, and 0.416, respectively. Based on the relevant experimental basis and data indicators, this study analyzed the impact of tourism disturbance on the soil of Wugong Mountain meadowland and made an objective evaluation. At present, though limited in terms of timescale and research scope, the research results can be used as an essential reference for short-term scientific research and productive work. A long-term study with a more extensive range and including more indicators is required to further optimize and improve the evaluation method and system for the analysis of subtropical mountain meadow soils.

Author Contributions: Conceptualization, Z.L. (Zhi Li) and X.G. (Xiaomin Guo); methodology, Z.L. (Zhi Li) and X.G. (Xiaomin Guo); software, Z.L. (Zhi Li) and S.R.; validation, Z.L. (Zhi Li), S.R. and Z.X.; formal analysis, Z.L. (Zhi Li), S.R. and Z.X.; investigation, Z.L. (Zhi Li) and X.G. (Xiaomin Guo); resources, Z.L. (Zhi Li) and X.G. (Xiaomin Guo); data curation, Z.L. (Zhi Li), S.R. and X.G. (Xiaomin Guo); writing—original draft preparation, Z.L. (Zhi Li), Z.X., R.S.J., J.F., H.Z., T.Z. and M.L.; writing—review and editing, Z.L. (Zhi Li), S.R., Z.L. (Zhen Liu), Y.W., X.G. (Xiaodong Geng) and Q.C.; visualization, Z.L. (Zhi Li) and S.R.; supervision, Z.L. (Zhi Li); project administration, Z.L. (Zhi Li) and X.G. (Xiaomin Guo); funding acquisition, Z.L. (Zhi Li), X.G. (Xiaomin Guo), Z.L. (Zhen Liu) and M.L. All authors have read and agreed to the published version of the manuscript.

Funding: This research was funded by the Henan Province Postdoctoral Research Project of China, grant number 2020053; the Key Forestry Science and Technology Promotion Project of China Central Government, grant number GTH[2020]17; the National Natural Science Foundation of China, grant numbers 32000267 and 31360177.

Institutional Review Board Statement: Not applicable.

Informed Consent Statement: Not applicable.

Data Availability Statement: The data sets generated during and/or analyzed during the current study are available from the corresponding author upon reasonable request.

Acknowledgments: We acknowledge support from the Wugong Mountain Industrial Company and the Pingxiang Forestry Research Institute. We thank the editor and reviewers for their constructive suggestions and insightful comments.

Conflicts of Interest: The authors declare no conflict of interest.

References

1. Gonzaga, M.I.S.; Bispo, M.V.C.; Da Silva, T.L.; Dos Santos, W.M.; De Santana, I.L. Atlantic Forest Soil as Reference in the Soil Quality Evaluation of Coconut Orchards (Cocos Nucífera L) under Different Management. *Semin. Agrar.* **2016**, *37*, 3847–3858. [CrossRef]
2. Paz-Ferreiro, J.; Fu, S. Biological Indices for Soil Quality Evaluation: Perspectives and Limitations. *Land Degrad. Dev.* **2016**, *27*, 14–25. [CrossRef]
3. Horváth, A.; Szűcs, P.; Bidló, A. Soil Condition and Pollution in Urban Soils: Evaluation of the Soil Quality in a Hungarian Town. *J. Soils Sediments* **2014**, *15*, 1825–1835. [CrossRef]
4. Shao, G.; Ai, J.; Sun, Q.; Hou, L.; Dong, Y. Soil Quality Assessment under Different Forest Types in the Mount Tai, Central Eastern China. *Ecol. Indic.* **2020**, *115*, 106439. [CrossRef]
5. Jiang, L.; Zhang, L.; Deng, B.; Liu, X.; Yi, H.; Xiang, H.; Li, Z.; Zhang, W.; Guo, X.; Niu, D. Alpine Meadow Restorations by Non-Dominant Species Increased Soil Nitrogen Transformation Rates but Decreased Their Sensitivity to Warming. *J. Soils Sediments* **2017**, *17*, 2329–2337. [CrossRef]
6. Huang, J.; Liu, W.; Yang, S.; Yang, L.; Peng, Z.; Deng, M.; Xu, S.; Zhang, B.; Ahirwal, J.; Liu, L. Plant Carbon Inputs through Shoot, Root, and Mycorrhizal Pathways Affect Soil Organic Carbon Turnover Differently. *Soil Biol. Biochem.* **2021**, *160*, 108322. [CrossRef]
7. Yang, Z.; Zhang, R.; Li, H.; Zhao, X.; Liu, X. Heavy Metal Pollution and Soil Quality Assessment under Different Land Uses in the Red Soil Region, Southern China. *Int. J. Environ. Res. Public Health* **2022**, *19*, 4125. [CrossRef]
8. Gao, R.; Ai, N.; Liu, G.; Liu, C.; Zhang, Z. Soil C:N:P Stoichiometric Characteristics and Soil Quality Evaluation under Different Restoration Modes in the Loess Region of Northern Shaanxi Province. *Forests* **2022**, *13*, 913. [CrossRef]
9. Mukhopadhyay, S.; Masto, R.E.; Yadav, A.; George, J.; Ram, L.C.; Shukla, S.P. Soil Quality Index for Evaluation of Reclaimed Coal Mine Spoil. *Sci. Total Environ.* **2016**, *542*, 540–550. [CrossRef]
10. Fan, S.; Zhao, J.; Su, W.; Yu, L.; Yan, Y. Comprehensive Evaluation of Soil Quality in Phyllostachys Edulis Stands of Different Stocking Stocking Densities. *Sci. Silvae Sin.* **2015**, *51*, 1–9.
11. Raiesi, F. A Minimum Data Set and Soil Quality Index to Quantify the Effect of Land Use Conversion on Soil Quality and Degradation in Native Rangelands of Upland Arid and Semiarid Regions. *Ecol. Indic.* **2017**, *75*, 307–320. [CrossRef]
12. Deng, S.; Zeng, L.; Guan, Q.; Li, P.; Liu, M.; Li, H. Minimum Dataset-Based Soil Quality Assessment of Waterlogged Paddy Field in South China. *Acta Pdeologica Sin.* **2016**, *53*, 1326–1333.
13. Wang, F.; Li, Q.H.; Lin, C.; He, C.M.; Lin, X.J. Establishing a Minimum Data Set of Soil Quality Assessment for Cold-Waterlogged Paddy Field in Fujian Province, China. *Chin. J. Appl. Ecol.* **2015**, *26*, 1461–1468.
14. Askari, M.S.; Holden, N.M. Indices for Quantitative Evaluation of Soil Quality under Grassland Management. *Geoderma* **2014**, *230–231*, 131–142. [CrossRef]
15. Li, Z.; Xiang, Z.; Niu, D.; Guo, X.; Xie, B.; Zhang, X. Hyperspectral Characteristics of Main Community Types in Wugong Mountain Meadow. *Pratacultural Sci.* **2016**, *33*, 1492–1501.
16. Dong, L.; Huang, P.; Hu, M. Tourist Garbage Treatment Under Festival Activity Backgroundin Mountain ScenicAreas: A Case Study of International Tent Festival of Wugong Mountain. *Biol. Disaster Sci.* **2016**, *39*, 67–73.
17. Li, Z.; Siemann, E.; Deng, B.; Wang, S.; Gao, Y.; Liu, X.; Zhang, X.; Guo, X.; Zhang, L. Soil Microbial Community Responses to Soil Chemistry Modifications in Alpine Meadows Following Human Trampling. *Catena* **2020**, *194*, 104717. [CrossRef]
18. Deng, B.; Yuan, Z.; Guo, X. Microelement Distributions and Responses to Human Disturbance in Meadow Soil of Wugong Mountain. *Pratacultural Sci.* **2015**, *32*, 1555–1560.
19. Deng, B.; Li, Z.; Zhang, L.; Ma, Y.; Li, Z.; Zhang, W.; Guo, X.; Niu, D.; Siemann, E. Increases in Soil CO_2 and N_2O Emissions with Warming Depend on Plant Species in Restored Alpine Meadows of Wugong Mountain, China. *J. Soils Sediments* **2016**, *16*, 777–784. [CrossRef]
20. Zhao, X.; Guo, X.; Zhang, J.; Niu, D.; Shan, L.; Zhang, W.; Wei, X.; Chen, F.; Huang, S.; Li, Z.; et al. Vertical Distribution Character of Soil Inorganic Phosphorus in Mountain Meadow System of WuGong Mountain. *Pratacultural Sci.* **2014**, *31*, 1610–1617.
21. Yuan, Z.; Deng, B.; Li, Z. Study on the Factors Affecting the Distribution of Alkali Hydrolyzable Nitrogen in Meadow Soil in Wugong Mountain. *Jiangsu Agric. Sci.* **2015**, *43*, 318–320.
22. Li, Z.; Yuan, Y.; Zhang, X.; Niu, D.; Zhang, W.; Hu, D.; Guo, X. Distribution Characteristics of Soil Mechanical Composition in Wugong Mountain Mountainous Meadow with Different Disturbance Degree. *J. Green Sci. Technol.* **2015**, *11*, 27–28.
23. Liu, J.; Ye, H.; Xie, G.; Yang, G.; Zhang, W. Re-Os Dating of Molybdenite from the Hukeng Tungsten Deposit in the Wugongshan Area, Jiangxi Province, and Its Geological Implications. *Acta Geol. Sin.* **2008**, *82*, 1572–1579.

24. Cheng, X. *Studies on Plant Community Characteristics and Spatial Distribution Attern in Subalpine Meadows of Wugong Mountain*; Jiangxi Agricultural University: Nanchang, China, 2014.
25. Gao, Y.; Wu, P. Effects of Alpine Meadow Degradation on Soil Insect Diversity in the Qinghai-Tibetan Plateau. *Acta Ecol. Sin.* **2016**, *36*, 2327–2336.
26. Liu, S.; Li, L.; Zhang, F.; Du, Y.; Li, Y.; Guo, X.; Ouyang, J.; Cao, G. Effects of Grazing Season and Degradation Degree on the Soil Organic Carbon in Alpine Meadow. *Pratacultural Sci.* **2016**, *33*, 11–18.
27. General Administration of Quality Supervision, Inspection and Quarantine. *Parameters for Degradation, Sanctification and Salification of Rangelands*; Standards Press of China: Beijing, China, 2004.
28. Klimkowska, A.; Kotowski, W.; Van Diggelen, R.; Grootjans, A.P.; Dzierża, P.; Brzezińska, K. Vegetation Re-Development After Fen Meadow Restoration by Topsoil Removal and Hay Transfer. *Restor. Ecol.* **2010**, *18*, 924–933. [CrossRef]
29. *Institute of Forestry Determination of Forest Soil Percolation Rate*; Standards Press of China: Beijing, China, 1999.
30. *Institute of Forestry Determination of Forest Soil Water-Physical Properties*; Standards Press of China: Beijing, China, 1999.
31. Bao, S. *Agriculture Soil Chemical Analysis*; Science Press: Beijing, China, 2000.
32. Guan, S. *Soil Enzyme and Its Research Methods*; China Agriculture Press: Beijing, China, 1986.
33. Chu, H.; Sun, H.; Tripathi, B.M.; Adams, J.M.; Huang, R.; Zhang, Y.; Shi, Y. Bacterial Community Dissimilarity between the Surface and Subsurface Soils Equals Horizontal Differences over Several Kilometers in the Western Tibetan Plateau. *Environ. Microbiol.* **2016**, *18*, 1523–1533. [CrossRef]
34. He, F.Y.; He, F.; Da Wu, Z.; Liu, S.R.; Liu, X.L.; Peng, P.H. Change of Soil Physical Properties along Elevation Gradients in A Bies f Axoniana Natural Forest. *J. Northwest Norm. Univ.* **2015**, *51*, 92–98.
35. Gong, L.; Zhang, X.; Ran, Q. Quality Assessment of Oasis Soil in the Upper Reaches of Tarim River Based on Minimum Data Set. *Acta Pedol. Sin.* **2015**, *57*, 682–689.
36. Cao, L. Effects of Tourist Disturbance on Soil Quality in Yuntai Mountain Scenic Area, He'nan Province. *Res. Soil Water Conserv.* **2015**, *22*, 67–71.
37. Li, L.; Zhang, Y.; Jiang, H.; Xie, Y.; Du, H.; Zhou, Y.; Zhang, H. Effects of Tourist Activities on Soil Quality of Wuyishan Scenery District. *Res. Soil Water Conserv.* **2009**, *16*, 56–62.
38. Zhou, L.; Zhu, H.; Zhong, H.; Yang, H.; Suo, F.; Shao, X.; Zhou, X. Spatial Analysis of Soil Bulk Density in Yili, Xinjiang Uygur Autonomous Region, China. *Acta Prataculturae Sin.* **2016**, *25*, 64–75.
39. Tang, F.; Deng, Y.; Zheng, M.; Guo, H.; Cao, F.; Wu, L. Soil Quality Evaluation in Rocky Desertification of Northwest Hunan Province Based on Gray Correlation Analysis. *J. Cent. South Univ. For. Technol.* **2016**, *36*, 36–43.
40. Zhang, L.; Lai, G.; Kong, Y.; Sun, C. Evaluation of Soil Quality in Beijing Jiulong Mountain Based on Factor Analysis Method. *J. Northwest For. Univ.* **2016**, *31*, 7–14.
41. Yuan, Z.; Deng, B.; Guo, X.; Niu, D.; Hu, Y.-W.; Wang, J.; Zhao, Z.; Liu, Y.; Zhang, W. Soil Total NPK's Distribution Pattern and Response to Different Degradation Degrees in Wugong Mountain Meadow. *J. Northwest For. Univ.* **2015**, *30*, 14–20.
42. Deng, B.; Yuan, Z.; Wen, W.; Fan, N.; Li, Y.; Li, Z.J.; Guo, X.; Zhang, W. Distribution Patterns and Relationships of Soil Organic Matter, Total Nitrogen and Alkali Hydrolyzable Nitrogen in Wugong Mountain Meadow. *J. Jiangsu Agric. Sci.* **2015**, *43*, 414–417.
43. Yuan, Y.; Li, Z.; Zhang, W.; Niu, D.; Guo, X. Soil Nitrogen Distribution and Its Correlation with Soil Physical Properties in Different Altitude in Mountain Meadow of Wugong Mountain. *J. Cent. South Univ. For. Technol.* **2016**, *36*, 108–113.

MDPI

St. Alban-Anlage 66

4052 Basel

Switzerland

Tel. +41 61 683 77 34

Fax +41 61 302 89 18

www.mdpi.com

Life Editorial Office

E-mail: life@mdpi.com

www.mdpi.com/journal/life